ARTIFICIAL INTELLIGENCE FOR NEUROLOGICAL DISORDERS

ARTIFICIAL INTELLIGENCE FOR NEUROLOGICAL DISORDERS

Edited by

AJITH ABRAHAM, PhD

Machine Intelligence Research Labs (MIR Labs), Scientific Network for Innovation and Research Excellence, Auburn, WA, United States

SUJATA DASH

Department of Computer Application, Maharaja Srirama Chandra BhanjaDeo University (erstwhile North Orissa University), Baripada, Mayurbhanj, Odisha, India

SUBHENDU KUMAR PANI

Krupajala Group of Institutions, BijuPatnaik University of Technology, Odisha, India

LAURA GARCÍA-HERNÁNDEZ

University of Córdoba, Spain

ELSEVIER

ACADEMIC PRESS

An imprint of Elsevier

Academic Press is an imprint of Elsevier
125 London Wall, London EC2Y 5AS, United Kingdom
525 B Street, Suite 1650, San Diego, CA 92101, United States
50 Hampshire Street, 5th Floor, Cambridge, MA 02139, United States
The Boulevard, Langford Lane, Kidlington, Oxford OX5 1GB, United Kingdom

ISBN 978-0-323-90277-9

For information on all Academic Press publications
visit our website at https://www.elsevier.com/books-and-journals

Publisher: Nikki Levy
Acquisitions Editor: Joslyn Chaiprasert-Paguio
Editorial Project Manager: Kristi Anderson
Production Project Manager: Sreejith Viswanathan
Cover Designer: Miles Hitchen

Typeset by STRAIVE, India

Dedication

We dedicate the book to our family members, contributing authors, and the staff at Elsevier, especially Ms. Kristi Anderson. Without their patience, understanding, support, and most of all love, the completion of this book would not have been possible.

Ajith Abraham, Editor

Sujata Dash, Editor

Subhendu Kumar Pani, Editor

Laura García-Hernández, Editor

Contents

Contributors

Mayowa J. Adeniyi
Department of Physiology, Edo State University Uzairue, Iyamho, Nigeria

Charles O. Adetunji
Department of Microbiology, Applied Microbiology, Biotechnology and Nanotechnology Laboratory, Edo State University Uzairue, Iyamho, Nigeria

Olorunsola Adeyomoye
Department of Physiology, University of Medical Sciences, Ondo, Nigeria

Rishabh Anand
Service Deliver Manager, HCL Technologies Limited, New Delhi, India

Korhan Cengiz
College of Information Technology, University of Fujiarah, Fujairah, United Arab Emirates

Ayobami Dare
Department of Physiology, School of Laboratory Medicine and Medical Sciences, College of Health Sciences, Westville Campus, University of KwaZulu-Natal, Durban, South Africa

Sujata Dash
Department of Computer Application, Maharaja Sriram Chandra Bhanja Deo University, Baripada, Odisha, India

Pijush Dutta
Department of Electronics and Communication Engineering, Greater Kolkata College of Engineering and Management, West Bengal, India

Alex Enoch
Department of Human Physiology, Ahmadu Bello University Zaria, Zaria, Nigeria

M.A. Jabbar
Department of CSE (AI & ML), Vardhaman College of Engineering, Hyderabad, India

Maheshkumar H. Kolekar
Department of Electrical Engineering, Indian Institute of Technology Patna, Bihta, Patna, Bihar, India

Asok Kumar
Student Welfare Department, Vidyasagar University, Medinipur, West Bengal, India

Hima Bindu Maringanti
MSCB University, Baripada, Odisha, India

Minati Mishra
P.G. Department of Computer Science, Fakir Mohan University, Balasore, Odisha, India

Ricky Mohanty
School of Information System, ASBM University, Bhubaneswar, India

Nihar Ranjan Nayak
Department of Computer Science and Engineering, Presidency University, Bangalore, India

Olugbemi T. Olaniyan
Department of Physiology, Laboratory for Reproductive Biology and Developmental Programming, Rhema University, Aba, Nigeria

Sagar Dhanraj Pande
Intelligent System, School of Computer Science and Engineering, Lovely Professional University, Phagwara, Punjab, India

Subhendu Kumar Pani
Krupajal Computer Academy, Bhubaneswar, India

Shobhandeb Paul
Department of Electronics and Communication Engineering, Guru Nanak Institute of Technology, Kolkata, West Bengal, India

Subhransu Pradhan
Directorate of Health Services, Bhubaneswar, Odisha, India

Syed Saba Raoof
Department of CSE, VIT University, Vellore, India

V. Selvakumar
Department of Maths and Statistics, Bhavan's Vivekananda College of Science, Humanities and Commerce, Hyderabad, Telangana, India

Neelam Sharma
Electrical and Instrumentation Engineering Department, Thapar Institute of Engineering and Technology, Patiala, Punjab, India

K. Rupabanta Singh
Department of Computer Application, Maharaja Sriram Chandra Bhanja Deo University, Baripada, Odisha, India

Brijesh K. Soni
Department of Computer Science and Technology, AKS University, Satna, Madhya Pradesh, India

Mukesh Soni
Department of Computer Science and Engineering, Jagran Lakecity University, Bhopal, India

Akhilesh A. Waoo
Department of Computer Science and Technology, AKS University, Satna, Madhya Pradesh, India

About the editors

Dr. Abraham is the director of Machine Intelligence Research Labs (MIR Labs), a Not-for-Profit Scientific Network for Innovation and Research Excellence connecting Industry and Academia. The Network with HQ in Seattle, USA has currently more than 1500 scientific members from over 105 countries. As an Investigator/Co-Investigator, he has won research grants worth over 100+ million US$. Since 2021, he holds two University Professorial appointments. He works as a Professor of Artificial Intelligence in Innopolis University, Russia and is the Yayasan Tun Ismail Mohamed Ali Professorial Chair in Artificial Intelligence of UCSI, Malaysia. Dr. Abraham works in a multidisciplinary environment and he has authored/co-authored more than 1400+ research publications, out of which there are 100+ books covering various aspects of Computer Science. One of his books was translated to Japanese and a few other articles were translated to Russian and Chinese. Dr. Abraham has more than 46,000+ academic citations (h-index of 102+ as per google scholar). He has given more than 150 plenary lectures and conference tutorials (in 20+ countries).

Dr. Abraham was the Chair of IEEE Systems Man and Cybernetics Society Technical Committee on Soft Computing (which has over 200+ members) during 2008–2021 and served as a Distinguished Lecturer of IEEE Computer Society representing Europe (2011–2013). Dr. Abraham was the editor-in-chief of Engineering Applications of Artificial Intelligence (EAAI) during 2016–2021 and serves/served on the editorial board of over 15 international journals indexed by Thomson ISI. Dr. Abraham received a Ph.D. degree in Computer Science from Monash University, Melbourne, Australia (2001) and a Master of Science Degree from Nanyang Technological University, Singapore (1998).

Sujata Dash is currently an associate professor in the Department of Computer Application, North Orissa University, India. She has been teaching and guiding students for more than three decades. She is a recipient of the prestigious Titular Fellowship from the Association of Commonwealth Universities, United Kingdom. She has worked as a visiting professor in the Computer Science Department, University of Manitoba, Canada. She has published more than 170 technical papers in international journals and proceedings of international conferences. She has also published several textbooks, monographs, edited books, and book chapters several book chapters. She has 11 patents to her credit. She is a member of international professional associations including Association for Computing Machinery (ACM), International Rough Set Society (IRSS), Computer Society of India (CSI), Indian Meteorological Society (IMS), Orissa Information & Technological Society (OITS), Orissa Mathematical Society (OMS), International Association of Computer Science and Information Technology (IACSIT), Indian Society for Technical Education (ISTE), Institution of Electrical & Electronics Engineers (IEEE). She is a reviewer and associate editor of 15 international journals and on the editorial boards of 10 others. She has received many national and international awards. She has visited many countries and delivered keynotes, invited speeches, and chaired many special sessions at conferences in India and abroad. Her current research interests include machine learning, data science, Big Data analytics, bioinformatics, and intelligent agents.

Subhendu Kumar Pani received his PhD from Utkal University, India, in 2013. He is currently a s research coordinator and professor in the Department of Computer Science, Orissa Engineering College (OEC), India. He has more than 15 years of teaching and research experience. His research interests include data mining, Big Data analysis, Web data analytics, fuzzy decision making, and computational intelligence. He is the recipient of 5 researcher awards. In addition to research, he has guided 2 PhD students and 31 MTech students. He has published 51 international journal papers. His professional activities include roles as associate editor, editorial board member, and/or reviewer of various international journals. He is a member of several conference societies. He has more than 100 international publications, 5 authored books, 15 edited and upcoming books, and 10 book chapters to his credit. He is a fellow in Scientific Society of Advanced Research and Social Change (SSARSC), Institute of Engineers (IE), Indian Society of Technical Education (ISTE), Indian Science Congress Association (ISCA), Odisha Bigyan Academy (OBA), Orissa Mathematical Society (OMS), Senior Member International Association of Computer Science and Information Technology (SMIACSIT), Senior Member Universal Association of Computer and Electronics Engineers (SMUACEE), Computer Society of India (CSI).

Laura García-Hernández received an MSc in Computer Science from the Universitat Oberta de Catalunya, Spain, in 2007, and a PhD in Engineering from the University of Córdoba, Spain, and the Institut Français de Mécanique Avancée, France, in 2011. She was an invited professor for a semester at the Institut Français de Mécanique Avancée. She is currently Associate Professor of Project Management at the University of Córdoba. Her primary areas of research are engineering design optimization, intelligent systems, machine learning, user adaptive systems, interactive evolutionary computation, project management, risk prevention in automatic systems, and educational technology. In these fields, she has authored or co-authored more than 70 international research publications. She has given several invited talks in different countries. She has realized several postdoctoral internships in different countries with a duration of more than two years. She received the prestigious National Government Research Grant "José Castillejo" for supporting their post-doc research during 6 months at the University of Algarve, Portugal. She has been an Principal Investigator in two Spanish research projects and has also been a Collaborator Investigator in some research contracts and projects. She is an expert member of the ISO/TC 184/SC working team and the National Standards Institute of Spain (UNE). Moreover, she is a member of the Spanish Association of

Engineering Projects (IPMA Spain). She received the Young Researcher Award from the Spanish Association of Engineering Projects (IPMA), Spain, in 2015. Additionally, she received the General Council of Official Colleges Award in 2017 and 2018. She is the co-editor-in-chief of the *Journal of Information Assurance and Security*. She is also an associate editor for *Applied Soft Computing, Complex & Intelligent Systems*, and *Journal of Intelligent Manufacturing*.

Preface

Overview

The novel applications of artificial intelligence (AI) and Big Data analytics for neurological disease research is an emerging field in computer science, medicine, biology application, and biomedical engineering. During the last 20 years, various AI techniques and data mining approaches have been successfully applied in different fields, particularly in medicine. Medicine is one of the most appropriate application domains for these techniques since they facilitate model diagnostic information based on causal and/or statistical data and therefore reveal hidden dependencies between symptoms and illnesses. Healthcare informatics and neurological analytics bring tremendous opportunities and challenges due to the availability of huge amounts of neurological data for analysis. Deep learning (DL) has rapidly developed into the de-facto approach, presenting tremendous improvement in accuracy, robustness, and cross-language generalizability over conventional approaches. The aim of healthcare informatics is to ensure high-quality and efficient health care, enhanced treatment, and improved quality of life via efficient analysis of abundant biomedical and healthcare data, including patient data, electronic health records (EHRs), and lifestyle models. While doctors and physicians are more than capable of detecting diseases of the brain, the most agile human mind cannot compete with the processing power of modern technology. Utilizing algorithmic systems in health care in this way may provide a way to treat neurological diseases before they occur. With the influence and the development of the Internet of Things (IoT), the need for AI techniques has become more significant than ever. The aim of these techniques is to detect neurological diseases at a faster rate than can be accomplished by normal means.

This book is written for researchers and practitioners in biomedical engineering, health informatics, bio-electronics, medical electronics, physiology, neurosciences, life sciences, and physics. It is a collection of chapters discussing the use of AI, DL, Big Data analytics, and machine learning (ML) in diagnosing and treating neurological diseases.

Objective

AI techniques for early detection of neurological disease can reveal useful information to specialists. As such, this book examines numerous ML techniques in biomedicine for detecting neurological diseases at the cellular level.

It discusses applications for image segmentation, classification, and image indexing using fuzzy approaches, neural networks, and a variety of computational intelligence and image processing approaches. It also discusses DL techniques for noise diminution and filtering brain MRI images. It outlines more challenging applications, using imaging techniques (EEG, MEG, fMRI, fNIRS, PET), including brain-computer interfacing, the prediction and detection of epilepsy, and the rehabilitation of patients with neuromuscular impairment.

We hope this book encourages an even wider adoption of these concepts and methods to assist physicians in problem-solving and stimulate research that may lead to additional innovations in this area.

Organization

Artificial Intelligence for Neurological Disorders consists of 22 edited chapters.

Chapter 1, "Early detection of neurological diseases using machine learning and deep learning techniques: A review," presents an overview of the application of hand-crafted ML algorithms and DL techniques for diagnosis of both Alzheimer's and Parkinson's. In addition, the chapter authors propose an edge-based healthcare system architecture for monitoring AD patients by using the accelerometer sensor of a smartphone.

Chapter 2, "A predictive method for emotional sentiment analysis by deep learning from EEG of brainwave data," recommends long short-term memory (LSTM) and multilayer perceptron neural networks (MLPNNs) to improve classification of EEG data obtained via a Muse EEG-powered headband with a global EEG position standard.

Chapter 3, "Machine learning and deep learning models for early-stage detection of Alzheimer disease and its proliferation in human brain," brings together medical and computational domains to discuss the use of DL and ML in the early detection of Alzheimer's disease.

Chapter 4, "Convolutional neural network model for identifying neurological visual disorder," proposes a convolutional neural network model called the Neurological Visual Disorder Identifying Model (NVDIM) for identifying neurological disorders within the human visual system.

Chapter 5, "Recurrent neural network model for identifying neurological auditory disorder," discusses neurological disorders related to the auditory cortex of the human brain and uses a recurrent neural network to recognize phonetic disorders or dysfunctions with the help of an artificial DL model.

Chapter 6, "Recurrent neural network model for identifying epilepsy-based neurological auditory disorder," develops a model for successful features and classifiers for epilepsy detection and identification. To enhance epilepsy identification, a multiview epilepsy detection approach is presented that blends deep and surface characteristics.

Chapter 7, "Dementia diagnosis with EEG using machine learning," summarizes all the work done to date for dementia diagnosis based on EEG and cognitive task data and compares various ML approaches used in this regard. It also summarizes different ML approaches with advanced EEG signal processing that can guide future researchers, practitioner, and technicians.

Chapter 8, "Computational methods for translational brain-behavior analysis," explains the requirement of physiological data acquisition and biomarkers to understand the interplay between sensory input and motor output in the central nervous system for successfully analyzing behavioral patterns. It focuses on several computational approaches in translational brain behavioral analysis.

Chapter 9, "Clinical applications of deep learning in neurology and its enhancements with future directions," highlights the importance of AI in the healthcare industry for managing medical data, thus improving healthcare's data-driven environment, optimization, clinical pathways, preclinical research, patient care, and population-level applications. It describes several algorithms and techniques used to assist clinicians in making prompt and smart decisions, particularly for neurological disease treatment, monitoring, prediction, and analysis.

Chapter 10, "Ensemble sparse intelligent mining techniques for cognitive disease," outlines advances in ML research algorithms and their applications for cognitive disease diagnosis, prognosis, and forecasting. It also addresses issues related to classification of patients with neurocognitive diseases using ML.

Chapter 11, "Cognitive therapy for brain diseases using deep learning models," addresses demographics, environmental data, imaging, clinical factors, and genetics linked to the understanding of complex algorithms in cognitive therapy for brain diseases utilizing DL models for precision medicine.

Chapter 12, "Cognitive therapy for brain diseases using artificial intelligence models," gives a detailed account of different AI models for risk assessment, classification, segmentation tasks, prognosis, diagnosis, and prediction of therapy reaction in cognitive therapy.

Chapter 13, "Clinical applications of deep learning in neurology and its enhancements with future predictions," discusses advances in the adoption of AI applications to enhance neurological techniques in clinical settings, such as medical imaging, translational bioinformatics, cheminformatics, pervasive sensing, public health, and medical informatics. It also provides a critical analysis of potential pitfalls of these techniques.

Chapter 14, "An intelligent diagnostic approach for epileptic seizure detection and classification using machine learning," focuses on the utilization of machine learning and different algorithms that provide smart, powerful, real-time, and low-cost approaches for the classification and detection of different pathophysiological characteristics in epileptic seizure such as unusual behavior, loss of awareness, confusion, and memory and attention deficits utilizing vital information derived from ictal and interictal signals from the brain electroencephalogram waves using discrete wavelet transform.

Chapter 15, "Neural signaling and communication using machine learning," discusses advances in neural signaling, signal decoding, and communication, and the application of different biometric data derived from the neural cells.

Chapter 16 "Classification of neurodegenerative disorders using machine learning techniques," reviews the classification of neurodegenerative diseases, such as Huntington's disease, Parkinson's disease, progressive brain dysfunction, vascular dementia, cognitive impairment, Alzheimer's disease, and amyotrophic lateral sclerosis, utilizing ML algorithms.

Chapter 17, "New trends in deep learning for neuroimaging analysis and disease prediction," provides an overview of different approaches in DL for the prediction of major brain diseases such as mild cognitive impairment, Alzheimer's disease, brain tumors, schizophrenia, Parkinson's disease, autism spectrum disorder, and many more. It also discusses current challenges in using DL for assessing brain disorders via neuroimaging data.

Chapter 18, "Prevention and diagnosis of neurodegenerative diseases using machine learning models," elaborates on the importance of ML techniques for diagnosing and preventing neurodegenerative diseases.

Chapter 19, "Artificial intelligence-based early detection of neurological disease using noninvasive method based on speech analysis," describes the latest noninvasive techniques such as emotion recognition/intelligence, virtual environments, and behavioral analysis for the diagnosis, screening, evaluation, and early detection of common neurological diseases.

Chapter 20, "An insight into applications of deep learning to neuroimaging," augments the existing literature on DL techniques applied to neuroimaging. It presents and describes various DL models and DL-based automatic diagnosis approaches for neurological disorders. It also discusses challenges in applying DL techniques to neuroimaging as well as future directions in this domain.

Chapter 21, "Incremental variance learning-based ensemble classification model for neurological disorders," proposes a novel incremental variance learning-based ensemble classification model that can be applied to a multitude of neurological diseases. The proposed model uses an ensemble of support vector machine (SVM), naïve Bayes (NB), random forest (RF), and logistic regression (LR) classifiers to create a novel incremental learning model. The authors test the model on a variety of datasets and present results.

Finally, Chapter 22, "Early detection of Parkinson disease using adaptive machine learning techniques: A review," provides an overview of the application of ML techniques and introduces some key concepts for diagnosing Parkinson's disease. The system proposed in this chapter uses the accelerometer sensor of a smartphone to generate a sequence of data while the patient is moving. This data is then used to identify the stage of disease.

Ajith Abraham
Sujata Dash
Subhendu Kumar Pani
Laura García-Hernández

Acknowledgment

The field of diagnostic imaging has undergone major changes in recent years. Radiological and nuclear medicine modalities are the options of choice to investigate neurological diseases. The improving capabilities of imaging devices and the increasing availability of storing, sharing, and computing facilities have led to the generation of larger and larger amounts of data. Consequently, there has been increasing attention on the development of computational methods for the extraction of biomarkers capable of correlating with disease phenotype, clinical outcome, and/or response to treatment. The combined use of imaging data, biomarkers, and artificial intelligence techniques makes it possible to build powerful predictive models that can assist the physician in the management of patients with a wide range of neurological disorders, ultimately leading to personalized treatment and better clinical outcomes. However, there are still open challenges to address before these methods can be translated into clinical practice.

We are grateful to all the contributors who trusted us and supported our work. I hope they will be as proud of this book as we are. I would also like to thank Elsevier and all the editorial members involved in the publication process. Without their support, publishing this book during the COVID-19 pandemic would have been near impossible.

We would like to thank our colleagues, family members, and friends who extended immense support and encouragement during the development of this book.

Ajith Abraham
Sujata Dash
Subhendu Kumar Pani
Laura García-Hernández

CHAPTER 1

Early detection of neurological diseases using machine learning and deep learning techniques: A review

K. Rupabanta Singh and Sujata Dash
Department of Computer Application, Maharaja Sriram Chandra Bhanja Deo University, Baripada, Odisha, India

Introduction

Neurological diseases are disorders that affect the human brain and the central nervous system. There are hundreds of neurological disorders that cause neurological diseases. These diseases negatively affect a person's lifestyle due to disturbing their sense of perception, judgment, reasoning, and memory as well as their physical abilities. Alzheimer's disease (AD) and Parkinson's disease (PD) are two common neurological diseases that affect millions of people older than 65 years. AD affects 5.2 million Americans older than 65 years and thousands of people younger than 65 years (Johns Hopkins Medicine, 2022). AD is a common form of dementia that is usually asymptomatic in the early stages and because of this early diagnosis is very difficult. Thus, many researchers are studying AD in an attempt to develop successful methods of early diagnosis. Various methods have been proposed, including approaches based on machine learning (ML) and deep learning (DL) algorithms. PD is a result of lack of dopamine in the brain. It differs from AD in that patients with PD often have a better quality of life than patients with AD. For every 1000 people aged 60 years or older, 1–2 are affected by PD, and PD prevalence increases with age, affecting 1% of the population older than 60 years (Tysnes & Storstein, 2017). Traditionally the diagnosis of PD is based on clinical assessment of motor symptoms and other cardinal signs. As the applications of ML expand in the healthcare sector, several methods have been proposed to identify biomarkers for PD. Different ML algorithms can detect many relevant features of PD that aren't used in traditional clinical diagnosis and thus diagnose the disease in preclinical stages.

Artificial Intelligence for Neurological Disorders
https://doi.org/10.1016/B978-0-323-90277-9.00001-8

In this chapter, we review different methods for the early detection of AD and PD. Over the last few years, many researchers have contributed to the identification of potential AD biomarkers and prediction of progression of the disease using MRI data independently. Several studies preferred ML algorithms for feature extraction, which support the investigator in diagnosis and decision-making. Mild cognitive impairment (MCI) is characterized by changes in a person's perception, intelligence, reasoning, and so on. It may affect activities of daily life and progress to AD. ML algorithms have been applied to predict the changes that occur in the transition from MCI to AD. In one study, researchers proposed a method that uses supervised ML techniques to identify which subjects with MCI will progress to AD (Grassi et al., 2018). Today, neuroimaging has become very advanced and is being used to detect damaged areas of the human brain. Deep residual neural networks have been used to analyze brain MRIs and to predict the conversion process from MCI to AD. Another study implemented a modified form of deep residual neural networks to study the neuroanatomical alteration in structural MRI data (Abrol et al., 2020).

The classical way to diagnose PD is based on the evaluation of clinical images of motor and nonmotor syndromes. Deep neural networks (DNNs) may be used to predict the severity of PD as either severe or nonsevere in the output layer (Grover, Bhartia, Akshama, & Yadav, 2019). Some in vivo studies have used automatic behavioral monitoring systems to observe behavior changes in mice. The DNN algorithm has been used for classifying features of PD (Sutoko et al., 2021). A metaheuristic approach is also used for the early diagnosis of neurological diseases like PD. Kakulapati and TejaSanthosh (2020) studied PD and proposed a hybrid metaheuristic approach for a Random Multimodal Deep Learning (RMDL) method in the analysis of Digital Imaging and Communications in Medicine (DICOM) data. The features were classified by the RMDL method based on a subject with PD.

Many studies have used different methods for detecting neurological disorders. We discuss some of these methods in the sections that follow.

Support vector machine

The support vector machine (SVM) is a common supervised ML technique used for regression and classification tasks. The idea of SVM is to find a hyperplane in the *n*th number of features that distinctly classifies the data points. Many possible hyperplanes are available, but only one hyperplane is chosen to classify two classes of the data points. The orientation and the position of the hyperplane is influenced by some data points that are closer to the hyperplane. These data points are called the support vectors of the hyperplane. If we remove the support vectors, the orientation and the position of the hyperplane may change. Fig. 1 shows the possible hyperplanes, the selected hyperplane, and the support vectors. The output of the linear function has been taken in the linear regression and flattened the value within the range of 0–1. During flattening of the output value, a threshold value of 0.5 is used. If the flattened value is greater than the threshold value, it is assigned "label 1," otherwise it is assigned "label 2." These two labels are used to identify the two different classes.

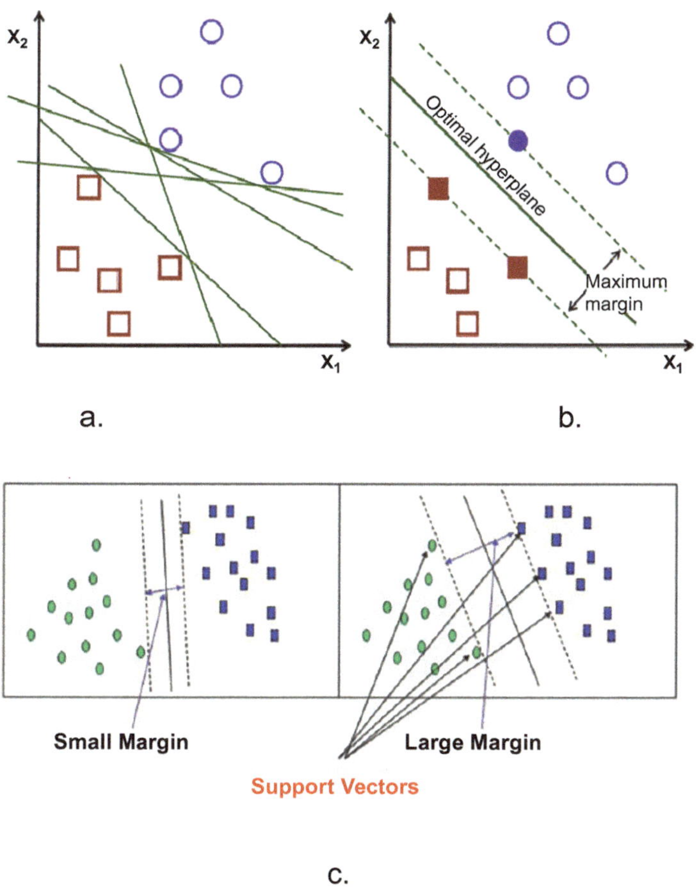

Fig. 1 (A) Possible hyperplane; (B) optimal hyperplane; (C) support vectors.

The margin between the data points and the hyperplane is maximized in the SVM algorithm with the help of a loss function (Eq. 1).

$$c(x, y, f(x)) = \begin{cases} 0, \text{ if } y \times f(x) \geq 1 \\ 1 - y \times f(x), \text{ else} \end{cases} \tag{1}$$

SVM with the feature selected by recursive feature elimination (RFE) has been used for the classification for identifying samples belong to AD and normal. This method can achieve a cross-validated accuracy of 0.961 (Grassi et al., 2018). SVM with optimizing gamma and C-constant parameters has been used to predict AD. It helps to screen subjects at high risk of preclinical disease. Different ML methods for the detection of AD were compared in a study by Kishore, Kumari, Kumar, and Pavani (2020), and the results showed that the SVM using a linear kernel achieved the greatest accuracy of 0.95.

Random forest

Random forest is one of the most popular ML classification algorithms. It is considered a supervised learning technique. In ML, random forest can be used for both classification and regression tasks. It includes many decision trees on different subsets of a dataset and takes the average output of the decision trees to enhance the accuracy of the model on the given dataset. To solve a problem, it combines the multiple classifiers, thus enhancing the performance of the model. The final output of the prediction will be based on majority votes of prediction from each tree. The level of accuracy of the prediction is fully dependent on the number of trees used. Also, it reduces the overfitting issue of the problem. Fig. 2 illustrates the process of random forest.

Classification via the random forest algorithm can be divided into five steps.

Step 1: select the *K* numbers of random data points from the given training dataset.

Step 2: make the decision trees of the corresponding selected *K* data points.

Step 3: choose the number *N* for the desired decision tree.

Step 4: repeat steps 1 and 2.

Step 5: check the majority votes and assign the category to the new data point.

The random forest method has been used in much research for the diagnosis and prediction of both AD and PD.

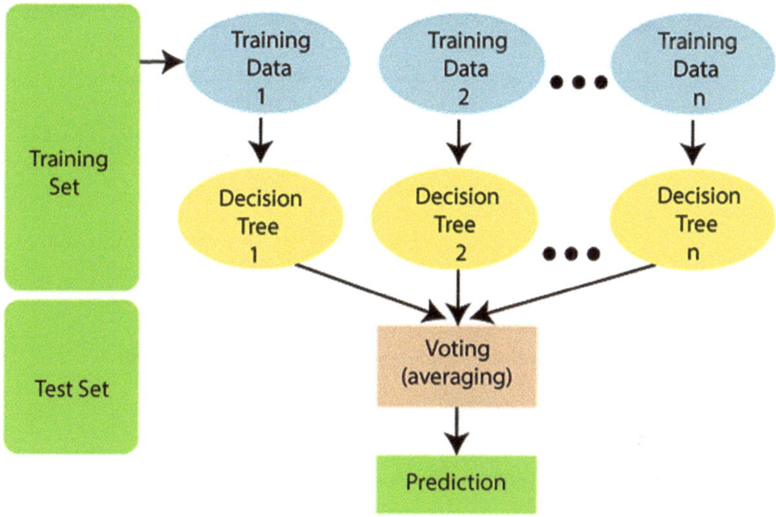

Fig. 2 Prediction process of random forest.

Logistic regression

Logistic regression is a supervised machine learning technique used to solve classification problems. It is one of the most common ML algorithms used for this purpose. The logistic regression predicts the categorical output (i.e., 1 or 0, Yes or No, True or False). The "S"-shaped logistic function called the sigmoid function is used instead of a regression line and this function can predict either 1 or 0 depending on the threshold value. The sigmoid function helps in mapping the predicted values to either 0 or 1 during classification. Fig. 3 shows the logistic function.

The logistic regression values must lie between 0 and 1; they cannot be beyond this range. Thus, the regression values form an S-shaped curve with a threshold value 0 or 1 for taking the decision.

If the regression value is greater than the threshold value, it is considered as 1; otherwise, it is considered as 0. The logistic regression equation is derived from the linear regression equation.

$$y = b_0 + b_1x_1 + b_2x_2 + b_3x_3 + \ldots + b_nx_n \tag{2}$$

The logistic regression value y is between 0 and 1.

Convolutional neural network

A convolutional neural network (CNN) is a supervised type of DL algorithm. It is preferred for image recognition and computer vision. A CNN

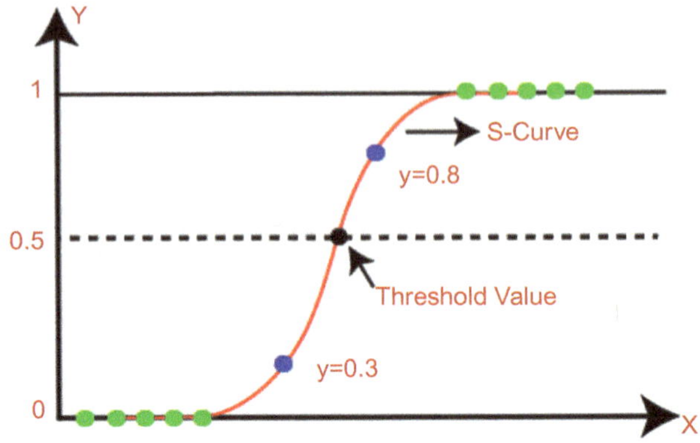

Fig. 3 Logistic function.

has multiple layers, and these layers work together to extract important features from a dataset. There are four types of layers in CNN architecture:

1. **Input layer**: In this layer, the data set is inputted for processing.
2. **Convolution layer**: This layer performs the convolution operation by using a filter. During the convolution process, the pixel values are multiplied with filter values and added up, then the filter will slide one pixel to the right and repeating this process.
3. **Pooling layer**: In this layer, the downsampling operation is performed to reduce the dimensions and computation; it also reduces the overfitting.
4. **Fully connected layer**: The different outputs of the pulling layers are inputted to the fully connected layer, which works as a feed-forward neural network. This layer can classify and recognize the image that has been inputted into the input layer.

Fig. 4 shows the architecture of a CNN.

The rest of this chapter is organized as follows: "Literature review" section presents a literature review, "Methodology and result analysis" section analyzes the different methods of ML and the DL techniques for diagnosis neurological disorders, "Proposed method" section defines our proposed method, and "Conclusion" section concludes with the feasibility of the proposed method for the prediction of AD.

Literature review

Several methods have been proposed to predict neurological disorders using ML and DL algorithms.

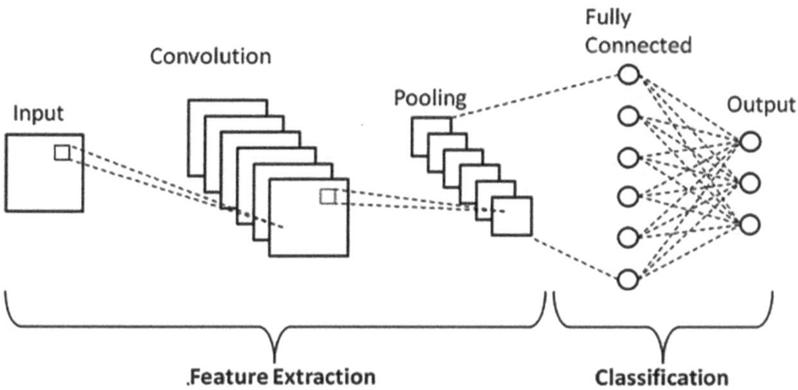

Fig. 4 CNN architecture.

Machine learning algorithms

In a recent study, researchers proposed a method using a supervised ML technique to predict conversion to AD in MCI and pre-MCI subjects (Grassi et al., 2018). The elastic net model with three-degree polynomial of the features was used to train the model on 90 subjects with MCI and 94 subjects with pre-MCI. Subjects with a Clinical Dementia Rating (CDR) score of 0.5 and memory complaints were considered to have MCI. Subjects with a CDR score of 0.5 with normal neuropsychological testing were considered to have pre-MCI. Two types of penalties, L1 norm (Eq. 3) and L2 norm (Eq. 4), were used during the training process of the elastic net model.

$$L1 = \lambda_1 \sum_{j=1}^{F} |\beta_j| \tag{3}$$

$$L2 = \lambda_2 \sum_{j=1}^{F} \beta_j^2 \tag{4}$$

where j is the features, β_j is the regression coefficient of the jth feature, and $\lambda 1$ and $\lambda 2$ are the amounts of penalization. For the classification of conversion to AD in MCI and normal subjects, the researchers used SVM with features selected by RFE. Random forest with RFE for recursive elimination was also used to compare performance with the SVM. During the training of the model, 36 features were included, and the model performed 23 h. of nonstop computation. The proposed method achieved a global cross-validated area under the curve (AUC) greater than 0.96. It achieved

AUC greater than 0.91 in the MCI subgroup. In another study, the echo state network (ESN) was implemented in place of traditional artificial neural network (ANN) and recurrent neural network (RNN) techniques to diagnose PD (Lacya, Smitha, & Lones, 2019). The proposed method determined the optimal configuration of ESN parameters for classification problems, and establishes whether it can directly modelling the separation waveform. During data collection, finger-taping protocols were used along with a pair of noninvasive electromagnetic sensors to record finger-tapping cycles. These sensors can record at a frequency of 60 Hz with a resolution of 1.52 mm. Training the model with ESN is quick and it allows the classifier to be more flexible to adapt to a new data source. In another study, researchers proposed an enhanced fuzzy min-max neural network (EFMM) with the OneR attribute evaluator (EFMM-OneR) as a hybrid model for diagnosing PD (Sayaydehaand & Mohammad, 2019). Performance evaluation was performed with 10-fold cross validation.

Gaubert, Ansart, Naccache, and Sitt (2021) proposed a prediction method using an ML algorithm with combined EEG, APOE4, neuropsychological, and MRI data. The model used EEG as the main information source. This study included 318 cognitively normal individuals in the age range of 70–85 years. The researchers used random forest, logistic regression, and linear SVM ML techniques and fivefold cross validation with 5 years of follow-up during performance evaluation. The logistic regression with the combination of EEG, APOE4, neuropsychological, and MRI data achieved 0.70 AUC, 83% specificity, 50% sensitivity, 97% negative predictive value (NPV), and 14% positive predictive value (PPV). In one recent study, the authors proposed a model for the detection of AD that improved efficiency and accuracy by using different ML algorithms (Kishore et al., 2020). The accuracy of the model increased because of the SVM algorithm with a linear kernel model. The linear kernel model helped to improve the accuracy of the earlier research model. Fivefold cross validation was performed. Individual models of each ML algorithm (logistic regression, decision tree, random forest, and Naive Bayes) and three different SVMs were developed and compared. The model using SVM with linear kernel had the greatest prediction accuracy of 0.95.

In another recent study, researchers proposed a method using an ML model to identify the implicit gene biomarkers of AD (Sharma & Dey, 2020). The expression datasets of two genes were merged for each brain region to increase sample size. Random forest for variable selection and the Least Absolute Shrinkage and Selection Operator (LASSO) regression method were used to select the important gene candidates. These two methods are the more efficient feature selection methods. LASSO is a regularization regression method that can fit a generalized linear model able to select an optimal feature. The three classification models of SVM, random forest, and elastic net classifier were developed and compared. Results showed that the elastic net classification model had the greatest prediction accuracy. Fivefold cross validation was performed. Yet another study proposed a method for early diagnosis of AD using different ML techniques, including SVM, import vector machine (IVM), and regularized extreme learning machine (RELM) (Sudharsan & Thailambal, 2020). Principal component analysis (PCA)-based feature selection was used to decrease the dimensionality of high-dimensional imaging. The SVM with kernel techniques separated the two groups during the training period with optimum margin from supports. Tenfold cross validation and LOOCV were used. In this study, the results showed that the RELM technique performed better than the SVM and IVM techniques.

Varrecchia et al. (2021) proposed an ML-based diagnostic algorithm to classify gait deficits in PD. The researchers used PCA analysis for feature selection and the selected features were used to diagnose PD using an ANN. The proposed model was trained on feed-forward neural networks. Various numbers of hidden layers with different numbers of neurons were used. In a hidden layer, the number of neurons may be either 20 or 50 and the number of nodes may be one-half or one-third of the neurons. Results showed that the proposed method can diagnose gait deficits in several symptomatic stages of PD. In another study, a predefined SVN was validated for the diagnosis of PD (Maass et al., 2019). During SVM model validation, 82 PD patients, 68 age-matched controls, and 7 additional normal pressure hydrocephalus (NPH) patients were included. Tenfold cross validation was used and achieved AUROC of 88. Another study used the ML methods of SVM, ANN, and classification and regression tree (CART) for early diagnosis of PD (Senturk, 2020). The dataset included 48 healthy phonetics and 147 PD phonetics. The CART decision tree produces either a regression tree or classification tree depending on the output variable. The proposed method achieved accuracy of 93.84%. In another study, researchers used a fuzzy classification algorithm for the diagnosis of PD (Dastjerd, Sert, Ozyer, & Alhajj, 2019). The proposed prediction model used a Parkinson's speech dataset from the UC Irvine machine learning repository (UCI) repository. Tenfold cross validation was performed and the proposed model achieved 100% accuracy. Celik and Omurca (2019) proposed a model for the prediction of PD using random forest, SVM, gradient boosting, and extra trees. A comparison of the different methods showed that SVM achieved the greatest accuracy of 75%.

UlHaq et al. (2019) used an L1-SVM for diagnosing PD because it can select highly related and suitable features for classifying PD patients and healthy controls. The researchers tested the model on 194 instances of a dataset from the repository of the University of Oxford. They performed 10-fold cross validation for selecting the optimal value. The L1-SVM gave the output with 99% accuracy. Another study used mel-frequency cepstral coefficients (MFCC), discrete wavelet transform (DWT), and a hybrid model of MFCC and DWT for feature extraction (Kuresan et al., 2017). The researchers applied a hidden Markov model (HMM) SVM for the classification of PD patients and healthy individuals. They used a dataset from the UCI machine learning repository and 20 voice samples from 40 people, including 20 pathological and 20 nonpathological to train the model. Results showed that the HMM classifier achieved 0.951 accuracy.

Other researchers used a metaheuristic optimization algorithm called the chaotic firefly algorithm to develop a model for diagnosing PD (Dash, Abraham, & Atta-Ur-Rahman, 2019; Dash, Thulasiram, & Thulasiraman, 2017). They used a voice measurement dataset from the UCI repository and stratified 10-fold cross validation for model validation. The model achieved accuracy of 89.326% (Table 1).

Deep learning algorithms

DL algorithms are some of the best algorithms for predicting neurological disease. In a recent study, researchers proposed a model using DL architecture based on magnetoencephalography (MEG) activity for the early detection of MCI (Lopez-Martin, Nevado, & Carro, 2020). 1D-CNN (5 filters, kernel size of 5 and stride of 5), 2D-CNN (2/3/4 filters, kernel size of 2/3, and stride of 1/2), and randomized 2D-CNN were used to make the prediction model. The proposed model performed the binary classification of Alzheimer's patients and healthy people. Two stages of nested cross validation were implemented. The 1D-CNN model had 140 trainable parameters and the 2D CNN model had 236 trainable parameters. In the experiment, randomized 2D-CNN gave the best result with 0.91 accuracy. Abrol et al. (2020) proposed a method to predict the progression from MCI to AD using deep residual neural networks. Only MCI individuals were used for training the proposed model. The diagnostic performance for the cognitively normal

Table 1 Different machine learning techniques and their outcomes.

Study	Neurological disease	Dataset	Machine learning technique	Cross validation	Accuracy
Grassi et al. (2018)	Alzheimer's disease	MCI = 90 PreMCI = 94	Elastic net, SVM, and random forest	Leave-pair-out-cross validation	AUC = 0.914 ACC = 0.874
Lacya et al. (2019)	Parkinson's disease	49 PD patients and 41 control subjects	Echo State Network (ESN)	–	AUC = 0.84
Sayaydehaand and Mohammad (2019)	Parkinson's disease	UCI machine learning repository	EFMM-OneR	10-fold cross validation	ACC = 0.94
Gaubert et al. (2021)	Alzheimer's disease	318 cognitively normal individuals	Random forest, logistic regression, and SVM	–	AUC = 0.68 ACC = 0.65
Kishore et al. (2020)	Alzheimer's disease	MRI data of 150 people	Logistic regression, decision tree, random forest, Naïve Bayes, and three different SVMs	Fivefold cross validation	ACC = 0.95
Sharma and Dey (2020)	Alzheimer's disease	NCBI-GEO database	SVM, random forest, elastic net classifiers, and regularized regression	fivefold cross validation	ACC = 0.99 (Average)
Sudharsan and Thailambal (2020)	Alzheimer's disease	ADNI repositories	SVM, IVM, and RELM	10-fold cross validation	ACC = 0.627
Varrecchia et al. (2021)	Parkinson's disease	PD = 76 HC = 67	ANN	–	AUC = 0.77

Continued

Table 1 Different machine learning techniques and their outcomes—cont'd

Study	Neurological disease	Dataset	Machine learning technique	Cross validation	Accuracy
Maass et al. (2019)	Parkinson's disease	82 PD patients, 68 age-matched controls and 7 additional normal pressure Hydrocephalus (NPH) patients	SVM	10-fold cross validation	AUROC = 0.88
Senturk (2020)	Parkinson's disease	UCI machine learning repository/48 healthy phonetics and 147 PD phonetics	SVM, ANN, and CART	–	ACC = 93.84%
Dastjerd et al. (2019)	Parkinson's disease	Parkinson speech dataset from UCI repository	Fuzzy classifier	10-fold cross validation	ACC = 1.00
Celik and Omurca (2019)	Parkinson's disease	1208 speech dataset from UCI repository	SVM, extra trees, gradient boosting, and random forest	–	ACC = 0.75
UlHaq et al. (2019)	Parkinson's disease	194 voice recording dataset from the repository of the University of Oxford	SVM-RBF and SVM-linear	10-fold cross validation	ACC = 0.99
Kuresan et al. (2017)	Parkinson's disease	20 voice samples from 40 peoples	HMM and SVM	–	ACC = 0.951
Dash, Abraham, and Atta-Ur-Rahman (2019)	Parkinson's disease	Voice dataset from UCI repository	Metaheuristic optimization algorithm	Stratified 10-fold cross validation	ACC = 89.326%

(CN) vs AD classification task was tested via ResNet architecture. Fivefold cross validation with 10 repetitions and 100 epochs was used for the validation task. In this experiment, the proposed model achieved accuracy of 0.91. In another study, the authors proposed a prediction model for the detection of PD severity (Jo, Nho, & Saykin, 2019). This study used several voice datasets of PD patients available from the UCI ML repository. A DNN was used to classify the two outputs: severe and nonsevere. The proposed method achieved an accuracy of 0.94 and 0.62 for training and testing, respectively. Another study developed a model to classify multiple diseases (Choi, Kim, Yoon, Lee, & Lee, 2019). The proposed model can predict AD and PD using a CNN with ReLU activation function. The model was trained with 636 PET images, including those of 243 AD patients and those of 393 normal people. It included 62 PD patient images to differentiate PD patients from those with dementia. The researchers selected 64 images for use during model validation. The proposed model achieved an accuracy of 0.94. A model with two CNNs was proposed for the prediction of AD (Zhang et al., 2019). The proposed model was trained by two independent CNNs and the two outputs were judged via correlation analysis. The result of the model were combined with the results of clinical neuropsychological diagnosis. The model was trained with a dataset from the Alzheimer's Disease Neuroimaging Initiative (ADNI), which included 186 AD subjects, 226 control normal (CN) subjects, and 395 MCI subjects. The proposed model was able to classify CN, MCI, and AD. In the convolution layer, the 1th convolution layers were calculated by the following equation:

$$x_j^1 = f\left(\sum_{i\epsilon M_j} x_i^{l-1} \times k_{ij}^l + b_j^l\right) \tag{5}$$

where x_i^{l-1} is the input feature map of the lth layer and the same is considered as the output of the previous layer. M_j represents the set of input feature maps, k_{ij}^l is the partial input feature map, and b_j^l is the bias offset of the jth feature map. The output feature of the lth pulling layer is calculated by the following equation:

$$x_j^l = f\left[\beta_j^l \cdot \text{down}\left(x_j^{l-1}\right) + b_j^l\right] \tag{6}$$

where β_j^l is the weight coefficients and b_j^l is the bias offset parameters. The output of the fully connected layer is collected by:

$$r^{l-1} = \text{Bernoulli}(p) \tag{7}$$

$$x^{\sim l-1} = r^{l-1} \times x^{l-1} \tag{8}$$

$$x^{\sim l} = f\left(w^l x^{l-1} + b^1\right) \tag{9}$$

where the Bernoulli function is the random generator function that generates a vector r^{l-1}, w^l is the weighting, and b^l is the offset parameter. The SoftMax function is used to calculate the result of the output layer.

$$x_i = \frac{e^{z_i}}{\sum_{j=1}^{k} e^{z_j}} \tag{10}$$

$$y_i = \frac{e^{z_i}}{\sum_{j=1}^{k} e^{z_j}} \tag{11}$$

where k represents the total number of categories and z_i is the ith dimension component in the kth dimension vector. X_i represents the probability of the ith class over k classes and y_i represents the probability of the ith class over k classes.

Some of the studies of the recent past illustrated in Table 2 is briefly explained here. In a recent study, a model based on CNN was proposed to identify the stage of AD (Bringas, Salomón, Duque, Lage, & Montaña, 2020). The proposed model was trained on mobility data obtained from the accelerometer sensors of 36 patients. The model was able to recognize patterns in patient movements and achieved an accuracy of 91%. The Iterative Sparse and Deep Learning (ISDL) model with a sparse regression module has been proposed for the diagnosis of AD and the prediction of conversion from MCI to AD (Chen & Xia, 2020). The proposed method joined deep feature extraction and critical cortical region identification to diagnose AD and MCI. The sparse regression model was designed to identify the critical cortical region and was merged with the deep feature extraction module. Fivefold cross validation was used for ISDL and the MCI conversion prediction. The proposed method achieved accuracy of 95.32%. To identify the stage of PD severity is challenging. In a recent study, a model using CNN was developed for predicting PD stage (Mozhdehfarahbakhsh et al., 2021). MRI data from a Parkinson's Progression Markers Initiative (PPMI) data repository was used to train the model. The proposed model was able to distinguish between healthy subjects and those with PD as well as identify PD stage. Tenfold cross validation was used to estimate the accuracy of the proposed model and results show the model achieved accuracy of 94%.

Table 2 Different deep learning techniques and their outcomes.

Study	Neurological disease	Dataset	Deep learning technique	Cross validation	Accuracy
Lopez-Martin et al. (2020)	Alzheimer's disease	78 MCI patients and 54 healthy aging participants	1D-CNN, 2D-CNN, and randomized 2D-CNN	Two-stage nested cross validation	ACC = 0.91
Abrol et al. (2020)	Alzheimer's disease	Alzheimer's disease neuroimaging initiative (ADNI) database	Modified ResNet	10-repeat, fivefold (stratified) cross validation	ACC = 0.91
Jo et al. (2019)	Parkinson's disease	UCI machine learning repository	DNN	–	ACC = 0.94
Choi et al. (2019)	Alzheimer's disease	PET images (243 AD, 62 PD, and 393 NC) from ADNI	CNN	–	AUC = 0.94
Zhang et al. (2019)	Alzheimer's disease	Dataset from ADNI database (186 AD subjects, 226 CN subjects, and 395 MCI subjects)	CNN	–	ACC = 0.9847 AUC = 0.9861
Bringas et al. (2020)	Alzheimer's disease	Data from 35 patients	CNN	–	ACC = 91% F1-score = 0.897

Continued

Table 2 Different deep learning techniques and their outcomes—cont'd

Study	Neurological disease	Dataset	Deep learning technique	Cross validation	Accuracy
Chen and Xia (2020)	Alzheimer's disease	T1-weighted sMRI data from ADNI	ISDL	Fivefold cross validation	ACC=77.60% SEN=71.62% SPE=75.85%
Grover et al. (2019)	Parkinson's disease	UCI's Parkinson's telemonitoring voice data set of patients	DNN	–	ACC=94.4422%
Mozhdehfarahbakhsh et al. (2021)	Parkinson's disease	MRI data from PMMI data repository	CNN	10-fold cross validation	ACC=0.94

Methodology and result analysis

Various ML and DL techniques have been used in diagnostic and early detection models of neurological disorders. The SVM with radial basis function (RBF) as kernel is a common ML technique used for the classification of neurological diseases illustrated in Table 3. It was used in one study to predict the conversion to AD in MCI and pre-MCI subjects (Grassi et al., 2018). RFE was used for feature selection. Scholkopf and Smola (2018) explain SVM and kernel models in detail. During cross validation, the proposed model achieved accuracy of 0.96. In the study by Gaubert et al. (2021), the SVM method was used with optimizing gamma and C-constant parameters. During performance evaluation, fivefold cross validation was used with 5-year follow-up. The accuracy of the detection model for AD is increased by using SVM with linear kernel (Kishore et al., 2020). The prediction model achieved accuracy of 0.95. An SVM method with varSelRF and LASSO performed well for predicting AD (Sharma & Dey, 2020), achieving an average accuracy of 0.99. In another study, SVM with leave-out cross validation gave good results for the prediction of AD (Sudharsan & Thailambal, 2020). A predefined SVM method has been validated for the diagnosis of PD, achieving accuracy of 0.88 (Maass et al., 2019). The SVM with RFE method for the diagnosis of PD achieved accuracy of 93.84% (Senturk, 2020). SVM with expanded features for the diagnosis of PD achieved accuracy of 75.49% (Celik & Omurca, 2019). The SVM method with 10 features for the diagnosis of PD achieved accuracy of 0.99 (UlHaq et al., 2019).

Most of the authors preferred the CNN model for the prediction of neurological diseases illustrated in Table 4. The randomized 2D-CNN gave a good result for the detection of AD in the early stage (Lopez-Martin et al., 2020), achieving accuracy of 0.91. The CNN model can also classify multiple diseases. A CNN model has been used to classify both AD and PD (Choi et al., 2019). To enhance the accuracy of AD diagnosis, two CNNs were used, both of which achieved accuracy of 98.47% (Zhang et al., 2019). A CNN model was deployed to detect the stage of AD (Bringas et al., 2020). This model can improve the classification result from the conventional feature base classifiers. It achieved accuracy of 90.91%. A CNN model was developed with five convolutional layers, a flatten layer, two fully connected layers, and the output layer (Mozhdehfarahbakhsh et al., 2021). It achieved accuracy of 94%.

Table 3 Different studies using SVM method.

Study	Neurological disease	Dataset	Year	Accuracy
Grassi et al. (2018)	Alzheimer's disease	MCI = 90 Pre-MCI = 94	2018	AUC = 0.914 ACC = 0.874
Gaubert et al. (2021)	Alzheimer's disease	EEG, APOE4, neuropsychological, and MRI data	2021	ACC = 0.65
Kishore et al. (2020)	Alzheimer's disease	MRI data of 150 people	2020	ACC = 0.95
Sharma and Dey (2020)	Alzheimer's disease	NCBI-GEO database	2020	ACC = 0.99 (average)
Sudharsan and Thailambal (2020)	Alzheimer's disease	ADNI repositories	2021	ACC = 79.03%
Maass et al. (2019)	Parkinson's disease	82 PD patients, 68 age-matched controls, and 7 additional NPH patients	2019	AUROC = 0.88
Senturk (2020)	Parkinson's disease	48 healthy phonetics and 147 PD phonetics	2020	ACC = 93.84%
Celik and Omurca (2019)	Parkinson's disease	1208 speech dataset from UCI repository	2019	ACC = 74.49%
UlHaq et al. (2019)	Parkinson's disease	Dataset from the repository of the University of Oxford	2019	ACC = 0.99

Table 4 Different studies using CNN method.

Study	Neurological disease	Dataset	Year	Accuracy
Lopez-Martin et al. (2020)	Alzheimer's disease	MCI = 78 Normal = 94	2020	AUC = 0.91
Choi et al. (2019)	Alzheimer's disease, Parkinson's disease	ADNI dataset	2019	ACC = 0.94
Zhang et al. (2019)	Alzheimer's disease	ADNI dataset	2019	ACC = 98.47%
Bringas et al. (2020)	Alzheimer's disease	Data from 35 patients	2020	ACC = 90.91%

Several ML techniques have been used for the prediction of neurological diseases. Most researchers prefer an SVM classifier because it outperforms other methods. There are many supporting techniques that have been used with SVM in prediction models to enhance model performance. These techniques include REF, kernel, varSelfRF, and LASSO. The prediction model for the detection of AD using SVM with varSelRF and LASSO achieved a good result (Sharma & Dey, 2020). The L1 Norm SVM was used to diagnose PD with an accuracy of 0.99.

DL-based prediction models for detecting neurological diseases also perform well. Many authors have used CNN techniques to develop prediction models for both AD and PD. The CNN technique is commonly used and achieves the best results. Two CNNs can be used to predict AD to enhance the performance of a single CNN model.

Proposed method

The smartphone has become an essential tool in daily life and plays a vital role in communication and computing architecture. Many researchers are applying mobile computing in the healthcare sector (Rahman et al., 2019; Rehman et al., 2018). A wide variety of wearable Internet of Things (IoT) devices are currently available in the market. These wearable computing devices make up a wireless network called a Body Sensor Network (BSN). The BSN can connect with smartphones or personal computers and can share data obtained by sensors in wearable devices. Smartphones have many sensors and these sensors can be used to collect a person's health information (Rupabanta Singh, Dash, Biswas, & Deka, 2021). Because a smartphone has enough processing capability, it can act as an edge node. It can perform some operation on the sensors' data and then transmit this data to the cloud for remote monitoring and processing. Nowadays, DL is commonly used for medical image processing (Dash, Biswas, Banerjee, & Rahman, 2019; Dash et al., 2020) and time series data analysis to diagnose disease. Very recently, edge computing has been applied in the IoT-based healthcare system to reduce system latency and allow performance of real-time operations. Edge computing extends healthcare services to users anywhere at any time. The main advantages of using edge computing in the healthcare system include reduced latency, real-time processing, and reduced patients' mobility. Fig. 5 shows the three-tier architecture of an edge computing-based healthcare system.

The system proposed in this chapter is designed to monitor AD patients by using the accelerometer sensor of a smartphone. The accelerometer sensor can generate a sequence of data depending on the patient's movements. These data will be used to detect the stage of AD. The Global Deterioration Scale (GDS) provides an overview of the stages of cognitive function for those suffering from primary degenerative dementia such as AD.

Stage 1: No cognitive decline.
Stage 2: Very mild cognitive decline.
Stage 3: Mild cognitive decline.
Stage 4: Moderate cognitive decline.

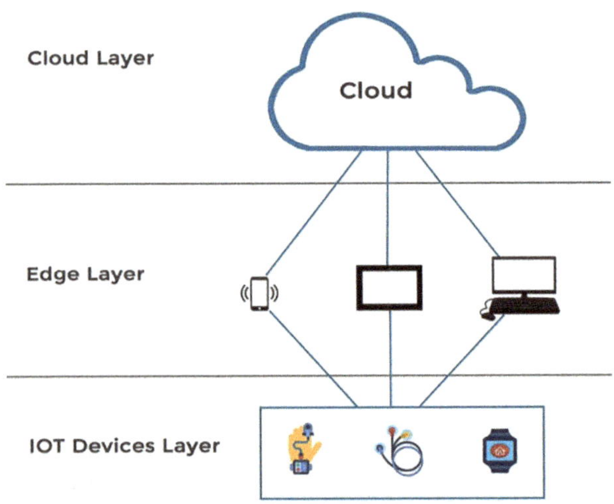

Fig. 5 Edge computing-based healthcare system architecture.

Stage 5: Moderately severe cognitive decline.
Stage 6: Severe cognitive decline.
Stage 7: Very severe cognitive decline.

AD stages can be categorized into three levels: early (stage 2 and stage 3), middle (stage 4 and stage 5), and late (stage 6 and stage 7) (Bringas et al., 2020).

Our proposed model has four phases:

1. **Data preprocessing**: Transform the data collected by the accelerometer sensor of the smartphone into same-length sequences of time series data.
2. **Build the prediction model**: Build a prediction model using the CNN method for the detection of AD stage.
3. **Develop the application**: Build an Android application that can implement the prediction model.
4. **Data uploading**: Upload the accelerometer sensor's data to cloud storage, so that it can be used for further analysis.

Fig. 6 shows our model's proposed system architecture. The model can monitor the AD patient continuously and identify whether the patient is improving or not. The edge node can check the status of the patient continuously. If the stage of the disease changes, then the edge node will send an

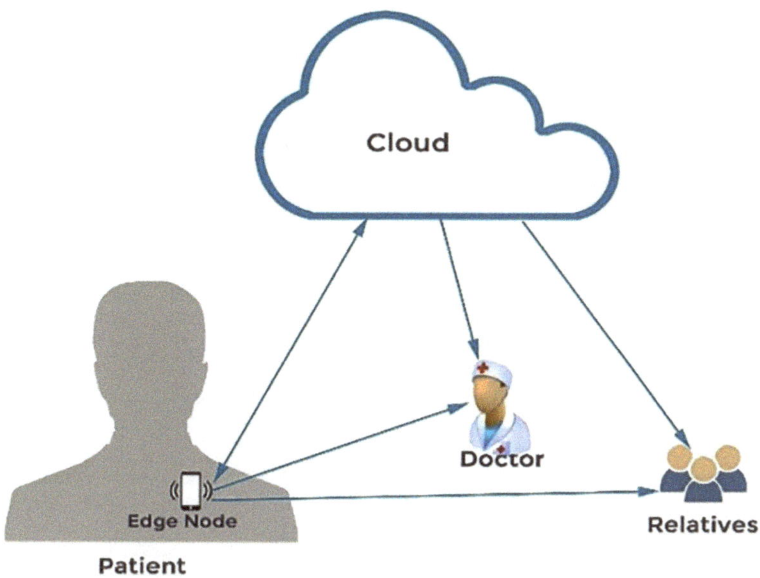

Fig. 6 Proposed system architecture.

alarm message. The proposed system will be helpful for patients as well as their healthcare providers and caregivers.

Conclusion

There are many ML and DL methods used for the early detection of neurological diseases like AD and PD. The SVM classifier is the most widely used to develop an ML-based prediction model for both diseases. As DL algorithms become more advanced, the CNN method is also widely used to develop prediction models for early diagnosis of neurological diseases. Nowadays, even smartphones are being used in the healthcare sector. For example, accelerometer sensors in smartphones can obtain data from a person's movements that can be used to diagnose and determine the stage of AD. In this example, the smartphone is used as the edge node of the proposed system. This proposed system can monitor the patient continuously and send alerts to the patient's doctor or relatives. The sensor's data can be uploaded to the cloud for storage and used for further analysis. This proposed system will benefit AD patients and make a valuable contribution to the healthcare system.

References

Abrol, A., Bhattarai, M., Fedorov, A., Du, Y., Plis, S., & Calhoun, V. (2020). Deep residual learning for neuroimaging: An application to predict progression to Alzheimer's disease. *Journal of Neuroscience Methods*. https://doi.org/10.1016/j.jneumeth.2020.108701. Elsevier.

Bringas, S., Salomón, S., Duque, R., Lage, C., & Montaña, J. L. (2020). Alzheimer's disease stage identification using deep learning models. *Journal of Biomedical Informatics*. Elsevier.

Celik, E., & Omurca, S. I. (2019). *Improving Parkinson's disease diagnosis with machine learning methods*. EBTT, IEEE. https://doi.org/10.1109/EBBT.2019.8742057,2019.

Chen, Y., & Xia, Y. (2020). Iterative sparse and deep learning for accurate diagnosis of Alzheimer's disease. *Pattern Recognition*. Elsevier.

Choi, H., Kim, Y. K., Yoon, E. J., Lee, J.-Y., & Lee, D. S. (2019). Cognitive signature of brain FDG PET based on deep learning: domain transfer from Alzheimer's disease to Parkinson's disease. *European Journal of Nuclear Medicine and Molecular Imaging*. https://doi.org/10.1007/s00259-019-04538-7. Springer.

Dash, S., Abraham, A., & Atta-Ur-Rahman. (2019). Kernel based chaotic firefly algorithm for diagnosing Parkinson's disease. In *International conference on hybrid intelligent* Springer. https://doi.org/10.1007/978-3-030-14347-3_18.

Dash, S., Biswas, S., Banerjee, D., & Rahman, A. (2019). Edge and fog computing in healthcare—A review. *Scalable Computing: Practice and Experience*, *20*(2), 191–205. http://www.scpe.org. 10.12694/scpe. v20i2.1504, ISSN 1895-1767.

Dash, S., Thulasiram, R., & Thulasiraman, P. (2017). An enhanced Chaos-based firefly model for Parkinson's disease diagnosis and classification. In *International conference on information technology*. https://doi.org/10.1109/ICIT.2017.43.

Dash, S., et al. (2020). *Deep learning techniques for biomedical and health informatics, studies in big data series. Vol. 68*. Springer Nature. https://doi.org/10.1007/978-3-030-33966-1. ISBN 978-3-030-33965-4.

Dastjerd, N. K., Sert, O. C., Ozyer, T., & Alhajj, R. (2019). Fuzzy classification methods based diagnosis of Parkinson's disease from speech test cases. In *Current aging science* Bentham Science Publishers.

Gaubert, S., Ansart, M., Naccache, L., & Sitt, J. D. (2021). A machine learning approach to screen for preclinical Alzheimer's disease. *Neurobiology of Aging*. https://doi.org/10.1016/j.neurobiolaging.2021.04.024. Elsevier.

Grassi, M., Perna, G., Caldirola, D., Schruers, K., Duara, R., & Loewenstein, D. A. (2018). A clinically-translatable machine learning algorithm for the prediction of Alzheimer's disease conversion in individuals with mild and Premild cognitive impairment. *Journal of Alzheimer's Disease*. https://doi.org/10.3233/JAD-170547. IOS Press.

Grover, S., Bhartia, S., Akshama, & Yadav, A. (2019). Predicting severity of Parkinson's disease using deep learning. *Computational Intelligence and Data Science*. https://doi.org/10.1016/j.procs.2018.05.154. Elsevier.

Jo, T., Nho, K., & Saykin, A. J. (2019). Predicting severity of Parkinson's disease using deep learning. In *International conference on computational intelligence and data science* Elsevier. https://doi.org/10.1016/j.procs.2018.05.154.

Johns Hopkins Medicine. (2022). *Alzheimer's disease*. Johns Hopkins Medicine. https://www.hopkinsmedicine.org/health/conditions-and-diseases/alzheimers-disease.

Kakulapati, V., & TejaSanthosh, D. (2020). Metaheuristic approach of RMDL: Classification of Parkinson's disease. In *Applications of hybrid metaheuristic algorithms for image processing* Springer. https://doi.org/10.1007/978-3-030-40977-7_17.

Kishore, P., Kumari, C. U., Kumar, S. S., & Pavani, T. (2020). Detection and analysis of Alzheimer's disease using various machine learning algorithms. *Materials Today: Proceedings*. https://doi.org/10.1016/j.matpr.2020.07.645. Elsevier.

Kuresan, H., et al. (2017). Fusion of WPT and MFCC feature extraction 1 in Parkinson's disease diagnosis. In *Technology and health care* IEEE. https://doi.org/10.3233/THC-181306.

Lacya, S. E., Smitha, S. L., & Lones, M. A. (2019). Using echo state networks for classification: A case study in Parkinson's disease diagnosis. In *Artificial Intelligence in Medicine* Elsevier. https://doi.org/10.1016/j.artmed.2018.02.002.

Lopez-Martin, M., Nevado, A., & Carro, B. (2020). Detection of early stages of Alzheimer's disease based on MEG activity with a randomized convolutional neural network. In *Artificial Intelligence In Medicine* Elsevier. https://doi.org/10.1016/j.artmed.2020.101924.

Maass, F., et al. (2019). Elemental fingerprint: Reassessment of a cerebrospinal fluid biomarker for Parkinson's disease. In *Neurobiology of disease* Elsevier. https://doi.org/10.1016/j.nbd.2019.104677.

Mozhdehfarahbakhsh, A., et al. (2021). *An MRI-based deep learning model to predict Parkinson's disease stages*. medRxiv. https://doi.org/10.1101/2021.02.19.21252081.

Rahman, A., et al. (2019). A Comprehensive study of mobile computing in Telemedicine. In A. K. Luhach, et al. (Eds.), *ICAICR 2018,CCIS 956* (pp. 413–425). Singapore: Springer. https://doi.org/10.1007/978-981-13-3143-5_34.

Rehman, A., et al. (2018). Management of resource usage in mobile cloud computing. *International Journal of Pure and Applied Mathematics*, *119*(16), 255–261. ISSN: 1314-3395 (online version) 2018.

Rupabanta Singh, K., Dash, S., Biswas, S., & Deka, B. (2021). *Mobile technology solutions for COVID-19*. Springer. https://doi.org/10.1007/978-3-030-60039-6_14.

Sayaydehaand, O. N. A., & Mohammad, M. F. (2019). Diagnosis of the Parkinson disease using enhanced fuzzy min-max neural network and OneR attribute evaluation method.

In *International conference on advanced science and engineering* IEEE. https://doi.org/10.1109/ICOASE.2019.8723870.

Scholkopf, B., & Smola, A. J. (2018). *Learning with kernels: Support vector machines, regularization, optimization, and beyond.* MIT Press.

Senturk, Z. K. (2020). Early diagnosis of Parkinson's disease using machine learning algorithms. In *Medical hypotheses* Elsevier.

Sharma, A., & Dey, P. (2020). A machine learning approach to unmask novel gene signatures and prediction of Alzheimer's disease within different brain regions. In *Genomics* Elsevier. https://doi.org/10.1016/j.matpr.2020.07.645.

Sudharsan, M., & Thailambal, G. (2020). Alzheimer's disease prediction using machine learning techniques and principal component analysis (PCA). *Materials Today: Proceedings.* https://doi.org/10.1016/j.matpr.2021.03.061. Elsevier.

Sutoko, S., et al. (2021). Early identification of Alzheimer's disease in mouse models: Application of deep neural network algorithm to cognitive behavioral parameters. *iScience, 24.* https://doi.org/10.1016/j.isci.2021.102198.

Tysnes, O., & Storstein, A. (2017). Epidemiology of Parkinson's disease. In *Neurology and preclinical neurological studies—Review article.* https://doi.org/10.1007/s00702-017-1686-y.

UlHaq, A., et al. (2019). *Feature selection based on L1-norm support vector machine and effective recognition system for Parkinson's disease using voice recordings.* IEEE. https://doi.org/10.1109/ACCESS.2019.2906350.

Varrecchia, T., et al. (2021). An artificial neural network approach to detect presence and severity of Parkinson's disease via gait parameters. *PLos One.* https://doi.org/10.1371/journal.pone.0244396.

Zhang, F., Li, Z., Zhang, B., Du, H., Wang, B., & Zhang, X. (2019). Multi-modal deep learning model for auxiliary diagnosis of Alzheimer's disease. *Neurocomputing*. Elsevier.

CHAPTER 2

A predictive method for emotional sentiment analysis by deep learning from EEG of brainwave dataset

Pijush Dutta[a], Shobhandeb Paul[b], Korhan Cengiz[c], Rishabh Anand[d], and Asok Kumar[e]

[a]Department of Electronics and Communication Engineering, Greater Kolkata College of Engineering and Management, West Bengal, India
[b]Department of Electronics and Communication Engineering, Guru Nanak Institute of Technology, Kolkata, West Bengal, India
[c]College of Information Technology, University of Fujiarah, Fujairah, United Arab Emirates
[d]Service Deliver Manager, HCL Technologies Limited, New Delhi, India
[e]Student Welfare Department, Vidyasagar University, Medinipur, West Bengal, India

Introduction

The brain-computer interface (BCI) utilizes technology to acquire human brainwaves to control external devices. For practical implementations, most BCI applications have been developed for noninvasive brain signal processing. The BCI is not only useful for mentally controlling devices, but is also useful for recognizing mental states and emotions. Automatic emotion recognition algorithms potentially bridge the gap between human and machine interactions.

Electroencephalography (EEG) is a very sensitive tool for measuring human emotions, but there are some drawbacks that limit its use in further analysis and processing of signals. Some of these include low spatial resolution, poor signal-to-noise ratio (SNR) (Bigdely-Shamlo, Mullen, Kothe, Su, & Robbins, 2015; Jas, Engemann, Bekhti, Raimondo, & Gramfort, 2017), and multiple sources of environmental, physiological, and activity-specific noise called artifacts. Due to high speed of electrical signal propagation, EEG can capture a signal within milliseconds. Characteristics of EEG signals vary, meaning they are non-stationary (Gramfort, Strohmeier, Haueisen, Hämäläinen, & Kowalski, 2013). Hence, a classifier never produces an accurate result for a trained model using limited data. Thus, it is a challenge to model various filtering techniques and noise reduction

Artificial Intelligence for Neurological Disorders
https://doi.org/10.1016/B978-0-323-90277-9.00002-X

processes to minimize the signal-to-noise effect and enhance the usefulness of EEG signals for further applications.

Various techniques have been examined for emotion acknowledgment using EEG signal investigation, including nonlinear analysis, time domain, frequency domain, time-frequency domain, nonlinear analysis, and other methods (Frantzidis et al., 2010). Differential asymmetry, rational asymmetry, and power spectral density (PSD) are extricated as highlights for emotional acknowledgment (Lin et al., 2010), introduction of higher-order crossing features (Petrantonakis & Hadjileontiadis, 2009), fractal measurement (FD) with respect to the evaluation of fundamental feelings (Liu, Sourina, & Nguyen, 2010), several entropy-based measurements of sign multifaceted nature have just been proposed for separating passionate states. applied two entropy measurements (estimated and wavelet entropy) to separate between two enthusiastic states (calm-neutral and negative-energized) (Hosseini & Naghibi-Sistani, 2011). Discrete wavelet transform (DWT) is used to partition the EEG signal into a few groups and determine highlights dependent on these groups (Murugappan, Ramachandran, & Sazali, 2010). Notwithstanding the promising advancements that have already been made, building up the best mix of highlight extraction and arrangement strategies requires further examination.

Better generalization and flexibility of the application of EEG signals is a challenge to improving EEG processing. Deep learning (DL) involves computational models that learn hierarchical representations of input data through non-linear transformations (Zeng et al., 2018). It encompasses (Zhao, Ouyang, Li, & Wang, 2015) data preprocessing, feature extraction, and classification modules. According to the problem DL is trying to solve, it can also be used to perform a competitive performance on the target datasets. Recently, DL has been applied in different domains such as images, text, and audio signals (LeCun, Bengio, & Hinton, 2015), large-scale visual recognition challenges (Przegalinska, Ciechanowski, Magnuski, & Gloor, 2018), and industrial applications.

Researchers have applied artificial intelligence (AI) to recognize human emotions using an emotional investigation dataset and the Muse headband, which is an EEG recording device with four terminals set on the TP9, AF7, AF8, and TP10 positions dependent on the worldwide EEG arrangement framework (Asif, Majid, & Anwar, 2019). To develop a device that shows these patients' feelings with straightforward and understandable pictures or charts so they can be perceived by the individuals who deal with them or potentially specialized instrument for the secured disorder

(Przegalinska et al., 2018), the chance of utilizing EEG signals as a sign of misdirection or untruths (Youssef, Ouda, & Azab, 2018), the value in estimating EEG sign of Muse is helpful in test work (Bansal & Mahajan, 2019).

After a brief introduction in Section "Introduction," the chapter discusses previous scientific work in the field of AI and EEG in Section "Literature review." Section "Materials and methods" presents techniques of feature selection and machine learning (ML) algorithms. Section "Result analysis" presents results and Section "Conclusion and discussion" concludes.

Literature review

This section highlights several scientific studies on machine learning (ML) and DL using EEG data. In one study, researchers combined multilayer extreme learning machine (MLELM) with extreme learning machine with kernel (KELM) to develop a deep extreme learning machine (DELM). They applied the DELM to EEG classification using MATLAB and two BCI datasets. Minimum error obtained from this technique was 39.44% (Ding, Zhang, Xu, Guo, & Zhang, 2015). Tabar and Halici (2016) performed a comparative study of a convolutional neural network (CNN) and stacked autoencoders used to classify datasets obtained from EEG motor imagery signals. The best model obtained a kappa value of 0.547. Researchers combined CNN architecture with a variation autoencoder to classify datasets obtained from motor imagery EEG signals (Dai, Zheng, Na, Wang, & Zhang, 2019). Results showed that average kappa values obtained from three-electrode and five-electrode EEGs were 0.568 and 0.603, respectively. Deep CNN and deep residual learning algorithms were applied to predict the mental states of drivers from EEG signal datasets (Zeng et al., 2018). Results showed that both the algorithms classified the mental states of the drivers with a high degree of accuracy. In another study, a multi-layer CNN (MCNN) was used to improve the classification accuracy of EEG motor imagery (Amin, Alsulaiman, Muhammad, Mekhtiche, & Hossain, 2019). Data classification was performed with the help of spatial and temporal features of EEG datasets. Analysis of results showed that the proposed MCNN exhibited 75.7% and 95.4% accuracy for the IV-2a dataset and the High Gamma Dataset, respectively. To classify EEG motor imagery datasets, researchers used the deep belief net (DBN) algorithm (An, Kuang, Guo, Zhao, & He, 2014) and obtained experimental results for eight hidden layers. The DBN outperformed a support vector machine

(SVM) and DBN classifier. A comparative study of a deep neural network (DNN) and a CNN was performed for measuring cybersickness from EEG data (Jeong, Yoo, & Yun, 2019). Classification was performed based on 360 video stream segments causing cybersickness. Outcomes were satisfactory. In another study, a CNN was applied for prediction of epileptiform discharges from scalp EEG recordings (Tjepkema-Cloostermans, de Carvalho, & van Putten, 2018). The proposed model produced an area under curve (AUC) of 0.94 for the test set and 47.4% sensitivity and 98.0% specificity for the validation model. The advanced ML algorithm of a stacked autoencoder with a softmax layer model was developed to predict drowsiness using EEG data (Rundo et al., 2019). The experiment was conducted on sixty-two volunteers and results showed the autoencoder achieved 100% accuracy in drowsy/wakeful discrimination.

Materials and methods

A BCI is an immediate communication pathway between the brain and an outside device (Kumar et al., 2020). There are three BCI methods for obtaining EEG signals.

(1) Invasive: In invasive methods, anodes are embedded into the dark matter of the cerebrum during neurosurgery.

(2) Partially invasive: BCI gadgets are embedded inside the skull yet rest outside the brain.

(3) Non-invasive: In this BCI system, cathodes are set externally on the skull. Non-invasive techniques are widely utilized in research. Magnetic resonance imaging (MRI) and EEG are examples of non-invasive techniques. EEG is used to estimate the electrical movement of the cerebrum, which is produced by billions of nerve cells called neurons. This EEG movement is recorded through terminals placed on the scalp (Sameer, Gupta, Chakraborty, & Gupta, 2019). EEG signals are ordered into various rhythm frequencies ranging from 0.5 Hz to 30 Hz. There are five types of EEG waves: Delta, Theta, Alpha, Beta, and Gamma waves. Table 1 lists their characteristics.

Fig. 1 shows the proposed model to classify EEG brainwaves collected from the Muse headband using DL algorithms.

IoT-based Muse headband

The Muse headband is an EEG recording device with four anodes on the TP9, AF7, AF8, and TP10 positions dependent on the worldwide EEG

Table 1 Different brain waves with ranges of frequency and emotions.

Serial No	Brain wave	Ranges of frequency (Hz)	Emotions
1	Delta	0.5–4	Deep sleep, coma
2	Theta	4–8	Trance, dreams
3	Alpha	8–13	Relaxation with eyes closed
4	Beta	13–20 for Beta 1 20–30 for Beta 2	Thinking, tension, anxiety, excitement
5	Gamma	26–70	High stress and disturbance

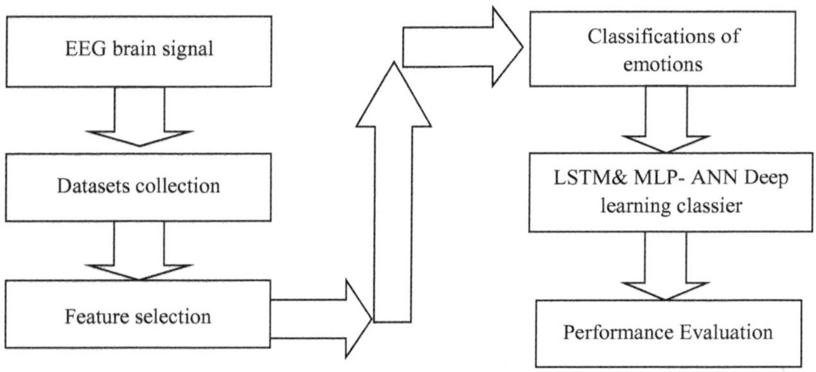

Fig. 1 Framework of the proposed model.

arrangement framework (Jasper, 1958). Since the signals are weak, signal commotion is a significant issue because it covers up the helpful data (Symeonidou, Nordin, Hairston, & Ferris, 2018). The overall process is done in three stages. In the first phase, the Muse headband collects the brain signals, processes them via ESP8266 Node MCU, and sends the EEG data to a computer or Wi-Fi module. The computing system then smooths the feature extraction and mental state fusion. In the second stage, comments by the user are clarified via natural language processing. In the final stage, the output from the headband is compared with the comments of the user to determine the sentimental score. The EEG headband utilizes different detachment methods to collect and maintain brainwave information and dispose of undesirable commotion (Oliveira, Schlink, Hairston, König, & Ferris, 2016). The Muse headband is being utilized alongside AI techniques to quantify various degrees of client/consumer satisfaction (Abujelala, Abellanoza, & Sharma, 2016; Plotnikov et al., 2012). Dream headbands are also used in neuroscience research because of their ease and simplicity of sending, just as its viability as far as to order and exactness (Krigolson, Williams, Norton, Hassall, & Colino, 2017).

Feature selection

Feature selection and characterization of EEG data is a central issue in the BCI domain. The signs are viewed as fixed uniquely inside short spans that are the reason the best practice is to apply a brief timeframe windowing method to meet this prerequisite. Feature selection (Dwivedi, Dey, Chakraborty, & Tiwari, 2021) can be performed by several statistical techniques like max-min features in temporal sequence, log-covariance, Shannon entropy, time-frequency based Fourier Transform, and so on. All highlights proposed to arrange the psychological states are registered as far as the worldly dissemination of the sign in a given time window. This sliding window is characterized as a time of 1 s at 250 Hz. A cover of 1 s is utilized while moving each of four windows. Another significant feature of the EEG Muse headband is that it offers five kinds of sign frequencies: Delta, Theta, Alpha, Beta, and Gamma (δ, θ, α, β, γ).

Statistical Features: To obtain a minimal portrayal of the sensor information in a given time range, several measurable features are demonstrated.

(1) Mean value

$$\mu = \frac{1}{N} \sum_{i}^{N} x_i \tag{1}$$

(2) The standard deviation

$$\sigma = \sqrt{\frac{1}{N} \sum_{i}^{N} (x_i - \mu)^2} \tag{2}$$

(3) Skewness and kurtosis

$$\text{Skewness} : \gamma = \frac{\mu^k}{\sigma^k} \tag{3}$$

$$\text{Kurtosis} : \mu^k = \frac{1}{N} \sum_{i}^{N} (x_i - \mu)^k \tag{4}$$

Log-covariance features: To attain a 12×12 square matrix to process the log-covariance from 150 temporal features, we disposed of 6 features.

$$\text{L}cM = (\log m(\text{cov}(M))) \tag{5}$$

where LcM is an upper triangular element of the matrix.

The covariance matrix is given by

$$(M) = \text{cov}\, ij = \frac{1}{N}\Sigma(x_{ik} - \mu_{\text{i}})\left(x_{kj} - \mu_j\right) \tag{6}$$

Shannon entropy and log-energy entropy: Shannon entropy is one type of non-linear analysis widely applicable in the field of signal processing and time series. In data hypothesis, the Shannon entropy is given by:

$$h = -\sum_j S_j \times \log\ (S_j) \tag{7}$$

where h is a feature computed every 1 s and Sj is each element (normalized) of this temporal window.

The log-energy entropy can be given by:

$$\log e = \sum_i \log\left(s_i^2\right) + \sum_j \log\left(s_j^2\right) \tag{8}$$

where i and j represent an index for the elements of the first and second sub window (0–0.5 s).

Frequency domain: A frequency domain method to analyze the time series spectrum can be represented by:

$$X(k) = \sum_{n=0}^{N-1} x(n).e^{-\frac{j(2\pi)*nk}{N}} \text{ where,} k = 0,1,2,3\ldots(N-1) \tag{9}$$

Datasets

Feature selection algorithms

Highlight choice intends to eliminate information that has no valuable application and just serves to unneeded increment the interest for assets (Witten & Frank, 2002). Five datasets were created utilizing various calculations. Each dataset has a similar focus, however, which had a diminished number of characteristics, chose by the calculation. The evaluators utilized were:

(1) One R: helps to calculate the error rate of every forecast and selects the lowest risk classification (Faria, Premebida, Manso, Ribeiro, & Núñez, 2018).

(2) **Information Gain**: estimating concerning the class (Jadhav, He, & Jenkins, 2018).
(3) **Correlation**: measures concerning with the help of attribute and class by means their Pearson coefficient (Jain, Jain, & Jain, 2018).
(4) **Symmetrical Uncertainty**: measures concerning the class (Saikhu, Arifin, Nopember, Fatichah, & Nopember, 2019).
(5) **Evolutionary Algorithm**: measure concerning subsets and positions their viability with a fitness capacity to gauge their prescient capacity of the class (Dutta, Biswas, Biswas, & Majumder, 2021; Dutta & Kumar, 2020; Dutta, Majumder, & Kumar, 2021a, 2021b; Li, He, Liang, & Quan, 2020).

Symmetric uncertainty

When a lot of information is amassed it becomes daunting to identify significant patterns. Noise in the information additionally sabotages the capacities of the main classification methods. As such, feature determination assumes a fundamental job. This chapter has presented a productive technique for feature selection. As referenced above for a filter-based feature selection mechanism their requirements to have some proxy classifier measure that gauges the value of a chose highlight subset (Ali & Shahzad, 2012; Dutta, Paul, Obaid, Pal, & Mukhopadhyay, 2021). The expression for symmetric uncertainty is given by

$$\mathrm{SU}(X, Y) = 2\left[\frac{IG\left(\frac{X}{Y}\right)}{H(x) + H(Y)}\right] \tag{10}$$

where IG $\left(\frac{X}{Y}\right)$ is the information gain of feature X concerning independent attribute Y, and H(X) and H(Y) are the entropy of feature X and Y, respectively.

Deep learning model

As a benchmark, every dataset is set to run by a symmetric uncertainty classifier. This oversimplified classifier picks one single class to apply to the entirety of the information to diminish off-base arrangements. It is normal to achieve accuracy of 33% with a reasonable appropriation of the three mental states. AI alludes to the utilization of computational algorithms that can figure out how to perform given errands from model information without the need directions. AI applied advanced statistical methods to remove prescient or unfair samples to play out the most precise forecasts on new information. AI based algorithms are obvious for cracking the book from the historical data to fit out the conjectures for the future outcomes. Machine learning (ML) and deep learning (DL) algorithms both are subsets of the AI. There are sanguine unconformities between machine learning (Dutta, Paul, & Kumar, 2021; Dutta, Paul, & Majumder, 2021; Dutta, Sen, Saha, & Chaki, 2021; Dutta, Shaw, Das, & Ghosh, 2021) and deep learning. Unraveling up growths in big data requires larger and deeper networks, having provisions for computers to learn, keep one's eye on the applicability and have the vibes to complex situations faster than humans. Here we proposed DL algorithms as an AI tool in the field of cardiovascular diagnosis for a non-expert audience. In this research under the Python platform, distinctive prescient calculations were picked to assemble the model, namely: Long short-term memory (LSTM) & multilayer perception neural network (MLP-ANN). Accordingly, we chose the most ideal model to accomplish the best on clinical datasets and permit viable data classification.

The dataset included 2132 records. On analyzing the label column of the dataset, a classification problem was identified and thus classification algorithms were used to train the models to obtain accuracy.

We assessed the results of the models based on the metrics of exactness, recall, accuracy, and F1-score.

In this consideration, we evolved and figured out the clinical predictive models to wind up the COVID-19 infection with laboratory outcomes. To conduct the study, we trained six different model types: Artificial Neural Network (ANN) and LSTM. ANN is an approach which can be disposed of the propaganda that is inspired by the biological nervous system of human brain. It is unruffled with neurons, activation functions, input, output, and hidden layers. LSTM is the refitted version of the RNN. In the LSTM, it is effortless to conjure up the past data in the memory. Table 4 lists the parameters of each classifier. To dig it up with the predictive performance of each of the refined predictive models, we kept the tab for their performance in terms of accuracy, f1-score, precision, recall, and training time. To analyze the results, we used 10-fold cross validation and an 80:20 train-test split approach.

LSTM classification

LSTM is based on recurrent neural network (RNN), which is a DL algorithm. An RNN is comprised of recurrent structures that provincially regale the firing strength thereby external registers or memories are not needed for

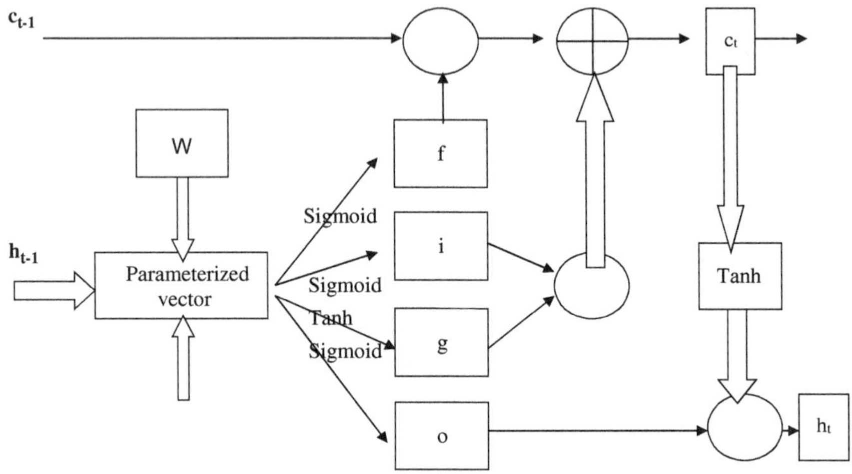

Fig. 2 Basic diagram of an LSTM.

storing previous outputs. Computational complexity is low in LSTM due to recurrent structures applied in RNN (Fig. 2).

Structure learning and parameter learning are carried out in two phases in an RNN. An LSTM has four different gates and each of the gates has different purposes:

(1) Forget Gate (f): This gate predicts what amount of previous data can be forgotten.

(2) Input Gate (i): This gate predicts what amount of new data can be written onto the internal cell state.

(3) Input Modulation Gate (g): This is a sub-part of the input gate used to modulate the information that the input gate will write onto the internal cell state by adding non-linearity to the information and making the information *zero-mean*. This is done to reduce the learning time as zero-mean input has faster convergence. Although this gate's actions are less important than the others, it is good practice to include this gate in the structure of the LSTM unit.

(4) Output Gate (o): This gate determines what output will be generating from the current internal cell state.

Result analysis

The dataset used in this study included 2132 records. On analyzing the label column of the dataset, a classification problem was identified and thus classification algorithms were used to train the models to achieve accuracy. The proposed method uses a stacked LSTM comprised of four bidirectional LSTM layers and four dense layers. This stacked LSTM model was built using 64, 32, and 3 nodes in each layer. Each of these is followed by dropout regularization of 20%. The dense layers were stacked by incorporating three nodes. All the LSTM layers and the final dense layers were activated using five different combinations of activation functions: ReLU-Sigmoid, ReLU-ReLU, ReLU-SELU, ReLU-Softmax, and ReLU-Softplus. Finally, the layers were compiled using Adam optimizer and binary cross entropy and logcoh loss function. Construction of this model was dependent on 100 epochs with a batch size of 32. Fine-tuning the hyper-parameters assisted in obtaining the best problem-solving approach. Once this model was constructed, it was fitted with the training data. Tables 2–5 provide a detailed description of the layers, including activation function used, output shape produced by each layer, and number of parameters accepted by each layer. Two different DL algorithms with two different loss models were developed as a classifier. Table 6 shows the evaluation results for the LSTM and MLP-ANN models using a 10-fold cross-validation approach.

In terms of predictive performance, the LSTM model achieved the best overall accuracy, F1 score, precision, recall, and MSE. LSTM performs better for data analysis when it contains time series. Figs. 9 and 10 show the evaluation results for the LSTM and MLP-ANN models.

We also tested the performance of the algorithms using an 80:20 train-test split approach. Although k-fold cross validation is frequently used in AI, especially in health studies in cases of relatively small samples, it generates less clearly the results in clinical applications. The clinical predictive performance of all algorithms was better in comparison with 10-fold cross-validation strategies with an accuracy of 97.189, F1-score of 95.560, precision of 97.55, and recall of 95.25 for the best-performing algorithm, which was the LSTM hybrid model shown in Table 7. The MLP-ANN showed a lower MSE than LSTM models during the train-test split approach.

Figs. 3–8 show that as the number of epochs increases, the accuracy increases gradually and reaches around a value of 96%. In contrast, the loss gradually decreases and attains lowest value around 0.18. Once the training process is done (i.e., after completing 100 epochs), accuracy, precision, F1 score, recall, and MSE value for test dataset.

Table 2 Description of LSTM, Adam optimizer, and binary cross-entropy loss models after 100 epochs.

Layer	Input shape	o/p shape	No. of parameters received	Activation function	Training time	Accuracy (%)
LSTM	(1, 2548)	(None,1,64)	668928	ReLU-Sigmoid	27.157 s	96.25
Dropout	(1, 2548)	(None,1,64)	0			
LSTM_1	(1, 2548)	(None,32)	12416			
Dropout_1	(1, 2548)	(None,32)	0			
Dense	(1, 2548)	(None,3)	99			
LSTM_2	(1, 2548)	(None,1,64)	668928	ReLU-ReLU	27.21 s	96.48
Dropout_2	(1, 2548)	(None,1,64)	0			
LSTM_3	(1, 2548)	(None,32)	12416			
Dropout_3	(1, 2548)	(None,32)	0			
Dense_1	(1, 2548)	(None,3)	99			
LSTM_4	(1, 2548)	(None,1,64)	668928	ReLU-SELU	26.489 s	96.487
Dropout_4	(1, 2548)	(None,1,64)	0			
LSTM_5	(1, 2548)	(None,32)	12416			
Dropout_5	(1, 2548)	(None,32)	0			
Dense_2	(1, 2548)	(None,3)	99			
LSTM_6	(1, 2548)	(None,1,64)	668928	ReLU-Softmax	27.16sc	97.189
Dropout_6	(1, 2548)	(None,1,64)	0			
LSTM_7	(1, 2548)	(None,32)	12416			
Dropout_7	(1, 2548)	(None,32)	0			
Dense_3	(1, 2548)	(None,3)	99			
LSTM_8	(1, 2548)	(None,1,64)	668928	ReLU-Softplus	26.48 s	97.189
Dropout_8	(1, 2548)	(None,1,64)	0			
LSTM_9	(1, 2548)	(None,32)	12416			
Dropout_9	(1, 2548)	(None,32)	0			
Dense_4	(1, 2548)	(None,3)	99			

Table 3 Description of LSTM, Adam optimizer, and logcosh loss models after 100 epochs.

Layer	I/p shape	o/p shape	No. of parameters received	Activation function	Training time	Accuracy (%)
LSTM_10	(1, 2548)	(None,1,64)	668928	ReLU-Sigmoid	26.98 s	96.252
Dropout_10	(1, 2548)	(None,1,64)	0			
LSTM_11	(1, 2548)	(None,32)	12416			
Dropout_11	(1, 2548)	(None,32)	0			
Dense_5	(1, 2548)	(None,3)	99			
LSTM_12	(1, 2548)	(None,1,64)	668928	ReLU-ReLU	26.52 s	96.721
Dropout_12	(1, 2548)	(None,1,64)	0			
LSTM_13	(1, 2548)	(None,32)	12416			
Dropout_13	(1, 2548)	(None,32)	0			
Dense_6	(1, 2548)	(None,3)	99			
LSTM_14	(1, 2548)	(None,1,64)	668928	ReLU-SELU	25.78 s	96.487
Dropout_14	(1, 2548)	(None,1,64)	0			
LSTM_15	(1, 2548)	(None,32)	12416			
Dropout_15	(1, 2548)	(None,32)	0			
Dense_6	(1, 2548)	(None,3)	99			
LSTM_16	(1, 2548)	(None,1,64)	668928	ReLU-Softmax	27.45 s	94.847
Dropout_16	(1, 2548)	(None,1,64)	0			
LSTM_17	(1, 2548)	(None,32)	12416			
Dropout_17	(1, 2548)	(None,32)	0			
Dense_7	(1, 2548)	(None,3)	99			
LSTM_18	(1, 2548)	(None,1,64)	668928	ReLU-Softplus	29.65 s	97.423
Dropout_18	(1, 2548)	(None,1,64)	0			
LSTM_19	(1, 2548)	(None,32)	12416			
Dropout_19	(1, 2548)	(None,32)	0			
Dense_8	(1, 2548)	(None,3)	99			

Table 4 Description of MLP-ANN, Adam optimizer, and binary cross-entropy loss models after 100 epochs and 10-fold cross validation.

Layer	I/p shape	o/p shape	No. of parameters received	Activation function	Training time	Accuracy (%)
Dense_180	(1, 2548)	(None,1,32)	81,568	ReLU-Sigmoid	2 min 59 s	96.610
Dense_181	(1, 2548)	(None,1,32)	1056			
Dense_182	(1, 2548)	(None,1, 32)	1056			
Dense_183	(1, 2548)	(None,1, 3)	99			
Dense_180	(1, 2548)	(None,1,32)	81,568	ReLU-ReLU	1 min 34 s	97.245
Dense_181	(1, 2548)	(None,1,32)	1056			
Dense_182	(1, 2548)	(None,1, 32)	1056			
Dense_183	(1, 2548)	(None,1, 3)	99			
Dense_184	(1, 2548)	(None,1,32)	128			
Dense_185	(1, 2548)	(None,1,32)	1056			
Dense_186	(1, 2548)	(None,1, 32)	1056			
Dense_187	(1, 2548)	(None,1, 3)	99			
Dense_188	(1, 2548)	(None,1,32)	128	ReLU-SELU	1 min 58 s	62.288
Dense_189	(1, 2548)	(None,1,32)	1056			
Dense_190	(1, 2548)	(None,1, 32)	1056			
Dense_191	(1, 2548)	(None,1, 3)	99			
Dense_190	(1, 2548)	(None,1, 32)	1056	ReLU-Softmax	2 min 15 s	99.576
Dense_191	(1, 2548)	(None,1, 3)	99			
Dense_192	(1, 2548)	(None,1,32)	128			
Dense_193	(1, 2548)	(None,1,32)	1056			
Dense_194	(1, 2548)	(None,1, 32)	1056			
Dense_195	(1, 2548)	(None,1, 3)	99			
Dense_196	(1, 2548)	(None,1,32)	128	ReLU-Softplus	1 min 54 s	99.576
Dense_197	(1, 2548)	(None,1,32)	1056			
Dense_198	(1, 2548)	(None,1, 32)	1056			
Dense_199	(1, 2548)	(None,1, 3)	99			

Table 5 Description of MLP-ANN, Adam optimizer, and logcosh loss models after 100 epochs and 10-fold cross validation.

Layer	I/p shape	o/p shape	No. of parameters received	Activation function	Training time	Accuracy (%)
Dense_17	(1, 2548)	(None,1,32)	1056	ReLU-Sigmoid	1 min 32 s	36.238
Dense_18	(1, 2548)	(None,1, 32)	1056			
Dense_19	(1, 2548)	(None,1, 3)	99			
Dense_20	(1, 2548)	(None,1,32)	128			
Dense_21	(1, 2548)	(None,1,32)	1056			
Dense_22	(1, 2548)	(None,1, 32)	1056			
Dense_23	(1, 2548)	(None,1, 3)	99			
Dense_20	(1, 2548)	(None,1,32)	128	ReLU-ReLU	1 min 22 s	29.872
Dense_21	(1, 2548)	(None,1,32)	1056			
Dense_22	(1, 2548)	(None,1, 32)	1056			
Dense_23	(1, 2548)	(None,1, 3)	99			
Dense_24	(1, 2548)	(None,1,32)	128			
Dense_25	(1, 2548)	(None,1,32)	1056			
Dense_26	(1, 2548)	(None,1, 32)	1056			
Dense_27	(1, 2548)	(None,1, 3)	99			
Dense_27	(1, 2548)	(None,1, 3)	99	ReLU-SELU	1 min 56 s	29.872
Dense_28	(1, 2548)	(None,1,32)	128			
Dense_29	(1, 2548)	(None,1,32)	1056			
Dense_30	(1, 2548)	(None,1, 32)	1056			
Dense_31	(1, 2548)	(None,1, 3)	99			
Dense_30	(1, 2548)	(None,1, 32)	1056	ReLU-Softmax	1 min 37 s	29.872
Dense_31	(1, 2548)	(None,1, 3)	99			
Dense_32	(1, 2548)	(None,1,32)	128			
Dense_33	(1, 2548)	(None,1,32)	1056			
Dense_34	(1, 2548)	(None,1, 32)	1056			
Dense_35	(1, 2548)	(None,1, 3)	99			

Continued

Table 5 Description of MLP-ANN, Adam optimizer, and logcosh loss models after 100 epochs and 10-fold cross validation—cont'd

Layer	I/p shape	o/p shape	No. of parameters received	Activation function	Training time	Accuracy (%)
Sense_31	(1, 2548)	(None,1, 3)	99	ReLU-Softplus	1 min 49 s	33.898
Dense_32	(1, 2548)	(None,1,32)	128			
Dense_33	(1, 2548)	(None,1,32)	1056			
Dense_34	(1, 2548)	(None,1, 32)	1056			
Dense_35	(1, 2548)	(None,1, 3)	99			
Dense_36	(1, 2548)	(None,1,32)	128			
Dense_37	(1, 2548)	(None,1,32)	1056			
Dense_38	(1, 2548)	(None,1, 32)	1056			
Dense_39	(1, 2548)	(None,1, 3)	99			

Table 6 Evaluation of results of LSTM and MLP-ANN with 10-fold cross-validation approach.

Name of the DL	Accuracy	F1 Score	Recall	Precision	MSE
LSTM	96.45	92.66	97.02	94.75	0.0923
MLP-ANN	92.12	88.45	93.56	91.89	0.187

Table 7 Evaluation result of LSTM and MLP-ANN for train-test approach.

Name of the DL	Accuracy	F1 Score	Recall	Precision	MSE
LSTM	97.189	95.56	97.55	95.25	0.0867
MLP-ANN	93.21	93.21	94.0	92.28	0.0756

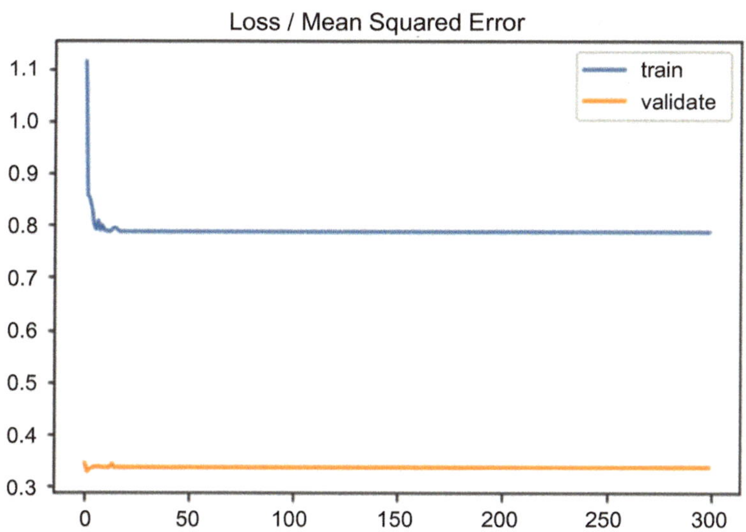

Fig. 3 ReLU-Sigmoid for binary cross entropy.

Conclusion and discussion

In this chapter, we presented our study on using DL models to predict EEG brainwaves obtained from sensors. We collected 2549 datasets dependent on time-frequency domain statistical features from the Kaggle "EEG Brainwave Dataset: Feeling Emotions" (Kaggle, 2019) The study was performed in two stages. In the first stage, we chose 640 datasets for further classification. At this stage, the data were consistent and used as inputs

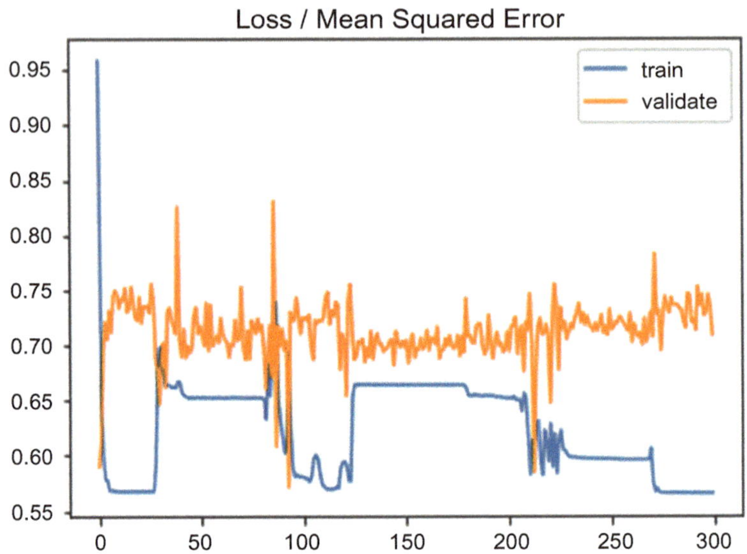

Fig. 4 ReLU-ReLU for binary cross entropy.

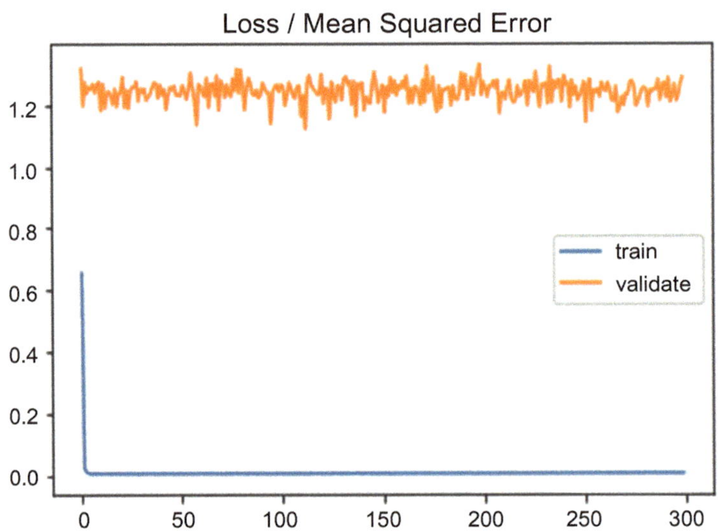

Fig. 5 ReLU-SELU for binary cross entropy.

for the DL models. In the second stage, classification was performed and the models were analyzed using 10-fold cross validations and a train-test split approach to determine their precision, recall, accuracy, MSE, and F1-scores. In the 10-fold cross-validation strategy, the best outcomes were observed from the LSTM model with accuracy of 96.45%, recall of

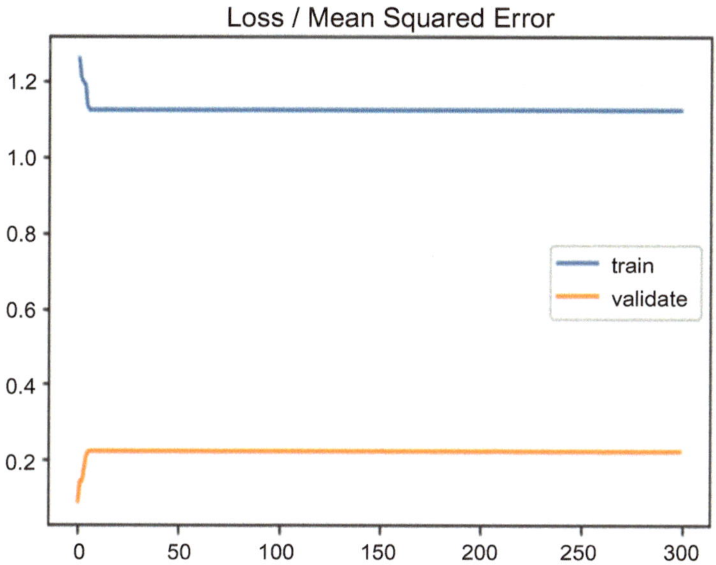

Fig. 6 ReLU-Softmax for binary cross entropy.

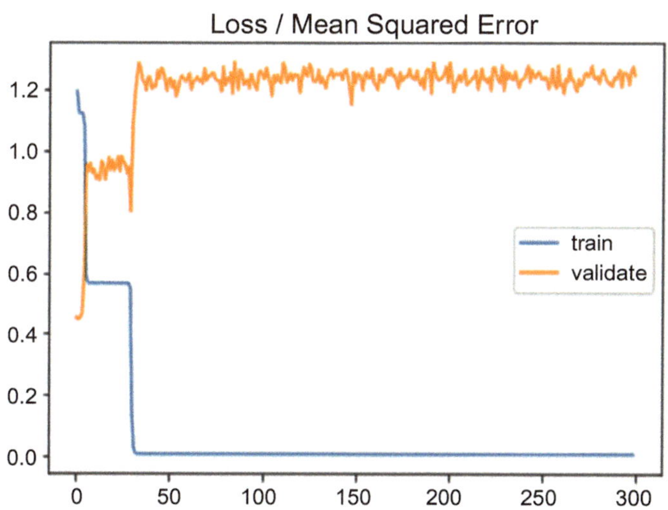

Fig. 7 ReLU-Softplus for binary cross entropy.

97.02%, F1 score of 92.66, precision of 94.75%, and MSE of 0.0923. In the train-test split approach, the LSTM model achieved accuracy of 97.189%, recall of 97.55%, and MSE of 0.0867. All DL models in the study achieved accuracy greater than 84%. Similar inferences can be made for precision and recall values.

Fig. 8 Different activation functioned logcosh model.

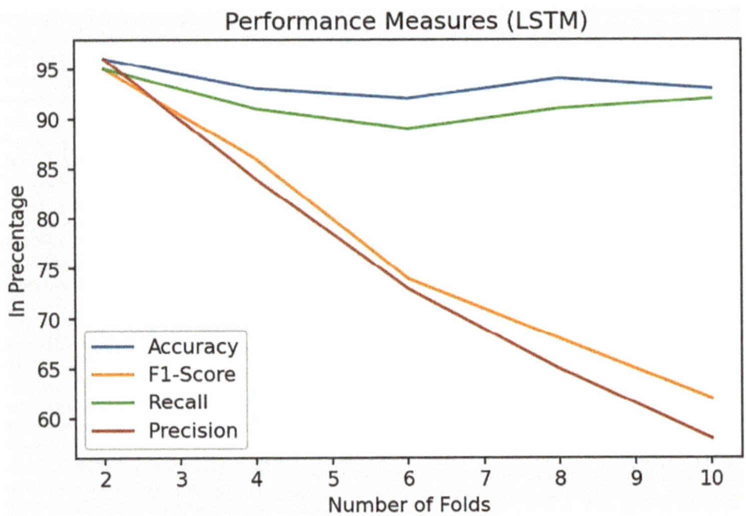

Fig. 9 Evaluation parameters for LSTM.

The major consideration in this study is the size of the data. Among 2549 datasets, only 640 datasets were utilized due to the disproportionate nature of the datasets. However, in a quantifiable population range, the prediction accuracy ranged between 92% and 97%. Hence the pursuance of these models can be elevated with a larger dataset.

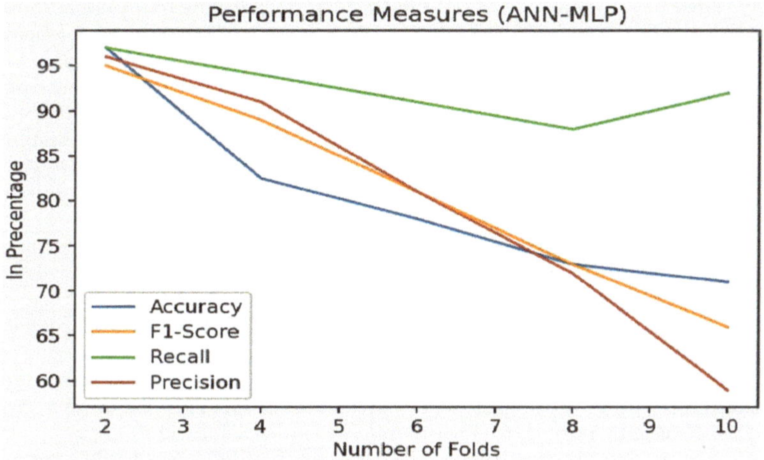

Fig. 10 Evaluation parameters for MLP-ANN.

Our results suggest that DL models can be applied to predict emotions from data collected from EEG sensors. Detecting positive and negative moods in real time can be useful for various applications, particularly in mental health care.

References

Abujelala, M., Abellanoza, C., & Sharma, A. (2016). *Brain-EE: Brain enjoyment evaluation using commercial EEG headband.* https://doi.org/10.1145/2910674.2910691.

Ali, S. I., & Shahzad, W. (2012). A feature subset selection method based on symmetric uncertainty and ant colony optimization. In *2012 International conference on emerging technologies* (pp. 1–6).

Amin, S. U., Alsulaiman, M., Muhammad, G., Mekhtiche, M. A., & Hossain, M. S. (2019). Deep Learning for EEG motor imagery classification based on multi-layer CNNs feature fusion. *Future Generation Computer Systems, 101*, 542–554.

An, X., Kuang, D., Guo, X., Zhao, Y., & He, L. (2014). A deep learning method for classification of EEG data based on motor imagery. In *International conference on intelligent computing* (pp. 203–210).

Asif, A., Majid, M., & Anwar, S. M. (2019). Human stress classification using EEG signals in response to music tracks. *Computers in Biology and Medicine, 107*, 182–196.

Bansal, D., & Mahajan, R. (2019). *EEG-based brain-computer interfaces: Cognitive analysis and control applications.* Academic Press.

Bigdely-Shamlo, N., Mullen, T., Kothe, C., Su, K.-M., & Robbins, K. A. (2015). The PREP pipeline: Standardized preprocessing for large-scale EEG analysis. *Frontiers in Neuroinformatics, 9*, 16.

Dai, M., Zheng, D., Na, R., Wang, S., & Zhang, S. (2019). EEG classification of motor imagery using a novel deep learning framework. *Sensors, 19*(3), 551.

Ding, S., Zhang, N., Xu, X., Guo, L., & Zhang, J. (2015). Deep extreme learning machine and its application in EEG classification. *Mathematical Problems in Engineering, 2015.*

Dutta, P., Biswas, S. K., Biswas, S., & Majumder, M. (2021). Parametric optimization of solar parabolic collector using metaheuristic optimization. *Computational Intelligence and Machine Learning, 2*, 7.

Dutta, P., & Kumar, A. (2020). Modelling of liquid flow control system using optimized genetic algorithm. *Statistics, Optimization & Information Computing, 8*(2), 565–582.

Dutta, P., Majumder, M., & Kumar, A. (2021a). An improved grey wolf optimization algorithm for liquid flow control system. *IJ Engineering and Manufacturing, 4*, 10–21.

Dutta, P., Majumder, M., & Kumar, A. (2021b). *Parametric optimization of liquid flow process by ANOVA optimized DE, PSO & GA algorithms.*

Dutta, P., Paul, S., & Kumar, A. (2021). Comparative analysis of various supervised machine learning techniques for diagnosis of COVID-19. In *Electronic devices, circuits, and systems for biomedical applications* (pp. 521–540). Elsevier.

Dutta, P., Paul, S., & Majumder, M. (2021). *An efficient SMOTE based machine learning classification for prediction & detection of PCOS.*

Dutta, P., Paul, S., Obaid, A. J., Pal, S., & Mukhopadhyay, K. (2021). Feature selection based artificial intelligence techniques for the prediction of COVID like diseases. *Journal of Physics: Conference Series, 1963*(1), 012167.

Dutta, P., Sen, S., Saha, S., & Chaki, S. (2021). Analysis of PCA based AdaBoost machine learning model for predict mid-term weather forecasting. *Computational Intelligence and Machine Learning, 2*(2), 41–52.

Dutta, P., Shaw, N., Das, K., & Ghosh, L. (2021). Early & accurate forecasting of mid term wind energy based on PCA empowered supervised regression model. *Computational Intelligence and Machine Learning, 2*(2), 53–64.

Dwivedi, R., Dey, S., Chakraborty, C., & Tiwari, S. (2021). Grape disease detection network based on multi-task learning and attention features. *IEEE Sensors Journal, 21*.

Faria, D. R., Premebida, C., Manso, L. J., Ribeiro, E. P., & Núñez, P. (2018). Multimodal Bayesian network for artificial perception. In *Bayesian networks-advances and novel applications* IntechOpen.

Frantzidis, C. A., Bratsas, C., Papadelis, C. L., Konstantinidis, E., Pappas, C., & Bamidis, P. D. (2010). Toward emotion aware computing: An integrated approach using multichannel neurophysiological recordings and affective visual stimuli. *IEEE Transactions on Information Technology in Biomedicine, 14*(3), 589–597.

Gramfort, A., Strohmeier, D., Haueisen, J., Hämäläinen, M. S., & Kowalski, M. (2013). Time-frequency mixed-norm estimates: Sparse M/EEG imaging with non-stationary source activations. *NeuroImage, 70*, 410–422.

Hosseini, S. A., & Naghibi-Sistani, M. B. (2011). Emotion recognition method using entropy analysis of EEG signals. *International Journal of Image, Graphics and Signal Processing, 3*(5), 30.

Jadhav, S., He, H., & Jenkins, K. (2018). Information gain directed genetic algorithm wrapper feature selection for credit rating. *Applied Soft Computing, 69*, 541–553.

Jain, I., Jain, V. K., & Jain, R. (2018). Correlation feature selection based improved-binary particle swarm optimization for gene selection and cancer classification. *Applied Soft Computing, 62*, 203–215.

Jas, M., Engemann, D. A., Bekhti, Y., Raimondo, F., & Gramfort, A. (2017). Autoreject: Automated artifact rejection for MEG and EEG data. *NeuroImage, 159*, 417–429.

Jasper, H. H. (1958). The ten-twenty electrode system of the International Federation. *Electroencephalography and Clinical Neurophysiology, 10*, 370–375.

Jeong, D., Yoo, S., & Yun, J. (2019). Cybersickness analysis with EEG using deep learning algorithms. In *2019 IEEE conference on virtual reality and 3D user interfaces (VR)* (pp. 827–835).

Kaggle. (2019). *EEG brainwave dataset: Feeling emotions.* https://www.kaggle.com/birdy654/eeg-brainwave-dataset-feeling-emotions.

Krigolson, O. E., Williams, C. C., Norton, A., Hassall, C. D., & Colino, F. L. (2017). Choosing MUSE: Validation of a low-cost, portable EEG system for ERP research. *Frontiers in Neuroscience*, *11*, 109.

Kumar, R. S., Misritha, K., Gupta, B., Peddi, A., Srinivas, K. K., & Chakraborty, C. (2020). A survey on recent trends in brain computer interface classification and applications. *Journal of Critical Reviews*, 7(11), 650–658.

LeCun, Y., Bengio, Y., & Hinton, G. (2015). Deep learning. *Nature*, *521*(7553), 436–444.

Li, H., He, F., Liang, Y., & Quan, Q. (2020). A dividing-based many-objective evolutionary algorithm for large-scale feature selection. *Soft Computing*, *24*(9), 6851–6870.

Lin, Y.-P., Wang, C.-H., Jung, T.-P., Wu, T.-L., Jeng, S.-K., Duann, J.-R., et al. (2010). EEG-based emotion recognition in music listening. *IEEE Transactions on Biomedical Engineering*, *57*(7), 1798–1806.

Liu, Y., Sourina, O., & Nguyen, M. K. (2010). Real-time EEG-based human emotion recognition and visualization. In *2010 International conference on cyberworlds* (pp. 262–269).

Murugappan, M., Ramachandran, N., & Sazali, Y. (2010). Classification of human emotion from EEG using discrete wavelet transform. *Journal of Biomedical Science and Engineering*, *3*(04), 390.

Oliveira, A. S., Schlink, B. R., Hairston, W. D., König, P., & Ferris, D. P. (2016). Induction and separation of motion artifacts in EEG data using a mobile phantom head device. *Journal of Neural Engineering*, *13*(3), 036014.

Petrantonakis, P. C., & Hadjileontiadis, L. J. (2009). Emotion recognition from EEG using higher order crossings. *IEEE Transactions on Information Technology in Biomedicine*, *14*(2), 186–197.

Plotnikov, A., Stakheika, N., De Gloria, A., Schatten, C., Bellotti, F., Berta, R., et al. (2012). Exploiting real-time EEG analysis for assessing flow in games. In *2012 IEEE 12th international conference on advanced learning technologies* (pp. 688–689).

Przegalinska, A., Ciechanowski, L., Magnuski, M., & Gloor, P. (2018). Muse headband: Measuring tool or a collaborative gadget? In *Collaborative innovation networks* (pp. 93–101). Springer.

Rundo, F., Rinella, S., Massimino, S., Coco, M., Fallica, G., Parenti, R., et al. (2019). An innovative deep learning algorithm for drowsiness detection from EEG signal. *Computation*, 7(1), 13.

Saikhu, A., Arifin, A., Nopember, I. T. S., Fatichah, C., & Nopember, I. T. S. (2019). Correlation and symmetrical uncertainty-based feature selection for multivariate time series classification. *International Journal of Intelligent Engineering and Systems*, *12*(3), 129–137. https://doi.org/10.22266/ijies2019.0630.14.

Sameer, M., Gupta, A. K., Chakraborty, C., & Gupta, B. (2019). Epileptical seizure detection: performance analysis of gamma band in EEG signal using short-time Fourier transform. In *2019 22nd international symposium on wireless personal multimedia communications (WPMC)* (pp. 1–6).

Symeonidou, E.-R., Nordin, A. D., Hairston, W. D., & Ferris, D. P. (2018). Effects of cable sway, electrode surface area, and electrode mass on electroencephalography signal quality during motion. *Sensors*, *18*(4), 1073.

Tabar, Y. R., & Halici, U. (2016). A novel deep learning approach for classification of EEG motor imagery signals. *Journal of Neural Engineering*, *14*(1), 016003.

Tjepkema-Cloostermans, M. C., de Carvalho, R. C., & van Putten, M. J. (2018). Deep learning for detection of focal epileptiform discharges from scalp EEG recordings. *Clinical Neurophysiology*, *129*(10), 2191–2196.

Witten, I. H., & Frank, E. (2002). Data mining: Practical machine learning tools and techniques with Java implementations. *ACM SIGMOD Record*, *31*(1), 76–77.

Youssef, A. E., Ouda, H. T., & Azab, M. (2018). MUSE: A portable cost-efficient lie detector. In *2018 IEEE 9th Annual Information Technology, Electronics and Mobile Communication Conference (IEMCON)* (pp. 242–246).

Zeng, H., Yang, C., Dai, G., Qin, F., Zhang, J., & Kong, W. (2018). EEG classification of driver mental states by deep learning. *Cognitive Neurodynamics*, *12*(6), 597–606.

Zhao, R., Ouyang, W., Li, H., & Wang, X. (2015). Saliency detection by multi-context deep learning. *Proceedings of the IEEE Conference on Computer Vision and Pattern Recognition*, 1265–1274.

CHAPTER 3

Machine learning and deep learning models for early-stage detection of Alzheimer's disease and its proliferation in human brain

Hima Bindu Maringanti[a], Minati Mishra[b], and Subhransu Pradhan[c]
[a]MSCB University, Baripada, Odisha, India
[b]P.G. Department of Computer Science, Fakir Mohan University, Balasore, Odisha, India
[c]Directorate of Health Services, Bhubaneswar, Odisha, India

Introduction

Named after the German psychiatrist and pathologist Alois Alzheimer, Alzheimer's disease (AD) is a chronic neurodegenerative disease (NDD) that generally affects the elderly population (Kang, Lin, Zhang, Shen, & Wu, 2021). It starts around the age of 65 years and contributes to as much as 60%–70% cases of dementia. Sometimes for a rare 10% of cases, it can even start at an early age of 30 years. Often confused with the normal symptoms of aging, the early symptom of AD is troubled recall of current memory. Symptoms like confusion, lack of motivation, mood swings, difficulties with language, partial loss of body functions, and withdrawal from society or even from family can be seen as the symptoms of disease severity (Dementia, 2021). The disease can be caused due to several reasons such as age-related changes in brain cells, sedentary lifestyle, environmental and genetic factors, and so on. Fig. 1 shows the different factors that contribute to the disease.

An inherent difficulty in diagnosing AD is that patients go through a long asymptomatic phase spanning approximately 10–20 years (Sperling et al., 2011) before showing clinical symptoms. To provide a biological assessment of the disease, AD has been recently defined as a pathology characterized by three main categories of biomarkers: amyloid, tau, and neurodegeneration (DeTure & Dickson, 2019). These three biomarkers can be measured using magnetic resonance imaging (MRI) and positron emission tomography (PET), which is an expensive technique, or in the case of amyloid and

Artificial Intelligence for Neurological Disorders
https://doi.org/10.1016/B978-0-323-90277-9.00024-9

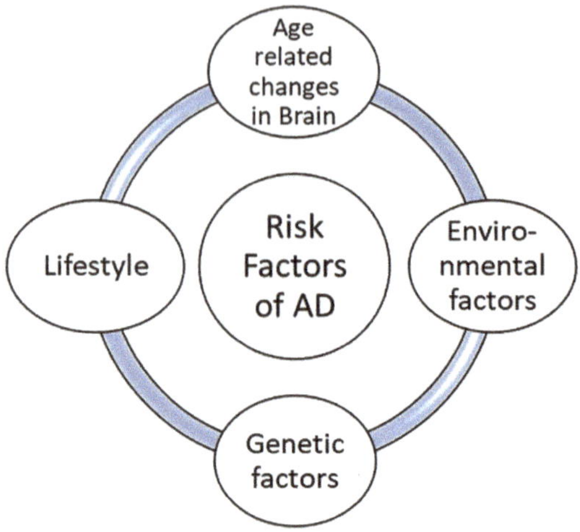

Fig. 1 Risk factors for Alzheimer's disease.

tau, also by lumbar puncture and blood collection. Monitoring these biomarkers is important to track disease progression and to potentially facilitate prevention or assess drug efficacy (Nader Clément et al., 2022).

How does AD affect the patient's life and normal functioning?

AD is a progressive NDD that causes impairments in cognitive, motor, and behavioral patterns. Though basic auditory ability is not impaired, greater auditory dysfunction is usually observed resulting in degraded speech. Circadian disruption and sleep disturbances are common in patients with AD and often occur early in the disease, which in turn impairs long-term memory establishment and hastens cognitive decline. In fact, sleep disturbances facilitate the progression of the disorder.

Primary symptoms involve memory recall (e.g., spatial, semantic, implicit, and episodic memory), diminished speech and linguistic abilities, and problems with executive functions, whereas secondary symptoms involve behavior and psychology of dementia such as depression, personality changes (e.g., irritation, aggression), presence of hallucinations, and delusional thinking. These symptomatic behaviors seriously disrupt daily activities and negatively affect quality of life. The prevalence of dementia is predicted to double by 2030 and even triple by 2050, with greater prevalence in low- and middle-income countries (Sutoko et al., 2021; Dementia, 2021).

Can AD onset be avoided or at least be delayed?

There is still a hope that onset of AD can be delayed and its progression arrested through exposure to bright sunlight, physical exercise, mental activities like solving constraint satisfaction problems, puzzles like crosswords and Sudoku, in addition to creative writing, recalling early childhood and writing short stories/ episodes about those times, and so on. Alternative therapies like yoga and meditation and mindfulness activities also help in arresting and avoiding the disorder to a large extent. A low-carb, sugar-free diet with extra servings of fresh fruits and vegetables may also help slow down the cognitive aging process.

Symptoms

Decline in mental ability in AD patients generally starts with loss of memory followed by decline in language and visuospatial abilities, though detection of memory loss is often delayed until daily life activities are affected severely. In almost 20% of cases, patients complain of difficulty in word-finding, organization, and navigation. Inability to make a decision, confusion, lack of concentration, aphasia, and apraxia are among the major symptoms of disease progression. In the final stage of AD, due to partial loss of body functions, patients become rigid, silent, and even bedbound in some cases requiring assistance for simple tasks such as eating, dressing, and toilet function. Often, death results from malnutrition, secondary infections, pulmonary emboli, heart disease, or, most commonly, aspiration (Kasper et al., 2016). We discuss the symptoms of AD through various stages in the sections that follow.

Early stage: Though difficulty in learning and loss of memory are the major symptoms of AD, older memories such as semantic and implicit memories are not affected much in comparison to recent memories. Difficulties with language such as reduced vocabulary, agnosia, and apraxia are more prominent than memory problems. Mild and unnoticeable difficulty in planning (apraxia), dressing, and writing may also be among the early-stage symptoms of the disease (Alzheimer's disease fact sheet, n.d.).

Middle stage: At this stage, due to severe memory problems, shrinking vocabulary, and paraphasias or unintended utterance of words, reading, writing, and speech difficulties become apparent. Increased risk of falling due to loss of body balance, failure to recognize close relatives, impairment of long-term memory, behavioral and neuropsychiatric changes, irritability, and aggression become evident (Alzheimer's disease fact sheet, n.d.).

Late stage: Complete loss of speech, complete dependence on caregivers, and indifference are the symptoms of the last stage of AD, eventually rendering the patient incapable of completing even simple tasks such as eating, drinking, and using the toilet independently. Mobility may be affected to such an extent that the patient becomes bedbound. The cause of death is usually an external factor, such as bedsores or pneumonia, not the disease itself (Alzheimer's disease fact sheet, n.d.).

Pathophysiology of AD

It is not clear how the disturbances in synthesis and accumulation of the beta-amyloid (Aβ) peptide give rise to the pathology of AD. Basically, calcium ion homeostasis is interrupted by accumulation of amyloid fibrils, which results in cell death. Further, glucose uptake by neurons is hampered due to deposits of Aβ protein in the mitochondria in the cells of persons suffering from AD. Cytokines and other inflammatory markers also play a role in the pathology of AD because of tissue damage or due to immunological response. Aging leads to mitochondrial dysfunction, and if the function of mitochondria can be improved then age-related oxidative stress can be reduced, thus delaying the process of AD. According to researchers, long-term physical activity can reduce reactive oxygen species (ROS) and improve cognitive health (Ionescu-Tucker & Cotman, 2021).

According to research, as little as 2% of AD cases have a genetic component, and these inherited forms of the disease generally occur at an early age. Though not confirmed and the reasons unknown, apolipoprotein E gene (APOE4) is believed to be a genetic risk factor for AD. Not all of the 25% of total population carrying one copy of APOE4 and the 2%–3% carrying two copies of the gene are at the risk of developing AD. However, recent research has discovered that the impaired ability of brain cells to process fats or lipids plays an important role in AD and the APOE protein helps the bloodstream in carrying cholesterol and other types of fats. Accumulation of excess Aβ in the forms of amyloid plaques, tau proteins, or neurofibrillary tangles in the brain reduces brain functioning leading to AD and other forms of NDD. The cases of AD that are not inherited are the result of environmental and genetic variations. Unlike the genetic cases that mostly show an early onset, 95% of sporadic cases start around the age of 65 years (Study reveals how APOE4 gene may increase risk for dementia, 2021).

Management of AD

Since AD is incurable, as of now, the only way to deal with the disease is to adapt measures to either prevent its onset or slow its progression. Even though there is no evidence available to support any particular prevention measure, and studies on measures to prevent the onset or progression have inconsistent results, epidemiological studies have proposed relationships between an individual's likelihood of developing AD and modifiable factors such as medications, lifestyle, and diet. Intellectual and physical activities such as meditation, regular exercise, playing chess, or regular social interactions have been linked to a reduced risk of AD. Since there is no cure for AD, its progression can only be managed by adopting the following measures.

Primary prevention: Primary prevention can be defined as intervention taken prior to the onset of the disease. It includes health promotion and specific protections such as health education, diet, exercise, lifestyle, social environment, and behavioral changes. Physical and mental exercises, meditation, educational and occupational achievements, controlled diabetes and hypertension, and so on can be seen as lifestyle activities alleviating the progression of AD, whereas sedentary lifestyle, stress, sleeplessness, smoking, and other unhealthy behaviors may be seen as the factors elevating the risk.

Secondary prevention: Secondary prevention is intervention that restricts the disease process at an early stage to prevent complications such as early diagnosis and adequate treatment.

Diagnosis: The AD brain is marked by excess ROS and bioactive metals such as copper, iron, zinc, and magnesium, which can promote the aggregation of Aβ into plaques and hyperphosphorylated tau into neurofibrillary tangles. Disease diagnosis involves study of the medical history of the person, determining the genetic linkage from the history of relatives, and observing the behavioral pattern of the person supported by neurological and neuropsychiatric conditions. Advanced medical imaging techniques such as magnetic resonance imaging (MRI), positron emission tomography (PET), computed tomography (CT), and single-photon emission computed tomography (SPECT) help differentiate AD from other biochemical and pathological cerebral abnormalities and certain types of dementia. A clear anatomical structure of the human brain can be seen and the change in size of the brain can be precisely measured using MRI. Since AD causes shrinking of the brain, an AD brain can be detected through MRI. The diffusion pattern of water at the microstructure level can be measured using diffusion tensor imaging (DTI); if there are any uncommon diffusion statistics seen in DTI then AD can be confirmed. A quantitative analysis of the pathological and metabolic changes can be studied via PET, leading to diagnosis of AD (Gao & Lima, 2022).

Treatment/medication: AD is incurable with no effective medicines available. The disease can be managed by careful use of cholinesterase inhibitor drugs, symptomatic administration of behavioral problems, and building rapport with the patient, family members, and other caregivers. Donepezil, rivastigmine, galantamine, and memantine are approved by the FDA for treatment of AD. If used before the onset of the disease, nonsteroidal anti-inflammatory drugs (NSAIDs) and statins may prevent dementia but cannot be used to treat AD. Antidepressants or cholinesterase inhibitors can be administered at early stages to handle depression. Because of their low anti-cholinergic side effects, selective serotonin reuptake inhibitors (SSRIs) are often used in treatment. Lifestyle diseases such as hypertension, diabetes, smoking, and hypercholesterolemia are factors that can cause or worsen AD, hence the use of anticholesterol, antidiabetic, and antihypertensive drugs can be used to reduce the risk (Kasper et al., 2016).

Tertiary prevention: Tertiary prevention involves actions taken after the early stage of AD, including limiting progress of the disease and rehabilitation through caregiving, continuation of treatment, and psychosocial interventions. Since AD is noncurable and makes the person suffering from the disease completely dependent upon others for normal day-to-day activities, caregiving is an important part of treatment and needs to be carefully managed over the course of the disease. Diet is an adjustable risk factor for the development of AD. The Mediterranean diet and the DASH diet are considered to support mental health, thus the authors propose the MIND diet plan, which combines the Mediterranean and DASH diets (Dominguez & Barbagallo, 2018). Fig. 2 shows tertiary prevention methods.

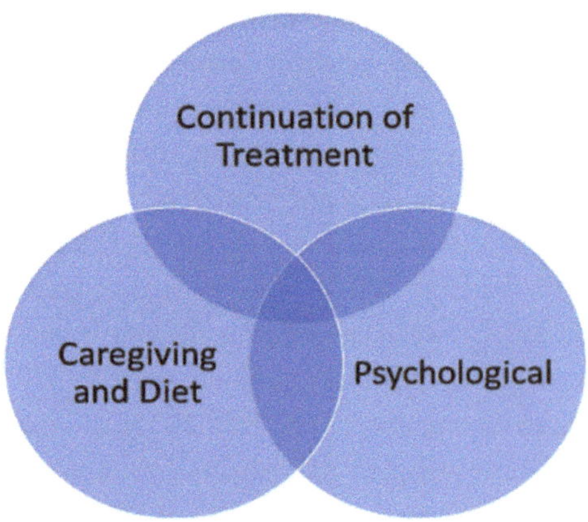

Fig. 2 Tertiary prevention in AD management.

Introduction to machine learning and deep learning and their suitability to AD diagnosis

Artificial neural networks (ANNs) are mathematical models used for solving all types of artificial intelligence (AI) problems, including recognition and identification tasks, prediction, pattern matching, and discovering or proving/disproving hypotheses about various problem domains, especially medical diagnosis. Such machine learning (ML) systems are being used by medical experts as digital tools functioning as expert systems.

Neural networks (NNs) have also been used to study the nonlinear course of AD. In a longitudinal study of 704 patients with AD, Grossi et al. found that a mixed-effects linear model provided the best results in the diagnosis of the disease. ANNs have been used for limited applications of modeling spatial characteristics of the human auditory system. Jenison and Fissell modeled a three-dimensional auditory space to simulate the virtual experience of free-field spatial hearing under headphones (Jenison & Fissel, 1996).

Contemporary models of ML are called deep learning (DL) models. These models consist of an increased number of hidden layers and introduce more complexity than simple ML algorithms. Convolutional neural networks (CNNs) and recurrent neural networks (RNNs) are the models currently in vogue, used to create a more realistic experience of the problem domain and its solution(s). In a recent paper (Ahmad & Pothuganti, 2020), about 5000 MRI images from the ADNI Dataset were used to run regional CNN and faster R-CNN models followed by a support vector machine (SVM) to identify abnormal areas in the MRI images (Ahmad & Pothuganti, 2020). Gao and Lima also in their paper have studied the applications of deep learning in detection of AD (Gao and Lima, 2022).

Ensemble learning architecture with a 2D-CNN comprising multiple models and multiple slices has been adopted to enhance AD diagnostics using a limited sample size (Kang et al., 2021). The experiment included 1.5TT1-weighted baseline sMRI scans from 798 participants in the ADNI-1 cohort. These subjects were divided into three groups (AD, mild cognitive impairment (MCI), and cognitively normal (CN)) according to their baseline diagnoses. VGG16, ResNet50, and the DCGAN discriminator were used as 2D slice classifiers. Two transfer strategies were implemented: domain transfer (i.e., fine-tuning pretrained classic CNNs with GMDM images) and task adaptation transfer (i.e., using an AD versus CN classifier as a baseline model for an MCI versus CN task) (Kang et al., 2021).

Disease progression modeling (DPM) is a challenging task for NNs, especially due to the incomplete and uncertainty of AD data, the disease's causes and effects on the brain functional areas or Broadman areas (BAs), and the predictability of disease progression. This ineffectiveness is probably due to noninclusion of temporal correlation between various biomarker parameters. Mehdipour Ghazi et al. (2019) in their work have applied an RNN to diagnose AD, with a preprocessing step to fill the gaps mentioned above. The different parameters considered by the researchers for building a long short-term memory (LSTM) model were the six volumetric MRI biomarkers: volumes of ventricles, hippocampus, whole brain, fusiform, middle temporal gyrus, and entorhinal cortex. The results were found to have a lower mean absolute error (MAE) compared to the regular regression model of prediction (Mehdipour Ghazi et al., 2019).

A dynamic model, SimulAD, on the independent GMC clinical cohort, a subset of ADNI dataset, was developed and experimented, resulting in the ability to differentiate subjects across clinical groups for both cohorts (Nader Clément et al., 2022).

Conceptual alignment deep neural networks (CADNNs) utilize some hidden neurons to learn human-interpretable concepts, while other neurons are trained freely. CADNNs achieve performance comparable to DNNs trained without interpretability constraints. Dai and Wang used a conceptual alignment deep auto-encoder to analyze tongue images that represent different body constituent types based on traditional Chinese medicine principles. Koh et al. introduced concept bottleneck models for osteoarthritis grading and used 10 clinical concepts such as joint space narrowing, bone spurs, calcification, and others. LaLonde et al. proposed a multitask capsule architecture called X-Caps for encoding high-level visual attributes within the capsule vectors to perform explainable diagnosis for lung nodule malignancy prediction.

State of the art/national and international status

The effects of oxidants and antioxidants in AD were studied by Goshtasbi et al. Alcohols, peroxides, ketones, and oxidized cholesterol are the byproducts of oxidative stress and are responsible for causing severe damage to neurological tissues and cells. Oxidative stress may be caused due to disproportionate cellular mechanisms and ROS, which may cause NDDs by impairing tissues and cells. According to the authors, clinical use of antioxidants can reduce age-related NDD by reducing oxidative stress (Goshtasbi et al., 2022).

Goenka and Tiwari proposed a three-class CNN through three neuroimaging computational methods for classification of AD. They experimentally proved that their proposed model has 97.48%, 96.62%, and 86.49% classification accuracies for large, medium, and small patches, respectively (Goenka & Tiwari, 2022).

According to Kang et al., because AD is basically a nontreatable NDD, monitoring and prediction of the disease are of paramount importance. In their papers, the authors proposed an ensemble learning model based on a 2D CNN. Due to unavailability of a large dataset, this model was found to be more efficient in comparison to the 3D CNN models. In stage one of this three-stage model, multislice predictions were combined with complementary results using a majority voting ensemble. An ensemble classifier using three generative adversarial network discriminators, such as VGG16, ResNet50, and DCGAN, were used in stage two for robust classification of AD from non-AD normal cognitive brain images. The learning was transferred using the domain and task adaptation transfer techniques. Results showed that the proposed ensemble learning model outperformed the single classifiers (Kang et al., 2021).

Ionescu-Tucker and Cotman studied the effect of oxidative stress on AD and brain aging. According to the authors, mitochondrial oxidative stress is a leading factor of AD and because there are no effective remedies available for the disease, long-term physical activities are considered to be the most promising among all therapeutic methods for delaying the process of brain aging. Long-term physical activities improve cognitive health by reducing the level of ROS (Ionescu-Tucker & Cotman, 2021).

Since AD is asymptomatic, it is a difficult task to detect the disease at an early stage. Sutoko et al. used DNNs and optimized feature selection and cross-validation to detect AD symptoms in mice. According to the authors, the model detects AD risk based on compulsive and learning behaviors with an accuracy of 89.3%–9.8% at a very early age of 8–12 months in older mice with underdeveloped symptoms (Sutoko et al., 2021).

Ahmad and Pothuganti compared ML techniques such as SVM, fast RCNN, and faster RCNN to detect AD. Study results showed that among the three models, faster RCNN is more efficient in terms of training and testing times (Ahmad & Pothuganti, 2020).

Atri compared signs and symptoms of normal aging with those of AD based on several parameters such as memory, problem solving and planning ability, completion of tasks accustomed, recognition of time and location, understanding of pictures and spatial relationships, communication ability, recollection and usage of words and names during verbal and written communications, ability of step repetitions, behavior and temperament, social activities, interpersonal interactions, and decision making. According to the author, persons suffering from AD show excessive dependence on memory aids, fail to recollect recently learned information, important dates, and appointments, and may repeat the same questions that have already been answered. They find it difficult to start, develop, follow, and finish plans and projects. Working with numbers and paying bills seems to be a troublesome task for them. Often struggling with vocabulary and avoiding social interactions are some of the primary symptoms of AD. The author emphasized biomarker-based multitiered testing for the diagnosis of AD, as the clinical diagnosis of the disease is cumbersome and lengthy (Atri, 2019).

Krishnamurti et al. employed an NN model to measure auditory impairments in AD. They concluded that with age central auditory dysfunction occurs irrespective of changes in cognitive function. AD and non-AD groups exhibit similar types of auditory attention. However, persons with AD show less auditory processing abilities compared to their non-AD counterparts (Krishnamurti, Drake, & King, 2011).

Conclusion

AI is useful for clinicians, helping them with disease diagnosis, prediction, and treatment. NNs are approximate solutions to problems that have come about with the advent and availability of myriad data and datasets. It has become more interesting for researchers to test the efficacy of their algorithms on these datasets. As the number of hidden layers increases, as is the case with DNNs, the more complex the algorithm becomes and encryption of the processes lead to convergence of neural networks. The internal process is supposed to be hidden, known as a *black box*. This poses a challenge to interpreting the results and mapping the quantitative, symbolic outputs to medical terms, diagnosis, and grade of the disease, as clinicians would do. Accordingly, the inputs need to be preprocessed and smoothed, missing values in datasets need to be filled in, the target values need to be determined prior to designing NNs, the number of hidden layers needs to be decided, and so on.

The nature of the disease and the selection of the dataset, comparing it with real data collected from hospitals, affects the efficacy of the algorithms used for the diagnosis and prediction of AD.

References

Ahmad, I., & Pothuganti, K. (2020). Analysis of different convolution neural network models to diagnose Alzheimer's disease. *Materials Today: Proceedings*. https://doi.org/10.1016/j.matpr.2020.09.625.

Atri, A. (2019). The Alzheimer's disease clinical spectrum: Diagnosis and management. *Medical Clinics of North America*, *103*(2), 263–293.

Dementia. (2021). World Health Organization. https://www.who.int/news-room/fact-sheets/detail/dementia.

DeTure, M. A., & Dickson, D. W. (2019). The neuropathological diagnosis of Alzheimer's disease. *Molecular Neurodegeneration*, *14*, 32. https://doi.org/10.1186/s13024-019-0333-5.

Dominguez, L. J., & Barbagallo, M. (2018). Nutritional prevention of cognitive decline and dementia. *Acta bio-medica: Atenei Parmensis*, *89*(2), 276–290. https://doi.org/10.23750/abm.v89i2.7401.

Gao, S., & Lima, D. (2022). A review of the application of deep learning in the detection of Alzheimer'sdisease. *International Journal of Cognitive Computing in Engineering*, *3*, 1–8. https://doi.org/10.1016/j.ijcce.2021.12.002 (December 2021).

Goenka, N., & Tiwari, S. (2022). AlzVNet: A volumetric convolutional neural network for multiclass classification of Alzheimer's disease through multiple neuroimaging computational approaches. *Biomedical Signal Processing and Control*, *74*, 103500. https://doi.org/10.1016/j.bspc.2022.103500 (January).

Goshtasbi, H., Pakchin, P. S., Movafeghi, A., Barar, J., Castejon, A. M., Omidian, H., et al. (2022). Impacts of oxidants and antioxidants on the emergence and progression of Alzheimer's disease. *Neurochemistry International*, *153*, 105268. https://doi.org/10.1016/j.neuint.2021.105268 (August 2021).

Ionescu-Tucker, A., & Cotman, C. W. (2021). Emerging roles of oxidative stress in brain aging and Alzheimer's disease. *Neurobiology of Aging*, *107*, 86–95. https://doi.org/10.1016/j.neurobiolaging.2021.07.014.

Jenison, R. L., & Fissel, K. (1996). A spherical basis function neural network for modeling auditory space. *Neural Computation*, *8*, 115–128.

Kang, W., Lin, L., Zhang, B., Shen, X., & Wu, S. (2021). Multi-model and multi-slice ensemble learning architecture based on 2D convolutional neural networks for Alzheimer's disease diagnosis. *Computers in Biology and Medicine*, *136*. https://doi.org/10.1016/j.compbiomed.2021.104678 (July).

Kasper, D. L., Fauci, A. S., Hauser, S. L., Longo, D. L., Jameson, J. L., & Loscalzo, J. (Eds.). (2016). *Harrison's manual of medicine* McGraw Hill Education.

Krishnamurti, S., Drake, L., & King, J. (2011). Neural network modeling of central auditory dysfunction in Alzheimer's disease. *Neural Networks*, *24*(6), 646–651. https://doi.org/10.1016/j.neunet.2011.02.010.

Mehdipour Ghazi, M., Nielsen, M., Pai, A., Cardoso, M. J., Modat, M., Ourselin, S., et al. (2019). Training recurrent neural networks robust to incomplete data: Application to Alzheimer's disease progression modeling. *Medical Image Analysis*, *53*, 39–46. https://doi.org/10.1016/j.media.2019.01.004.

Nader Clément, A., Ribaldi, F., Frisoni, G. B., Garibotto, V., Robert, P., Ayache, N., et al. (2022). SimulAD: A dynamical model for personalized simulation and disease staging in Alzheimer's disease. *Neurobiology of Aging*. https://doi.org/10.1016/j.neurobiolaging.2021.12.015.

Study reveals how APOE4 gene may increase risk for dementia. (2021). National Institute on Aging. https://www.nia.nih.gov/news/study-reveals-how-apoe4-gene-may-increase-risk-dementia.

Sperling, R. A., Aisen, P. S., Beckett, L. A., Bennett, D. A., Craft, S., Fagan, A. M., et al. (2011). Toward defining the preclinical stages of Alzheimer's disease: Recommendations from the National Institute on Aging-Alzheimer's Association workgroups on diagnostic guidelines for Alzheimer's disease. *Alzheimer's & Dementia*, 7(3), 280–292.

Sutoko, S., Masuda, A., Kandori, A., Sasaguri, H., Saito, T., Saido, T. C., et al. (2021). Early identification of Alzheimer's disease in mouse models: Application of deep neural network algorithm to cognitive behavioral parameters. *iScience*, *24*(3), 102198. https://doi.org/10.1016/j.isci.2021.102198.

Alzheimer's disease fact sheet. (n.d.). National Institute on Aging. Retrieved April 9 from https://www.nia.nih.gov/health/alzheimers-disease-fact-sheet.

Further reading

Dai, Y., Wang, G., & Li, K. C. (2018). Conceptual alignment deep neural networks. *Journal of Intelligent Fuzzy Systems*, *34*, 1631–1642.

Koh, P. W., Nguyen, T., Tang, Y. S., Mussmann, S., Pierson, E., Kim, B., et al. (2020). Concept bottleneck models. In H. Daumé III,, & A. Singh (Eds.), *Proceedings of the 37th international conference on machine learning, PMLR* (pp. 5338–5348). http://proceedings.mlr.press/v119/koh20a.html.

LaLonde, R., Torigian, D., & Bagci, U. (2020). Encoding visual attributes in capsules for explainable medical diagnoses. In *Medical image computing and computer assisted intervention—MICCAI 2020—23rd international conference, proceedings* (pp. 294–304). https://doi.org/10.1007/978-3-030-59710-8_29.

CHAPTER 4

Convolutional neural network model for identifying neurological visual disorder

Brijesh K. Soni and Akhilesh A. Waoo
Department of Computer Science and Technology, AKS University, Satna, Madhya Pradesh, India

Introduction

The process humans use to acquire knowledge about their surroundings is called visual perception. It begins when external light enters the eye and induces electrical signals that are subsequently processed within the brain where an image is created. The process of knowledge acquisition is a cognitive process that is strictly distinct from optical mechanisms. The camera and eye have some optical similarities. A camera can capture images of the environment, but it does not have perceptual or cognitive abilities like an eye. The major components of the visual pathway are the eye, optic nerve, optic chiasma, optic tract, lateral geniculate body, optic radiation, visual cortex, and visual-association cortex. The visual cortex is the most prominent cortical region of the brain that is wholly responsible for receiving, integrating, and processing visual information at relay mode (LGN) from the retinas. This occurs in the occipital lobe of the cerebral cortex, which is the posterior region of the brain. The overall visual cortex is divided into five parts (V1–V5) based on physical structure and corresponding function. Visual information travels from the retina to the visual cortex. It passes through the thalamus area first and then synapses in the lateral geniculate nucleus (LGN) and travels to the first area of the visual cortex (V1). Each hemisphere has one visual cortex, which receives visual information from the collateral eye. That means the right cortical areas process information from the left eye and the left cortical areas process information from the right eye. The initial purpose of the visual cortex is to receive, segment, and integrate visual information. Processed visual information is subsequently sent to other regions of the brain from the visual cortex to be analyzed and utilized (Gupta, Ireland, & Bordoni, 2021). This process is specialized and enables the brain to recognize objects and patterns quickly without highly significant and

Artificial Intelligence for Neurological Disorders
https://doi.org/10.1016/B978-0-323-90277-9.00003-1

conscious effort. An advantage of this specialization process is that other cortical regions are mostly free to perform other important computational activities, such as those responsible for executive functioning and decision making. Because visual information processing is an unconscious process, frequently it leads to misinterpretations of visual information, which are demonstrated by the efficacy of visual illusions. Usually, people experience various types of visual problems in their lives (James, Zoccolan, & Rust, 2012).

Some people cannot see objects from far away and some people have problems reading small fonts. Usually, these types of conditions can be easily treated with eyeglasses, but sometimes parts of the eye or brain required to process images become damaged. In these cases, severe or total vision loss can occur and vision cannot be fully restored even with clinical treatment, medical surgery, or glasses. According to a report by the American Blind Foundation, approximately 10 million people in the United States are visually impaired. Visual impairmentis a common term frequently used to describe vision loss. It may be partial or full vision loss. Some people are fully blind, but many others are partially or "legally" blind. Cortical blindness is an important cause of blindness due to damage to the visual cortex in the occipital lobe and is associated with posterior circulation stroke. This condition requires proper diagnosis and management for improved outcomes. Cortical blindness is defined as vision loss without any logical ophthalmic causes and with minor pupillary light reflexes by bilateral lesions in the striate cortex in occipital lobes. Primarily, cortical blindness is a type of cerebral blindness defined as loss of vision that is secondary to damage to the visual pathways to the lateral geniculate nucleolus (Salvin, 2016). In the Roman era, the philosopher Seneca reported a case of a slave, despite her blindness, she didn't accept and kept constantly arguing about room darkness. Michel de Montaigne also reported a case of a patient who, despite obvious signs of blindness, didn't believe he was blind. Gabriel Anton described patients having bilateral occipital lobe lesions who were fully blind but were not aware of their blindness due to memory disorder (confabulation). French scientist Joseph François Babinski described this phenomenon as anosognosia (Sarkar & Tripathy, 2022).

Convolutional neural networks are a class ofneural networks that specialize in processing data that has a grid-like topology, such as an image. A digital image is a binary representation of visual data. It involves a series of pixels arranged in a grid manner that has values that denote the brightness and color of each pixel. In the brain, each neuron is working in its receptive field and connected to other neurons in a manner that covers the whole visual field.

Usually, each neuron gives a response to stimuli only in a restricted region of a visual field called the receptive field. In biological vision, each neuron in a convolutional neural network processes data only in its receptive field. All layers are arranged in a manner so that they detect normal patterns first and then detect complex patterns. Using a convolutional neural network can enable computers to see (Yamashita, Nishio, Do, & Togashi, 2018). Image recognition involves various methods for analyzing and detecting images to perform the automation of a specific task. It is a technology that can identify people, places, and objects within an image and draw by analyzing those objects. Image or video detection can be performed differently, depending on the type of data or concept find. If a model can recognize a specific element it can normally assign an image to large data. Image classification refers to extracting information classes from a multiband raster image. Raster view from input image classification is very useful to create thematic maps. Based on the interaction between the system analyst and computer system during the classification process, classification methods can be categorized as either supervised or unsupervised (de Mingo López, Lucas, Blas, & Ivanova, 2019).

ArcGIS including the Spatial Analyst module is a full application package in a multivariate toolset for performing supervised and unsupervised classification. Because classification is a multistep process, the Image Classification toolbar was developed to provide an integrated environment for performing classification with tools. The ArcGIS software also includes additional modules to analyze input data, create training samples and signature files, and determine the quality of training samples and signature files. In the era of instant image sharing, it is essential to incorporate the language of images. Although the brain can easily process what exactly an image means, and what it indicates and associates with, developing a machine to perform the same thing is a very complicated process. To decode an image, a computer views the image as a two-dimensional (2D) array of numbers. If the image includes various colors, it becomes a three-dimensional (3D) array where the last field signifies the corresponding RGB value. The computer's job is to act like a human brain and take the image as an input and provide a classified output. To accomplish this, convolutional neural networks were developed. Three-dimensional convolutional neural networks duplicate simple and complex cells of the brain, including receptive fields that humans experience through their senses.

This chapter describes convolutional neural networks, their structure and biological connection, and their functionality (Lundervoldab & Lundervoldacd, 2019). It also discusses discuss how we can identify and understand the various cortical vision disorders with the help of an artificial deep neural network model, specifically a convolutional neural network model called the Neurological Visual Disorder Identifying Model (NVDIM). The architectural modifications required in this model may be weight adjustment with backpropagation and increasing the number of layers for maintaining the receptive field, with the intent of modifying clear visual stimulus into blur scenes and vice versa. In the proposed model, we attempt to put an identical and enough amount of blood flow inside the visual cortex in the occipital lobe to the node weight applying in a convolutional neural network.

In a deep learning domain, the receptive field is defined as the size of the region in the input that produces a feature. It is a measure of the association of an output feature to the inputregion. A convolutional unit depends on a local region of input. Therefore, it never refers to the receptive field on fully/completely connected layers since each unit has access to the entire input region. Our aim is to provide insight into the concept for understanding and analyzing how deep convolutional networks perform with when light weight actions take place. An activation map is output activation for a given filter, and the definition is the same regardless of what layer it is on. It is a mapping that correlates to the activation of various parts of the image. It is also referred to as a feature map because it is also a mapping of where a certain kind of feature is found in an image. High activation means a certain feature was found. A rectified feature map is a feature map created using the rectified linear activation function (ReLU) function. Convolution uses a kernel to extract particular features from an input image. A kernel matrix is slid across the image and multiplied with input such that output is enhanced in a certain desirable manner. The filter certainly concatenation of multiple kernels, which are associated with a input channels. Filters must always be in one dimension, which is more than the kernels. Given it is a 2D convolution, filters are 3D matrices (which is a concatenation of the kernels); thus, for a convolutional neural network layer with kernel dimensions height × width and input channels (k), the filter dimensions are $k \times$ height × width. Commonly in a convolutional neural network, the convolution layer consists of multiple such types of filters, assuming the existence of only a single filter unless specified since the same behavior is reflected across all the filters (Ganesh, 2019).

Human visual system

The visual system associates both eyes and the brain. Light enters the eye and hits the retina, which enables light receptors to send an electrical pulse through the optic nerve, which travels to the back of the brain where the first step of visual perception takes place (Williams, 2018). The further brain sends filtered signals forward through the brain until they can be seen. In various ways, the visual system plays an important role in the brain.

Visual cortex

The visual cortex is a region of the cerebral cortex that is responsible for processing visual data. It is situated posterior in the occipital lobe. Principally, sensory input begins from the eyes and travels through the lateral geniculate nucleus and until it reaches the visual cortex. The area of the visual cortex that receives sensory input information from the lateral geniculate nucleus is referred to as the primary visual cortex or striate cortex. It is also popularly referred to as V1 and Brodmann area 17. Additionally, extrastriate areas include areas V2, V3, V4, and V5 or Brodmann areas 18 and 19. Symmetrically, both hemispheres of the brain have a visual cortex; the visual cortex in the left hemisphere takes signals from the right visual field and the visual cortex in the right hemisphere takes signals from the left visual field (Huff, Mahabadi, & Tadi, 2021).

Vision disorders

Vision disorder is a loss of sense of vision; it does not mean an eye disease. Although many other vision disorders do have their immediate effect inside the eye, some others may occur at other locations in the optic pathway (Brady, 2021).

Cortical blindness

Cortical blindness is a type of neurological visual impairment. Neurological visual impairment is broadly classified as cortical blindness, cortical visual impairment, and delayed visual maturation. Neurological visual impairment and ocular visual impairment must be distinguished according to their causes and structures, in the brain and eye respectively. The pupils of an individual with cortical blindness will give a response to light, whereas those of an individual with ocular visual impairment will not see. Cortical blindness is total or partial vision loss in the normal-appearing eye due to damage to the brain's occipital cortex (Shibuki, Wakui, Fujimura, Tomikawa, & Hasegawa, 2020). Cortical blindness is acquired or hereditary and may also be transient in some cases. Acquired cortical blindness is mostly due to blood flow deficiency to the visual cortex and blockage from unilateral/bilateral posterior cerebral artery or cardiac surgery. Sometimes full loss of vision does not exist lifelong, and patients can recover some level of vision. Congenital cortical blindness is most often caused due to perinatal ischemic stroke, meningitis, and encephalitis. A patient with acquired cortical blindness might deny their vision loss, which is known as Anton Babinski syndrome (Davis, 2021).

Acquired cortical blindness

Acquired cortical blindness mostly occurs due to blood flow deficiency to the visual cortex and blockage from unilateral/bilateral posterior cerebral artery or ischemic stroke and cardiac surgery.

Congenital cortical blindness

Congenital cortical blindness refers to a collection of medical conditions that occur in individuals younger than 14 years old. If not treated properly, it can lead to permanent blindness.

Transient cortical blindness

Transient cortical blindness is described by loss of acquired vision, normal fundus, normal pupillary light reflexes, and unaltered extraocular movements (8, 9), and has been reported following cerebral angiography (10, 11). Symptoms may occur from the start of the cerebral angiography procedure until 12 h afterward (5).

Convolutional neural network

A convolutional neural network is like an ordinary neural network. It is designed by neurons that have weights and biases. Every respective neuron receives some input, performs dot product, and optionally follows with nonlinearity. The whole network follows a differentiable function that takes the raw input pixels of an image on one side and produces a class score at the other side. A convolutional neural network still has a loss function on the last layer and all ways developed for learning of neural network still significant (Mishra, 2020).

Image recognition

In the first step of image recognition, the maximum number of characteristics, called features, are extracted from the image. The fundamental unit of an image object is referred to as a "pixel." A pixel is denoted by a unique value or a combination of values and the whole range of these values is known as color depth. This means that the color depth denotes the greatest number of colors that represent various regions of an image surface. In the case of an 8-bit grayscale image or black-and-white image, the pixel has one value that starts from 0 and ends with a 255-color code. However, nowadays images use a 24-bit color code system.

A three color image is made up of the combination of three colors (Red-Green-Blue) (Leban, 2020). Here every potential color will be any one of 0–255 values. That means this color-generator method uses a tri-color system (RGB). Further, a pixel identifies a combination of three consequent values (R-102, G-255, B-102) refers to colorcode #66ff66. For example, an image with a width of 900 pixels and a height of 300 pixels height contains $900 \times 300 = 270{,}000$ pixels = 0.27 megapixels. Further, if an image with resolution of 1024×768 is a matrix of 1024 and 768 columns and rows respectively, the image contains a total of 0.78 megapixels. If an image is converted into such type of labeled feature, we can easily use these labeled images to train a software model. Such images can be used for each class to represent distinct images so that the model can be trained in a better way to recognize an object whether it is a cat or a mouse. Even if it is already known to which specified class the images belong, we can still use them to train the model. This is known as the supervised or labeled training method. The matrix used here can be assumed as a filter. The specified images enter the input layer of the network and labels are put forth in the output layer. The ultimate objective here is to train neural network models that mean an image with its features coming from the input side will match the label on the right and output side. Finally, when a model or network is well trained, it can be efficiently used to recognize a randomly selected image. Again, the new image will also follow the whole feature recognition or detection procedure (Das, 2020).

Image classification

Connectivity pattern in the convolutional neural network comes from research regarding the organization of the visual cortex. In the eye, individual neurons respond to the visual stimuli only in the receptive field, which is a restricted region. Receptive fields of different regions partially overlap so the entire visual field is covered. At the first level, the input image is divided into pixels. In the case of black/white images, those pixels are interpreted in a 2D array, and every pixel has a unique value between 0 and 255, where 0 is black and 255 is white; grayscale color is available between 0 and 255.

Based on the information, the computer can start working on the data. For a color image, there is a 3D array with a blue layer, a green layer, and a red layer. Each color has a value between 0 and 255. The desired color may generate by combining values in all three layers (Brownlee, 2019).

Cognitive application

The concept of the convolutional neural network was inspired by neurons in human brains. Goodfellow defined three important benefits of the convolutional neural network: equivalent representations, sparse interactions, and parameter sharing. Normally, fully connected networks share their weights and local connections, but convolutional neural networks use 2D input data such as an image. This process employs very few parameters, which both simplifies the training process and improves network performance. This happens in the same way as in the biological visual cortex cells. Only small regions of a scene, rather than the whole scene, are observed by these cells (Yamashita et al., 2018).

Neurological visual disorder identifying model

In this section, we discuss how we can use the proposed NVDIM to identify and understand the various cortical vision disorders.

Receptive field

The visual receptive field in the context of a convolutional neural network is identified as a portion of the input size for a particular feature. A precisely receptive field of a feature can be referred to by its centralized area and size. However, not all pixels in the receptive field are equally important with their corresponding feature. In a receptive field, the closer a pixel is to the center of the field, the more it contributes to the calculation of the output feature. This means that the feature not only looks at a particular region input image but also focuses specifically on the middle of the region (Adaloglou, 2020) (Fig. 1).

A convolutional neural network modifies its kernel biases based on receptive field performance. Since modern convolutional neural networks are deep, they contain a stack of multiple convolutional layers; the receptive field at every layer is different. If the layer is a little deeper in the architecture, its receptive field will be larger because its input space is activation maps from the subsequent previous layer (i.e., already down sampled input image).

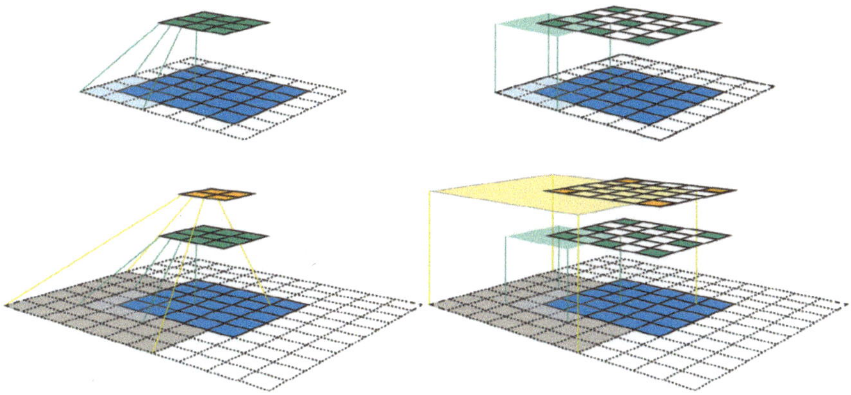

Fig. 1 Receptive field.

In the context of convolutional neural networks, the receptive field is defined as the size of the region at the input layer that produces the output features. It is a measure of output feature association at any layer to the input region (Baeldung, 2021).

The size of the receptive field is adequate if it covers a larger area than the largest flow amount of input data. If the receptive field in a convolutional neural network needs to be increased, this can be done in various ways. Convolutional neural networks have a unique layered architecture, specifically designed to incorporate the functionality of the biological nervous system. Having eachneuronin each layer associated with all neurons in the corresponding next layer, the neurons are gathered in the layered structure in such a manner to consider the spatial relationships among the various neurons concerning the basic dataset.

Convolutional neural networks were initially used in computer vision and digital image processing The specified data element that the neurons incorporate is usually an image object; each neuron in the input layer denotes a single pixel of that input image. The input layer contains all the input neurons, and neurons in the hidden layer will receive input from some of the remaining input neurons. Each neuron at the hidden layer does not have to take input from all specified neurons in the previous layer, even if it uses a receptive field in which each neuron takes input only from some selected neurons from the previous layer. Accordingly, the receptive field of a neuron in one of the input layers incorporates only a small portion of an input image, whereas the receptive field of an input neuron in the next layers contains receptive fields collectively from several input neurons in the layer before such that a neuron in a higher layer considered at a larger area of image than encompasses a neuron in a lower layer. Accordingly, each subsequent higher layer can easily learn the required features of the specified input image. The use of receptive fields in this way helps convolutional neural networks detect visual patterns more accurately than other artificial neural network models.

In convolutional neural networks, the receptive field is a region of input that influences a definite unit of the network. This input area can be not only the input of the network but also the output from other units in the network so that this receptive field can be evaluated on the basis of the given input value and it's another receiver unit. If required, the receptive field can be improved by using various techniques such as stacking, striding, pooling, and son. Traditionally when more layers are stacked, the receptive field increases linearly. However, not all pixels in the receptive field contribute similarly to the output unit's response. When working with forward pass, central receptive field pixels can propagate respective information to output using many different roots, as they are part of multiple output unit's calculations (Hien, 2017).

Activation map

An activation map or feature map is output activation for a given filter, and the definition is the same regardless of what layer it is on. A high-level activation means a case where a particular feature was found. Supervised learning and machine learning take data and results as inputs while training to generate rules or data patterns. Fundamental structures or patterns of input data produced by the convolutional neural network model completely represent how exactly the actual outcome is used to extract from the input data pattern. The convolutional neural network model takes a 2D image and the corresponding category of the input image, like a flower or a doll, as an input object. As a result of training, we get trained weights, which are data patterns or rules extracted from images. The image will be the only input passed to the trained model, and the trained model will output the class of the image. The specified class of input images will uniquely identify the learned patterns of data during the training process. We can generate feature maps or activation maps by applying filters to the input image and using the supportive ReLU activation function. Filters are the basic tools to detect various properties that existed in an input image such as lines, bends, edges, and so on. The pooling method is used over the activation maps for invariance translation. The pooling method is somehow based on the law; if we change the input pattern by a very nominal amount, the corresponding outputs remain unchanged. There are three pooling methods available: min pooling, max pooling, and average pooling. Max pooling is used most frequently because it gives better results compared to other pooling methods. By flattening all the input objects and passing these inputs to a convolutional neural network model, the model will provide the appropriate class of the given object. The outcome may be in the binary class classification like a flower or doll or it can be a multiclass system for categorizing various digits or other similar objects. A convolutional neural network generally looks like a black box and acquires features that are not easily interpretable. We can input an image and the model will process it internally and return the appropriate results. The method of feature visualization translates abstract features available in an input image into a visually understandable form with the help of computer vision. It is also helpful for identifying the learned features externally. First, it visualizes different filters or feature detectors that are applied to the input image and second, it visualizes feature maps or activation maps that are generated. Convolutional neural networks frequently use trained filters for convolving the feature maps from the specified previous layer (Fig. 2).

Filters are nothing but the 2D weights, and these associated weights pose a spatial relationship with the other weights. Artificial neural network models are frequently referred to as being opaque, which means they are not capable of explaining the reason why a particular decision or prediction was made (Sarkar, 2019). Convolutional neural network models are designed to process visual data, and their structure and function suggest that should be more interpretable than other types of neural networks. Models are evaluated along with small linear-type filters and consequent results to apply filters called feature maps or activation maps. Here, both filter and

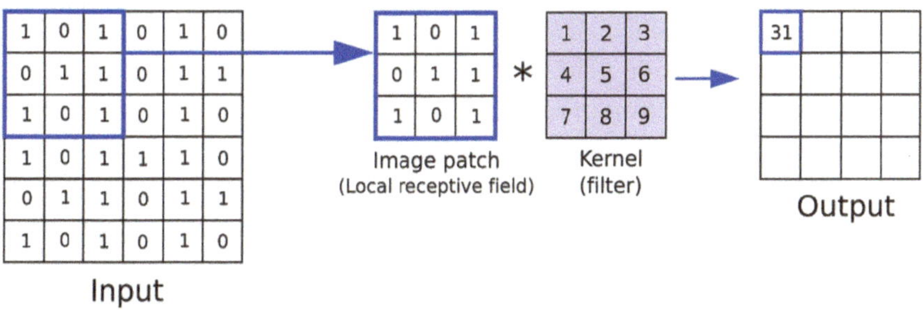

Fig. 2 Activation map.

feature maps can be visualized when required. Further, we can understand and design small filters, such as line detectors. It might be that visualizing a filter within a learned convolutional neural network will provide inside into how exactly the model works. The activation maps that generate from applying a filter to the input image and activation map output by the prior layer could provide insight into the internal representation that the model has a certain input at a given point in the model. As mentioned, deep learning models are opaque, which means they can make a useful prediction; however, it is not clear how or why the given prediction was made. Internal structures of convolutional neural networks are designed to operate 2D image data and preserve the spatial relationship for what was learned via the network model. The 2D filters learned by the model can also be inspected and visualized for discovering various features the model will detect and activation map output through convolutional layers can be inspected to understand which features were detected for a given input image (Choraria, 2020).

Kernel filter

A kernel is a matrix that slides across an image and multiplies with the input such that output is improved in a definite and desirable way. Convolution is using a kernel to extract certain features of an input image. A filter is a multiplication of multiple kernels, and each kernel is assigned to a particular channel of the input. A kernel is a matrix of weights, which is multiplied with the input value to find useful features. The number of dimensions of the kernel matrix determines the name of the convolution. For example, in 2D convolutions, the kernel is a 2D matrix (Ganesh, 2019) (Fig. 3).

Convolutional neural networks use kernels for detecting features in images. Every kernel looks at a certain part of the input image, performs multiplications, and then moves by a defined number of pixels or stride. The area where a particular convolutional neural networks kernel is looking is known as the receptive field. A deep neural network model is a stack of layers defined by the action of several filters on input; those filters considered as kernels. For example, the kernels in the convolutional layer are the convolutional filters. The kernel size is denoted by the width x height of the filter mask. The max pooling layer returns pixels with the highest value from pixels set within the kernel that swept across input subsampling. They can be thought themselves only as feature extractors. A kernel is a matrix mostly used to apply effects such as sharpening, outlining, blurring, or embossing.

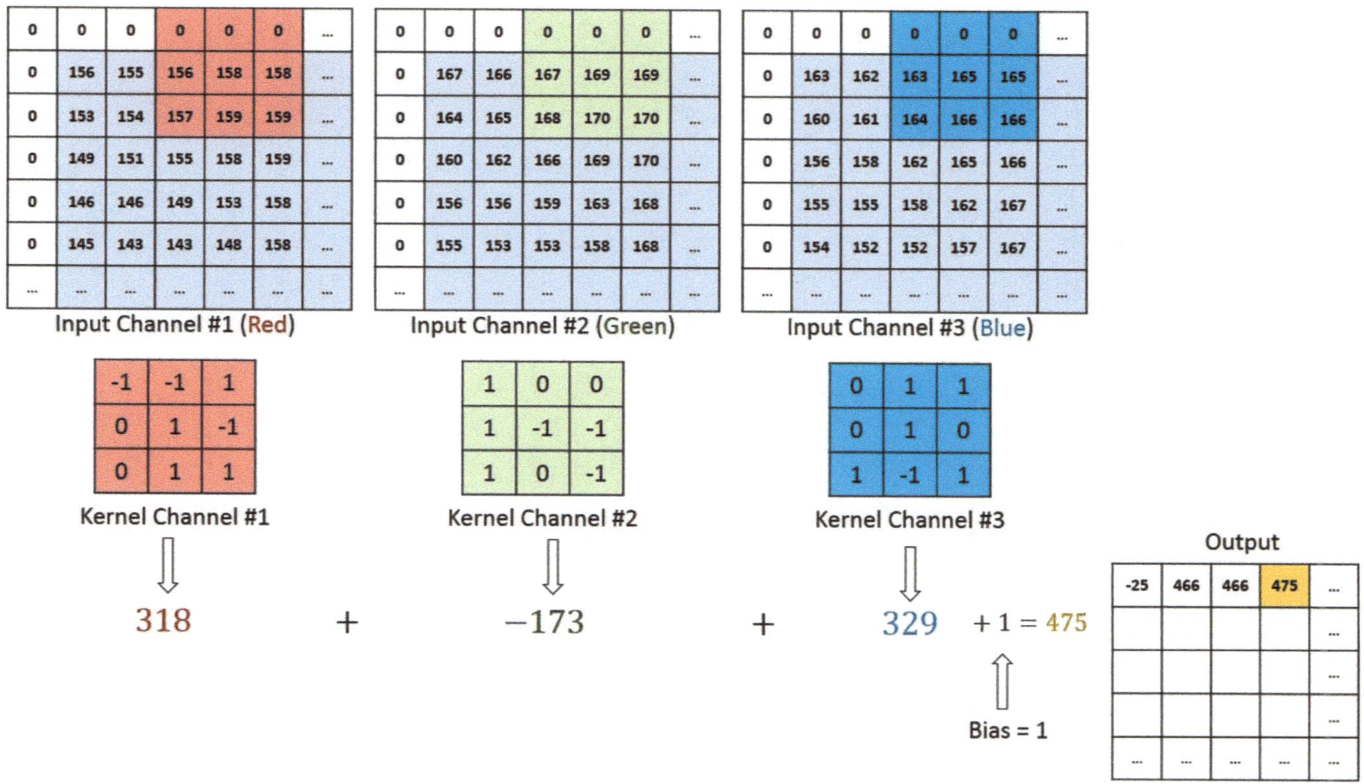

Fig. 3 Kernel filter.

They are also used for feature extraction, a technique to determine important parts of an image in a process called "convolution." In convolutional neural network models, there are more than 3-convolutional kernels, 16-kernels, or even 64-kernels in a single convolutional layer. These convolution kernels individually act as different filters creating feature maps representing something unique, such as kernels that can be filtering top-edges, bottom-edges, diagonal lines, and so on. In much deeper network models these kernels can also be filtering features such as an eye or wing. A greater number of convolutional kernels creates greater numbers of feature maps, increases the amount of data, and uses more memory. This means reflection padding is used; otherwise, it is slightly smaller (Kadam, 2021).

Conclusion

Deep learning is only the domain where we can solve most of the problems beyond the traditional computing limit in computational neuroscience. Convolutional neural networks are popular tools for solving such types of problems easily. They have strong capability to recognize and predict dysfunctions inside the human cortical regions, especially in the visual cortex.

This chapter depicted how convolutional neural networks are helpful for implementing fundamental mechanisms of image processing and recognizing visual disorders or dysfunctions with the help of an artificial deep learning model.

References

Adaloglou, N. (2020). *Understanding the receptive field of deep convolutional networks*. https://theaisummer.com/receptive-field/.

Baeldung. (2021). *How to calculate receptive field size in CNN*. https://www.baeldung.com/cs/cnn-receptive-field-size.

Brady, C. J. (2021). *Vision loss, sudden*. https://www.msdmanuals.com/en-in/home/eye-disorders/symptoms-of-eye-disorders/vision-loss-sudden.

Brownlee, J. (2019). *How to manually scale image pixel data for deep learning*. https://machinelearningmastery.com/how-to-manually-scale-image-pixel-data-for-deep-learning/.

Choraria, A. (2020). *Class activation mapping in deep learning*. https://www.loginradius.com/blog/engineering/class-activation-mapping/.

Das, A. (2020). *Convolution neural network for image processing—Using keras*. https://towardsdatascience.com/convolution-neural-network-for-image-processing-using-keras-dc3429056306.

Davis, C. P. (2021). *Definition of cortical blindness*. https://www.rxlist.com/cortical_blindness/definition.htm.

de Mingo López, L. F., Lucas, C. M., Blas, N. G., & Ivanova, K. (2019). *Pattern recognition with convolutional neural networks: Humpback whale tails*. https://www.preprints.org/manuscript/201902.0257/v1.

Ganesh, P. (2019). *Types of convolution kernels: Simplified*. https://towardsdatascience.com/types-of-convolution-kernels-simplified-f040cb307c37.

Gupta, M., Ireland, A. C., & Bordoni, B. (2021). *Neuroanatomy, visual pathway*. https://www.ncbi.nlm.nih.gov/books/NBK553189/.

Hien, D. H. T. (2017). *A guide to receptive field arithmetic for convolutional neural networks*. https://blog.mlreview.com/a-guide-to-receptive-field-arithmetic-for-convolutional-neural-networks-e0f514068807.

Huff, T., Mahabadi, N., & Tadi, P. (2021). *Neuroanatomy, visual cortex*. https://www.ncbi.nlm.nih.gov/books/NBK482504/.

James, J., Zoccolan, D., & Rust, N. C. (2012). *How does the brain solve visual object recognition?*. https://www.ncbi.nlm.nih.gov/pmc/articles/PMC3306444/.

Kadam, R. (2021). *Kernels (filters) in convolutional neural network (CNN), let's talk about them*. https://medium.com/codex/kernels-filters-in-convolutional-neural-network-cnn-lets-talk-about-them-ee4e94f3319.

Leban, J. (2020). *Image recognition with machine learning on python, image processing*. https://towardsdatascience.com/image-recognition-with-machine-learning-on-python-image-processing-3abe6b158e9a.

Lundervoldab, A. S., & Lundervoldacd, A. (2019). *An overview of deep learning in medical imaging focusing on MRI*. https://www.sciencedirect.com/science/article/pii/S0939388918301181.

Mishra, M. (2020). *Convolutional neural networks, explained*. https://towardsdatascience.com/convolutional-neural-networks-explained-9cc5188c4939.

Salvin, J. H. (2016). *Visual impairment*. https://kidshealth.org/en/teens/visual-impairment.html.

Sarkar, T. (2019). *Activation maps for deep learning models in a few lines of code*. https://towardsdatascience.com/activation-maps-for-deep-learning-models-in-a-few-lines-of-code-ed9ced1e8d21.

Sarkar, S., & Tripathy, K. (2022). *Cortical blindness*. https://www.ncbi.nlm.nih.gov/books/NBK560626/.

Shibuki, K., Wakui, I., Fujimura, T., Tomikawa, M., & Hasegawa, S. (2020). *Rapid recovery from cortical blindness caused by an old cerebral infarction*. https://www.frontiersin.org/articles/10.3389/fneur.2020.00069/full.

Williams, S. E. I. (2018). *The visual pathway*. https://www.sciencedirect.com/topics/neuroscience/visual-pathway.

Yamashita, R., Nishio, M., Do, R. K. G., & Togashi, K. (2018). *Convolutional neural networks: An overview and application in radiology*. https://insightsimaging.springeropen.com/articles/10.1007/s13244-018-0639-9.

CHAPTER 5

Recurrent neural network model for identifying neurological auditory disorder

Akhilesh A. Waoo and Brijesh K. Soni
Department of Computer Science and Technology, AKS University, Satna, Madhya Pradesh, India

Introduction

The human auditory system processes how we hear and understand sounds in the environment. It is comprised of both peripheral nerve structures and different brain regions. Auditory brain circuits encode frequency, attenuation, and location in space and time. Some nerve structures also function collectively on these contents to appreciate the one's understanding and how correctly he can predict auditory information. Processing of auditory information varies continuously in descending order; feedback of nerve structures based on a dynamic environment and perceive various environmental factors (Gray, 2021). Neurological auditory disorders present with multiple variants having complex symptoms. Focal neurological signs such as aphasia deficits altogether with often serious state of a neurocognitive patient often consider auditory symptoms into the background. In this chapter, we review literature on cortical auditory dysfunctions that may present in any general neurological dysfunction, such as cortical stroke, cochleovestibular nerve compression syndrome (CNCS), multiple sclerosis, epilepsy, cerebral palsy, brain tumors, and myasthenia gravis. We focus on the neural anatomical causes of auditory dysfunctions, their characteristics, and their severity (Peterson, Reddy, & Hamel, 2021).

The central auditory nervous system (CANS) includes nerve fibers and nuclei of the brainstem, midbrain, and cortical region. The core work of the CANS is to process specified input signals for auditory attention, sound localization, sound recognition, and pattern recognition. Like the other brain regions, the auditory nervous system regulates attention to target sounds in noisy backgrounds and the association of auditory signals with other various sensory signals. In the cortical regions, recognition and meaning are attached to the input sounds (Britannica, 2021). As people age, they

Artificial Intelligence for Neurological Disorders
https://doi.org/10.1016/B978-0-323-90277-9.00103-6

may experience changes in the CANS. Anatomical changes include the slowing of nerve conduction velocity and reduction of synchronous firing due to axonal thinning and loss of myelin sheath. Transmission between the neurons may slow down due to minimal release of neurotransmitters and fewer post-synaptic receptors. Efferent feedback efficiency may also be reduced. Changes in the CANS due to aging may result in central auditory processing deficits as opposed to simple loss of sensitivity. These changes likely contribute to the difficulty older people have in interpreting speech in abnormal listening environments. The auditory cortex has a major contribution to our perception and understanding of sound waves. There is a highly judgmental phenomenon to integrating the perception of the basic things of an acoustic stimulus, such as sound pitch. Although it's also significant in various other means of audio processing systems, such as determining the region from where exactly the sound originates, and associatively identifying what exactly produces the sound. Consequently, the auditory cortex is also involved in core and higher-level acoustic processing systems, such as aspect recognition of sound that is specific to speech. Any damage inside the auditory cortex may disturb facets of auditory perception. For instance, damage may reduce capability for detecting variation in sound pitch, sound localization, or speech interpretation (Jayakody, Friedland, Martins, & Sohrabi, 2018).

Recurrent neural networks (RNNs) are a special type of deep neural network (DNN) where the output from the previous layer is collected as an input for the current layer. In traditional neural network models, all the input and output data are independent of each other, but in special cases, for example when it is required to guess the next word in a sentence, the previous words are required and hence it is mandatory to remind previous words. Thus, RNNs were developed to solve this type of issue by using the middle layer or hidden layer. The most significant feature of an RNN is a hidden state, which reminds some important features of the sequential data. RNNs have memory that can store all the necessary information, intensively about what to be calculated. Further, this network uses the same number of input parameters for each data it performs the same work with all inputs, or hidden layers to produce the final output. It reduces the overall complexity of parameters used , unlike most other neural networks (Cheng & Zhao, 2011). Speech recognition is the capability of a machine or program to identify spoken words and convert them into readable text. Primarily, speech recognition software has limited vocabulary and may only identify words and phrases when spoken clearly. However, some stronger software tools can handle speech, accents, and different languages. The speech recognition process uses a range of research in linguistics and computer technology. Various modern robotic tools and text-focused software applications have speech recognition capability to enable easier, even hands-free use of the device. Auditory event-related potentials (AERPs) are measures of cortical activity related to auditory stimuli. They are neurophysiological correlates of sound discrimination and sound processing inside the cortical region and can be used for describing auditory cortical developmental plasticity. Sentence boundary detection is the technique of detecting where one sentence ends and another sentence starts. Although this may seem easy because of punctuation, this is not always the case, for example, in the case of acronyms with periods. Major analyses are required to be done at the sentence level; detecting the corner points or boundaries of the sentences is compulsory (Zwass, 2021).

In this chapter, we identify cortical processing ambiguities due to neurological disorder. RNNs such as gated recurrent units and long-short memory units with sequence controlling can efficiently learn various temporal as well as spatial relationships of speech. This is only possible by using the advanced techniques of sound and speech recognition that are fully adopted by the RNN in the broad domain of cognitive auditory information processing and natural language processing such as audio segmentation, phonetic recognition, attention mechanism, and so on. Audio signal segmentation denotes the category of the algorithms developed to categorize semantically relevant temporal data segments into audio signals, also known as auditory information. This auditory information can be calculated as equivalents of lines and paragraphs in textual mode and can act as inputs into audio classification processes in either a supervised or unsupervised manner. Along with these functionalities, meaningfully equivalent auditory information can be combined or labeled using semantic indexes to avail multi-level, non-linear, content-based access to large audio materials and collections (Gimeno, Viñals, Ortega, Miguel, & Lleida, 2020). One of the phonetic-classification schemes based on a feed-forward recurrent backpropagation the neural network working on the audio information is described. The speech's signals are processed using an auditory neural network model producing spectral-like matrices, measuring lip and jaw movement. Some responses will give for different speaker-dependent and independent phonetic-recognition activities regarding any plosive context. In human psychology, attention is the cognitive process of gradually concentrating on a few valuable things than avoiding other less important things. A neural network model is a tool to implement or develop human-brain functionalities effectively and simply. Attention mechanism is also a step to implement the cognitive behavior and functionalities in gradually concentrating on a few favorable and valuable things while avoiding others in the neural network model (Cosi, Dugatto, Ferrero, Magno Caldognetto, & Vagges, 1996).

Human auditory system

The human auditory system transforms a huge number of low mechanical signals into a complex series of electrical signals inside the central nervous system. Sound is a sequence of pressure alteration into the air that frequently changes in both intensity and frequency. The ear's auricle and external auditory meatus gather all incoming sound waves, alter them, and redirect them to the tympanic membrane. The corresponding movements of the eardrum are transmitted throughout the three middle ear bones, the ossicles malleus, incus and stapes, to the fluid of the inner ear. Normally the footplate of the stapes fits strongly into the oval window of the cochlea, now the inner ear part is filled with fluid which is not compressible, as the stapes part moves inside and outside there also needs a temporary movement on the opposite side. The round window membrane, situated under the oval window, moves in the opposite direction because the tympanic membrane has more area than the stapes; here, the footplate is the hydraulic role of sound pressure. In addition, the arm of the malleus to which the tympanic membrane is attached is longer than the arm of the incus to which the stapes is attached, thus there is a minor amplification of the sound pressure via lever activity (Gray, 2021). Two impedance matching processes effectively transmit air-generated sound waves into the existing fluid of the inner part of the ear. If the middle ear apparatus eardrum and ossicles are absent, sound waves reaching the oval window and round window would be largely affected.

Neurological auditory disorder

Neurological auditory processing disorders are a collection of various pathologies in which the ability to detect sounds, localize their supporting sources, or consider their identity and semantics are negatively affected due to disease, damage, or maldevelopment of the CANS. The major symptoms of these disorders range from subtle problems in understanding speech in noisy environments to complete incapability to extract semantics from speech, identify natural sounds, or appreciate music. Various parts of the body are involved in the hearing process and damage to any of these parts may result in some hearing degradation or even complete cortical deafness.

Hearing loss is a general problem and many people lose their hearing as they age; others may be genetically predisposed to hearing loss. About one-third of elderly people suffer from hearing loss. Some cases of hearing loss are temporary, whereas others are permanent. Sometimes people experience earwax accumulation and swelling of the outer ear, which can result in temporary hearing loss. In cases where the inner ear is damaged, permanent hearing loss may result and surgery may be required to restore hearing. Age-related hearing loss, referred to as presbycusis, often occurs gradually. Scientists believe that it occurs due to changes in the inner ear and the auditory nerves that occur over time as people age. This type of hearing problem usually affects both ears and can cause people to be overly sensitive to loud noises or to have difficulty discerning speech. Another common hearing problem in older people is tinnitus, which is a type of hearing problem that presents with a sensation of ringing in both ears (Ludwig et al., 2014). Sometimes tinnitus can also feel like clicking, roaring, hissing, or buzzing sounds. Generally, a variety of health problems including allergies, high blood pressure, diabetes, stroke, viruses, heart disease, and tumors can lead to tinnitus. A noisy environment may also cause hearing loss. Accordingly, loud music and gunshots are the most common reasons for hearing loss.

Central auditory nervous system

The hearing process involves the inner and outer parts of the ear. Sound waves travel toward the outer parts first. Then, the sound waves enter the inner ear. These vibrations stimulate nerves fibers inside the cortical region and generate a sound signal that contains valuable information. The sound signal is transmitted along with nerve fibers toward the next part of the brain, which enables hearing. Thus, the hearing process and nervous system function simultaneously to enable hearing. The three major parts of the ear are the outer part, the middle part, and the inner part. The outer part consists of those parts that are visible externally; these are the ear canal and eardrum. An auricle is an external curved structure that is fully outside of the head. Its outer and inner rims are known as the helix and antihelix. The curved structure of the auricle is naturally designed to receive and redirect sound waves inside the middle ear. The middle ear has four major components: tympanic cavity, ossicles, auditory tube, and oval and round-shaped window structures. Ossicles are tiny bones (malleus, incus, and stapes) in the middle ear that improve sound vibration and transmit that sound to the inner part of the ear. Sound waves are amplified up to 20% in the middle ear. The inner ear has a bony labyrinth with an oval vestibule. Both components are fully associated with the cochlea and semi-circular canal structure. The cochlea is a spiral-shaped bone that plays a key role in hearing. While sound vibrations are received at the outer part of the ear, there is a group of hairy cells inside the cochlea that make vibrations and translate sound waves into electrical pulses or signals. Further, these electrical signals are transmitted through the auditory nerve fibers, which are responsible for transmitting the necessary information to the cortical area. The auditory cortex allows for recognizing electrical signals in the form of sounds waves. It is located within the temporal lobe of the brain. There are some special types of neurons inside the auditory cortex that process certain frequencies of sound waves that objects perceive in the form of high or low sound pitches.

The midbrain and other parts of the brainstem are responsible for regulating some automatic reflex actions to particular types of sounds. Because humans and some other animals have ears on either or both sides of the head, there might be a particular distance size between the sounding object and corresponding receiver. This distance creates a normal variation in the amount of time it takes for certain sounds to reach either ear. The brain is capable of perceiving both sound inputs and identifying the location from which these sounds originate based on the temporal variation between when each ear picks up the sound. Most of the time a detector object cannot precisely notice this time variation because it is on the scale of microseconds, whereas hearing and the nervous system functioning together make this distinction possible. This is why a stereo system avails sound to feel like it is around the object rather than directly coming from the speaking object. If each speaker is placed correctly, both ears can hear the same sounds at slightly different times, and thus the brain senses the sounds being in different locations.

Cortical deafness

Extreme hearing loss is seen in patients with acquired bilateral lesions of the primary auditory cortex or the auditory radiations projecting along the Heschl gyri. Generally, unilateral lesions of the auditory cortex have a minimal effect on hearing. Only some patients with bilateral-type auditory cortex lesions become permanently deaf; most can hear pure sound but have dysfunction in higher-level acoustic sound processing such as recognition of meaningful sounds, temporal sequencing, and sound localization. As in the case of visual agnosia, cortical hearing deficits have less clearness than auditory agnosias. A patient with auditory agnosia can hear noises but cannot predict their meanings clearly, for example, recognizing animal sounds or the ring of a doorbell (Narayanan, Abdul Majeed, Subramaniam, Narayanan, & Navaf, 2017). Most of these patients cannot interpret speech contents or predict music. Auditory agnosia can be categorized as pure word deafness, pure auditory nonverbal-type agnosia, phonagnosia or the inability to recognize a person's voice, and pure amusia. Patients might have one or a combination of these types of deficits.

Recurrent neural network

In deep learning (DL) modeling when it comes to sequential or time-series data, traditional feed-forward networks cannot be used for learning and prediction. A technique is required that can uses historic information to forecast future values. RNN models are a variant of the conventional feed-forward ANNs that can deal with sequential data elements and can be trained to hold knowledge about historic data. An RNN is a special type of neural network model developed for temporal data or data having sequences or series. Elementary feed-forward neural network models only work for data points that are usually independent of each other. However, if we have data elements in a series where one data point is associated with the previous data point, we need to update the neural network model to inculcate the association between these data points. RNNs have memory that stores information from previous inputs to generate the next outputs of the sequential models (Alzubaidi et al., 2021).

Speech recognition

The process of speech recognition is the capability of any machine or software to classify spoken words and translate them into a readable text format. Speech recognition tools typically have limited vocabulary and may only recognize words or phrases when spoken exactly. However, highly trained software is capable of handling natural speech, different types of accents, and different types of languages. The speech recognition process uses a broad variety of innovations in linguistics and computer technology. Many modern robotic devices and text-focused applications have speech recognition functionality to enable easier or hands-free use of a robotic device or expert system (Hosom, 2003).

Auditory event-related potentials

AERPs are measures of cortical activities related to auditory stimuli. They are neurophysiological correlates of cortical sound discrimination and sound processing. They may also be useful for describing auditory system cortical plasticity development. AERP responses influence the sound detection process and transient-type sound encoding of physical features of auditory stimuli. They are highly correlated with the acoustic features of stimulus and integration of the primary cortical auditory pathway. In young children, AERPs contain a first positive peak of approximately 100 ms and a second negative peak of approximately 250 ms. Lower positive and negative amplitudes have been explored in children with cognitive-behavioral problems. In pre-term and full-term born children, lower positive and negative amplitudes are associated with less favorable outcomes in language domains. Mismatch negativity is a component of AERP caused by a modification in the auditory stimulus. It is a frontocentral distributed negativity peaking between 100 and 250 ms from the onset of a stimulus deviant. It is obtained by subtracting the average potential to a standard stimulus from that recorded after a random stimulus (Glennon, Svirsky, & Froemke, 2021). Mismatch negativity resulting sound discrimination classes, which are based on the automatic creation of neural memory traces for known auditory events and has concerned with pre-attentive cognitive type processes in addition, while it is mostly negative in adults, it may have positive in infants and young children. For slight sound changes, mismatch negativity is followed by a frontocentral P3a response, which is considered to show an involuntary switch in attention mechanism.

Various investigations have shown that AERPs are associated with neurocognitive impairments in infants and children, especially those born prematurely. The survival rate of extremely premature infants continues to improve and thus may be contributing to the increased prevalence of neurocognitive abnormalities. In premature infants with abnormally low birth weights, the development of AERPs during the first year of life differs from that of healthy full-term infants. Smaller positive and larger negative amplitudes have been recorded in children five years old born at 28 weeks or less (Joseph et al., 2016). In children born earlier than 32 weeks, mismatch negativity reflected abnormal phoneme discrimination, which was related to a slower native language acquisition.

Sentence boundary disambiguation

Sentence boundary disambiguation or detection is the technique of locating where one sentence ends and another sentence starts. This may seem like an easy task because of the presence of punctuation (periods, question marks, etc.), but this is not always the case. For example, acronyms may include full stops. Various types of analyses at the sentence level are required for identifying the boundaries of sentences. Sentence boundary disambiguation, also known as sentence segmentation or sentence boundary detection, is a well-understood and explored problem in general English text. In well-formed English text, most sentences are terminated by sentence-ending symbols. Ambiguities due to acronyms, quotations, abbreviations, and ordinal numbers are mostly controlled by existing rules or statistical processes. For example, maximum entropy classification of boundaries using these processes on English text provides an accuracy greater than 95 percent (Griffis, Shivade, Fosler-Lussier, & Lai, 2016). Sentence boundary symbols for detection on the whole text of clinical records are errors prone in detecting non-terminated sentence boundaries. Although DL is a popular approach for various machine learning (ML) problems, it remains underutilized in clinical applications.

Neurological auditory disorder identifying model

The process of identifying the cortical processing ambiguities occurred due to any neurological disorder. An RNN can efficiently learn various temporal as well as spatial relationships of speech by using some popular variants of the recurrent neural network like GRUs and LSTM with sequence controlling. This is only possible by using the advanced techniques of sound and speech recognition that are fully adopted by the RNN in the broad domain of cognitive auditory information processing and natural language processing such as audio segmentation, phonetic recognition, attention mechanism, and so on.

Audio segmentation

RNNs are especially useful when dealing with temporal sequences of acoustic information because they are capable of modeling temporal dependencies introducing a feedback loop between input and output of the ANN. LSTM networks are a special type of RNN that incorporate the availability of storage memory. This memory is capable of storing and forgetting data in long temporal dependencies. This capability makes LSTM networks powerful tools to carry out long and short term simultaneously. Two LSTMs are associated with a bidirectional LSTM to model special types of causal and anti-causal dependencies. One BLSTM processes the sequence in the forward direction, and the other BLSTM processes the sequence in the backward direction. Therefore, BLSTM networks have successfully associated with speech recognition techniques for sequence modeling tasks such as automated speech recognition, speaker verification, and language modeling (Brownlee, 2017). In the audio-segmentation tasks, mostly the LSTM-based systems deal with speech and non-speech classification. Some studies have presented both speech as well as music detection using the recurrent LSTM model.

Phonetic recognition

An RNN may learn the temporal relationship of speech data is capable of modeling time-dependent phonemes. The conventional neural network models of multi-layer perceptron are increasingly useful in both sound recognition and speech processing applications. LSTM-based RNN design makes more influential use of model parameters forlearning acoustic network modelsfor large vocabulary-based speech recognition. They have introduced a hybrid model that combines a phonetic model with an arbitrary frame-level acoustic model. The phonetic recognition task is one of the most critical problems in the domain of speech analysis. These popular applications of LSTM include dialect identification, mispronunciation detection, and spoken document retrieval. There are various ways to solve these problems such as improving the feature selection on input speech, applying DL techniques, or combining both having sequence data as phonetics, architecture which is based on recurrent neural network is an appropriate approach. It is even more powerful when combined with the improvement of features selection on input data elements. Mel-frequency cepstral coefficients (MFCCs) are used in automatic speech and speaker recognition that use some RNNs for phonetic classification. Previous experiments are implemented on the phone recognition dataset. Especially, data processing and features selection methods give consistently better results than another research using the same neural network model. They had achieved the lowest error test rate by using BLSTM, which is the best result in the dataset with a reduction of about 3.5% over the last best result. The motivation behind applying recurrent neural networks to the language processing domain is to take advantage of their capability to process short-term spectral features but yet respond to long-term temporal events. New research has proven that performance of the speaker recognition process increases as the duration of occurrence increases. RNNs exhibit better performance and learn in a shorter period than conventional feed-forward neural network models.

Recently, a simple model that has feedback connections of the self-loop type around hidden layer units has been proposed as a useful tool for recognizing speech contents, including voiced plosive sounds. This neural network has three layers: the input layer, hidden layer, and output layer. Each one of the output layers has a feedback connection with itself. A fully recurrent neural network model feed backs the hidden layer to itself. Generally, partially recurrent neural networks start with a fully recurrent neural network and add a feed-forward connection that passes the concept of recurrences, effectively treating the recurrent part in the state memory. An RNN can have limitless memory and thus find relationships through time as well space. Most real-world data have information in its time-based dimensions. RNNs are applicable in nonlinear time-series prediction, temporal pattern classification, and system identification.

Attention mechanism

Attention is proposed as a method to both align and translate information elements. Element alignment is the problem in machine translation that identifies which parts of the input sequence are relevant to each word in the output. Translation is a process of using the relevant information elements to select the suitable output. Commonly attention-based RNN models have successfully been assigned to multiple activities, for example, handwriting synthesis, machine translation, image caption generation, and visual objects classification. Such models iteratively process their input elements by choosing original content at each step. Fundamentally this idea makes applicability area of the training process, making it possible to construct connections with external memory. We introduce attention-based RNNs applicable for speech recognition. The learning recognizing speech may be considered as learning to generate a series forgiven another series (Patel, 2020). From this point of view, it is a little bit similar to machine-translation and handwriting-synthesis tasks, to which attention-based methods, are found more appropriate.

Although speech recognition and machine translation are conceptually different due to requesting longer input data sequences, which present a challenge of differentiating similar speech from another fragment in a single time. It also looks changed from handwritten text synthesizing, since the input sequence is much noisy and it doesn't have enough clear structure. Due to these causes, the speech recognition process is an interesting thing for developing a modern attention-based model capable of processing lengthy and noisy input elements. The application of the attention-based network model in the speech recognition process is also a major step for developing fully end-to-end modifiable recognition systems, which is a trending area in research.

The major approach is based on a hybrid model comprised of a DL-based acoustic information processing model, a triphone hidden Markov model, and a language processing model. This requires databases of human pronunciation techniques and phoneme lexicons, and a multi-level training process to enable the components to work together. Excellent results by a hidden Markov model less recognizer have recently been found, with the model consisting of a connectionist temporal classification trained neural network model and a communicative language model (Chorowski, 2015). Still, the language processing network-based model has connected at the last stage only in that work, which means have a question of how enough an auditory learning model can help to be aware of a language processing network model during the training and testing procedure.

Conclusion

DL is only the domain where we can solve most of the problems relaying beyond the traditional computing limit in computational neuroscience. Recurrent neural networks are also a major tool to solve such types of problems easily. They have strong capability to recognize and predict dysfunctions inside the human cortical regions, especially in the auditory cortex. In this chapter, we studied and depicted how a recurrent neural network helps to implement the fundamental mechanism of audio processing and recognizing phonetic disorders or dysfunctions with help of developing an artificial DL model.

References

Alzubaidi, L., Zhang, J., Humaidi, A. J., Al-Dujaili, A., Duan, Y., Al-Shamma, O., et al. (2021). *Review of deep learning: Concepts, CNN architectures, challenges, applications, future directions*. https://journalofbigdata.springeropen.com/articles/10.1186/s40537-021-00444-8. (Accessed June 2022).

Britannica. (2021). *Cochlear nerve and central auditory pathways*. https://www.britannica.com/science/ear/Cochlear-nerve-and-central-auditory-pathways.

Brownlee, J. (2017). *A gentle introduction to long short-term memory networks by the experts*. https://machinelearningmastery.com/gentle-introduction-long-short-term-memory-networks-experts/. (Accessed June 2022).

Cheng, F., & Zhao, J. (2011). *Recurrent neural network (RNN) is one of the most widely used NN to model dynamic processes*. https://www.sciencedirect.com/topics/chemical-engineering/recurrent-neural-networks.

Chorowski, J. (2015). *Attention-based models for speech recognition*. https://arxiv.org/pdf/1506.07503.pdf. (Accessed June 2022).

Cosi, P., Dugatto, M., Ferrero, F., Magno Caldognetto, E., & Vagges, K. (1996). *Phonetic recognition by recurrent neural networks working on audio and visual information*. https://www.istc.cnr.it/en/content/phonetic-recognition-recurrent-neural-networks-working-audio-and-visual-information.

Gimeno, P., Viñals, I., Ortega, A., Miguel, A., & Lleida, E. (2020). *Multiclass audio segmentation based on recurrent neural networks for broadcast domain data*. https://asmp-eurasipjournals.springeropen.com/articles/10.1186/s13636-020-00172-6.

Glennon, E., Svirsky, M. A., & Froemke, R. C. (2021). *Auditory cortical plasticity in cochlear implant users*. https://www.ncbi.nlm.nih.gov/pmc/articles/PMC7002179/. (Accessed June 2022).

Gray, L. (2021). *Auditory system: Structure and function*. https://nba.uth.tmc.edu/neuroscience/m/s2/chapter12.html.

Griffis, D., Shivade, C., Fosler-Lussier, E., & Lai, A. M. (2016). *A quantitative and qualitative evaluation of sentence boundary detection for the clinical domain*. https://www.ncbi.nlm.nih.gov/pmc/articles/PMC5001746/. (Accessed June 2022).

Hosom, J.-P. (2003). *Speech recognition*. https://www.sciencedirect.com/topics/computer-science/speech-recognition. (Accessed June 2022).

Jayakody, D. M. P., Friedland, P. L., Martins, R. N., & Sohrabi, H. R. (2018). *Impact of aging on the auditory system and related cognitive functions: A narrative review*. https://www.frontiersin.org/articles/10.3389/fnins.2018.00125/full.

Joseph, R. M., O'Shea, T. M., Allred, E. N., Heeren, T., Hirtz, D., Jara, H., et al. (2016). *Neurocognitive and academic outcomes at age 10 years of extremely preterm newborns*. https://www.ncbi.nlm.nih.gov/pmc/articles/PMC4811321/. (Accessed June 2022).

Ludwig, A. A., Fuchs, M., Kruse, E., Uhlig, B., Kotz, S. A., & Rübsamen, R. (2014). https://www.ncbi.nlm.nih.gov/pmc/articles/PMC4010590. (Accessed June 2022).

Narayanan, S., Abdul Majeed, K., Subramaniam, G., Narayanan, A., & Navaf, K. M. (2017). *A case of cortical deafness due to bilateral Heschl gyrus infarct*. https://www.ncbi.nlm.nih.gov/pmc/articles/PMC5368354/.

Patel, H. (2020). *Neural machine translation (NMT) with attention mechanism*. https://towardsdatascience.com/neural-machine-translation-nmt-with-attention-mechanism-5e59b57bd2ac. (Accessed June 2022).

Peterson, D. C., Reddy, V., & Hamel, R. N. (2021). *Neuroanatomy, auditory pathway*. https://www.ncbi.nlm.nih.gov/books/NBK532311/.

Zwass, V. (2021). *Speech recognition technology*. https://www.britannica.com/technology/speech-recognition.

CHAPTER 6

Recurrent neural network model for identifying epilepsy based neurological auditory disorder

Mukesh Soni[a], Nihar Ranjan Nayak[b], V. Selvakumar[c], and Sagar Dhanraj Pande[d]

[a]Department of Computer Science and Engineering, Jagran Lakecity University, Bhopal, India
[b]Department of Computer Science and Engineering, Presidency University, Bangalore, India
[c]Department of Maths and Statistics, Bhavan's Vivekananda College of Science, Humanities and Commerce, Hyderabad, Telangana, India
[d]Intelligent System, School of Computer Science and Engineering, Lovely Professional University, Phagwara, Punjab, India

Introduction

Epilepsy affects between 1% and 2% of the world's population. Most people who have epilepsy are just like people who don't have it. The unpredictability of epilepsy's rapid onset is the main reason people with epilepsy have problems and may even die. Electroencephalogram (EEG) signals are used to determine how many brain cells are moving around. It looks at changes in electrical waves in the cerebral cortex caused by electrophysiological markers, which act as a proxy for neuronal activity in the cerebral cortex. Electrical impulses are studied in biomedical research, and when they are, the EEG signal during epilepsy is different from the EEG signal in the normal state because neurons in the brain aren't firing normally. As a result, epilepsy can now be diagnosed with the help of EEG.

Support vector machines (SVMs), naive Bayes, neural networks, and fuzzy logic systems are some of the more intelligent algorithms that are being used to detect epilepsy in EEG signals as machine learning improves. Principal component analysis (PCA), wavelet packet decomposition (WPD), and higher-order crossings (HOC) are also being used to detect epilepsy in EEG signals. These strategies obtain features from original characteristics, use the added features to train a classification model, and then use the trained model to make predictions. Even though many algorithms have been used to find effective features and classify EEG epilepsy, finding enough discriminative information for successful testing is still a big problem.

Artificial Intelligence for Neurological Disorders
https://doi.org/10.1016/B978-0-323-90277-9.00023-7

Deep learning is a machine learning method that can be used for feature learning among other tasks. In each layer of the hierarchy, the features are changed in a way that makes them more likely to produce the desired outcome. For example, each layer improves learning of the input attributes, which leads to more discriminative features. EEG data processing has been done well with the help of deep learning technology in the last few years. Several studies have used different algorithms to extract features from EEG signal characteristics and then used convolutional neural networks (CNNs) to diagnose epilepsy based on the features.

When learning from multiview data, multiview learning algorithms take into account many different views (Haselsteiner & Pfurtscheller, 2000; Kostov & Polak, 2000). Based on how similar and different they are, multiview learning algorithms can be broken down into three learning groups: collaborative training, multikernel learning, and subspace learning. All three systems, no matter how different they are, require consensus or ideas that work together to make multiperspective learning work. When someone has epilepsy, multiview learning techniques are often used to look for EEG data. To get multiview data from EEG signals, several feature extraction methods are applied. Then, multiview learning algorithms are used to determine whether someone has epilepsy (Borujeny, Yazdi, Keshavarz-Haddad, & Borujeny, 2013; LeCun, Bottou, Bengio, & Haffner, 1998).

This chapter discusses a multiview classification strategy that includes both deep and shallow features. It uses deep learning and multiview learning techniques to use EEG data and generate useful features for diagnosing epilepsy. It takes many shallow features to obtain deep features. An EEG signal has many shallow properties, like those in the frequency domain, the time-frequency domain, and the nonlinear domain, which are all different. All the layer's properties are integrated to develop a dataset with different views. A multiview learning technique is used to group the features in a multiview set of features to detect epilepsy.

The algorithm proposed in this research can be broken down into three sections.

(1) Acquiring shallow EEG signal features: Constructing shallow characteristics of EEG signals can be done in a variety of ways. The shallow features built in various methods have distinct impacts and have their own sets of benefits and drawbacks. Multiple alternative shallow features are gathered for epilepsy detection to leverage the advantages of shallow features created by diverse methods. In this research, the shallow features in the frequency domain and the shallow features in the time-frequency domain of the EEG signal were obtained using fast Fourier transform (FFT) and WPD.

(2) Acquiring deep EEG signal features: To increase the efficiency of shallow EEG signal features and obtain features with greater generalization ability, this chapter analyzes shallow EEG signal characteristics using CNNs. Depp learning is used to construct deep features. Deep features of EEG signals have smaller data dimensions and higher classification effects than shallow features.

(3) Building a classifier and deciding on a classification: The EEG signal's shallow and deep features are referred to as a multiview feature set. A multiview classifier is built, and the output of the multiview classifier is used to determine whether someone has epilepsy.

The multiview classification method that mixes deep and shallow features gathers shallow and deep features obtained via deep learning technology and then makes judgments using multiview learning technology. Deep learning's deep structure optimizes feature representation; shallow features preserve the original information of EEG signals and reduce information loss; multiview technology effectively utilizes the independence of each view feature and the correlation between view features; and multiview technology effectively utilizes the independence of each view feature and the correlation between view features. The classifier's generalization ability can be improved by using a combination of shallow features, deep features, and multiview learning approaches to fully extract the useful information in an EEG signal.

Related research

This section provides a quick overview of the relevant work for the technique suggested in this research, including the importance of utilizing EEG data for epilepsy identification and the advancement of using machine learning, deep learning, and multiview learning technologies in EEG epilepsy detection.

Bioinformatics, medical image processing, and biological signal processing are all applications of intelligent technology in biomedicine. Bioinformatics studies protein and gene information; medical image processing mainly includes analysis of computed tomography (CT), nuclear magnetic resonance (NMR), and so on; biological signal processing is the study of electrical signals such as EEG and electrocardiogram (ECG). An EEG signal is the manifestation of neuronal activity in the human brain. It contains a lot of information about human physiological activities. EEG signals have been widely used in the field of epilepsy detection. Epilepsy detection generally refers to the use of automatic algorithms to analyze a patient's biological signals to determine whether the patient is having or has had seizures. An important goal of epilepsy detection is to do it as quickly and efficiently as possible. In recent years, a variety of algorithms for epilepsy detection (Muller-Putz, Scherer, Brauneis, & Pfurtscheller, 2005; Zheng et al., 2021) have been proposed.

Machine learning techniques such as random forest (RF), C4.5, SVM + RF, and SVM + C4.5 have been used to detect EEG epilepsy (Sharbrough et al., 1991). RF was determined to be the most effective method of categorization. The study authors identified EEG epilepsy using the estimated entropy and sample entropy from WPD. They classified objects using SVMs and extreme learning machines. The authors used WPD and KPCA to minimize the number of features (kernel principal component analysis). Additionally, the TSK fuzzy logic system was utilized to determine if someone had epilepsy. These techniques have been used to diagnose some forms of epilepsy.

All these techniques are applicable to biomedicine, owing to the growth of deep learning. They include stacked auto encoders (Aarabi, Fazel-Rezal, & Aghakhani, 2009; Fleishman et al., 2011; Harikumar & Narayanan, 2003) and deep belief networks (Kuhlman et al., 2003; Siegel et al., 2010), as well as CNNs and recurrent neural networks (RNNs) (Anfinsen, 1973; Anfinsen, Haber, Sela, & White, 1961; Hsu, Lander, & Zhang, 2014). In the past several years, many researchers have attempted to utilize CNNs to interpret EEG data with varying degrees of success. By combining raw EEG data with one-dimensional convolutions, CNNs may be used to predict seizures. This approach was employed by (Burkitt, Silberstein, Cadush, & Wood, 2000; Ren & Wu, 2014) to predict seizures. Fourier analysis was used to analyze the raw EEG data, which was then converted to frequency data. Following this, we employed CNNs to group the raw EEG data.

Multiperspective learning techniques

Multiview learning algorithms are machine learning algorithms that work with datasets that have features from a lot of different places. This is because the new multiview collaborative learning mechanism can use each viewpoint's own independence and the connections between different viewpoints to learn more effectively than a single-view dataset. Canonical correlation analysis (CCA) (Huang, Tsai, & Kuo, 2001; Makeig & Jung, 1996), co-training (Jung, Makeig, Stensmo, & Sejnowski, 1997; Van Orden, Jung, & Makeig, 2000), sparse multiview SVM (Makeig, Jung, & Sejnowski, 2000), and other common multiview algorithms are used in a wide range of applications. A single-view method hasn't worked as well as scenes have. Multiview technology has been used in several studies to help researchers detect epilepsy from EEG signals. These studies used tensor decomposition to obtain new features for detecting epilepsy from EEG multiview features, and they came up with new features for this. A new feature proposed by (Borujeny et al., 2013) can be used to detect epilepsy, and the research shows that these new features are better at classifying things. A method for detecting epilepsy from multiview EEG signals that is based on a fuzzy system and a multiview learning frame.

TSK fuzzy system

The fuzzy inference rules that define the TSK fuzzy system are as follows:

$$\begin{aligned} &\text{if } x_1 \, is \, A_1^k \wedge x_2 \, is \, A_2^k \wedge \ldots\ldots \wedge x_d \, is \, A_d^k \\ &\text{then} f^k(x) = p_0^k + p_1^k x_1 + \ldots\ldots\ldots + p_d^k x_d \\ &k = 1, 2, \ldots\ldots, K \end{aligned} \tag{1}$$

where K is the total number of fuzzy rules in the rule system, A_i^k is the fuzzy set of the kth rule of x_i, p_i^k ($i = 0, 1, \ldots, d$) is the coefficient of x_i on the kth rule in the linear function, and $\wedge$ is a fuzzy join operation. The kth rule maps the input vector $x = [x_1, x_2, \ldots, x_d]^T$ from the input space to the output $f^k(x)$, the output $f(x)$ of the TSK fuzzy system can be defined as:

$$f(x) \sum_{k=1}^{K} \frac{\mu^k(x)}{\sum_{k=1}^{K} \mu^{k'}(x)} f^k(x) = \sum_{k=1}^{K} \mu^{\sim k}(x) f^k(x) \tag{2}$$

where $\mu^k(x)$ is the fuzzy membership degree of the kth rule, which is obtained after normalization $\tilde{\mu}^k(x)$.

$$\mu^k(x) = \prod_{i=1}^{d} \mu_{A_i^k}(x_i)$$

Among them, $\mu_{A_i^k}(x_i)$ is the membership degree of x_i on the fuzzy set A_i^k, and the Gaussian membership function is widely used to calculate the fuzzy membership degree.

$$\mu_{A_i^k}(x_i) = \exp\left(\frac{-\left(x_i - c_i^k\right)^2}{2\delta_i^k}\right)$$

δ_i^k is the kernel width and c_i^k is the center parameter. The antecedent parameters c_i^k and δ_i^k specify the specific shape of the fuzzy set Aik, which can be derived using the clustering technique or the fuzzy space division method.

Proposed method

The approach suggested in this study consists of three major components: the acquisition of shallow EEG signal features, the acquisition of deep EEG signal features, and the learning of multiview classifiers. Fig. 1 depicts the fundamental framework of our strategy.

Shallow feature acquisition of EEG signals

It is a time series of varying frequency impulses. Despite its tremendous temporal capabilities, it lacks fundamental frequency properties. A Fourier transform is used to convert a signal's time domain to frequency domain. The Fourier transform may convert any periodic signal to the frequency domain, but it must be periodic. Longer EEG waves have smaller periodicity and less precise frequency domain characteristics. To acquire the instantaneous frequency at each time point, the wavelet transform converts the signal into the time-frequency domain for analysis, keeping both the time sequence of the

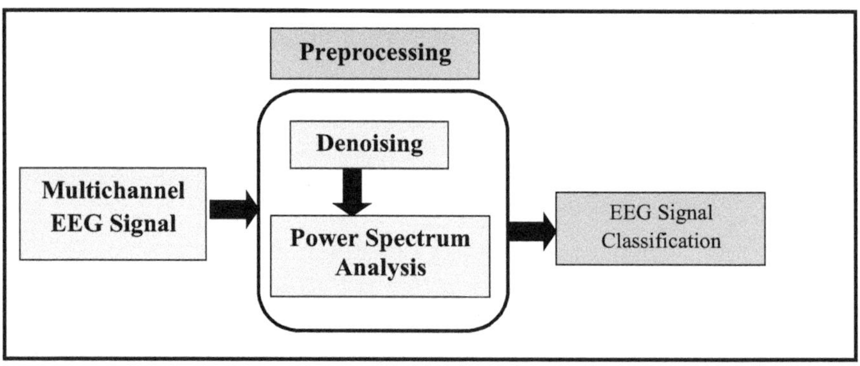

Fig. 1 Framework of proposed method.

signal and the resulting frequency. Using the previous research's conclusions, this study extracts shallow features from the original EEG signal in two dimensions: frequency and time-frequency.

Shallow feature construction in time-frequency domain

The signal's frequency domain properties are determined in addition to its time domain features. Wavelet decomposition is a typical way to convert signals to time-frequency domain. It converts the Fourier transform's trigonometric function basis into a wavelet function basis with two variables: scale and translation. The scale controls the wavelet transform's expansion and contraction, whereas the translation amount controls the wavelet transform's time. These two variables regulate the wavelet transform, which may enhance the signal at various scales to meet the needs of adaptive time-frequency signal analysis.

WPD (Mark & Aleksandar, 2003; Zhong & Joydeep, 2002) is a popular wavelet transform. Using WPD, Daubechies is used to generate shallow features in the time-frequency domain of EEG data (dbN). The dbN wavelet has good regularity. The smoothness and vanishing moment order both improve with order. The greater the frequency domain's ability to localize, the greater the effect. A large vanishing moment order would harm real-time performance by reducing temporal domain support and increasing effort. The wavelet function's order N is 4 in this work. Since the signal transformed to the time-frequency domain is three-dimensional data, the parameters differ from those in the frequency domain (channel frequency time). The sampling rate is 2 Hz. Set the decomposition level to six for wavelet transformations. Like the frequency domain, this research only looks at time-frequency features between 4 and 30 Hz. Fig. 2 shows the shallow characteristics in the time-frequency domain, where time is horizontal and frequency is vertical. The graphic depicts energy changes at different times and frequencies. The colored bars on the graph's right represent the control values for energies.

In this chapter, we use a CNN to automatically extract shallow characteristics from an EEG signal's frequency and time-frequency domains. Two separate CNNs are built for the shallow features of the two domains to automatically extract the shallow features in depth and obtain the corresponding deep features.

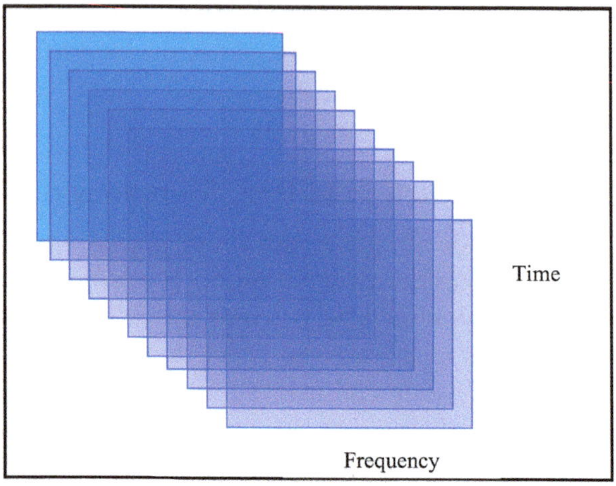

Fig. 2 Time-frequency domain shallow features of EEG signals.

Acquisition of deep features based on deep learning

The CNN calculates and distributes the error backwards during training by using the output of the previous layer to judge how bad things are. This is due to the fact that the penultimate layer's output feature vector only goes through one completely linked layer before reaching the output layer. When the network is trained on the output of the penultimate layer, its ability to expose features improves. After being joined in a nonlinear fashion, they have a greater potential to divide, which might make the next classification model more general and successful.

Frequency domain deep feature extraction network

Fig. 3 depicts the CNN architecture used for deep feature extraction in the frequency domain. The system includes two convolutional layers and three fully connected layers. The network's input is a two-dimensional matrix of 23×27 because the multichannel EEG frequency domain view feature is a two-dimensional matrix of frequency channel number. The network's first convolutional layer consists of 20 convolution kernels of 4×4 with a stride of 1, yielding 20 feature maps of 20×24; the network's second convolutional layer consists of 10 convolution kernels of 88 with a stride of 1, yielding $10 \times 13 \times 17$ feature maps. The network is then divided into three completely linked layers. These three completely linked layers transform the feature map into a 1×512 feature vector, then a 1×100 feature vector, and lastly a binary classification result.

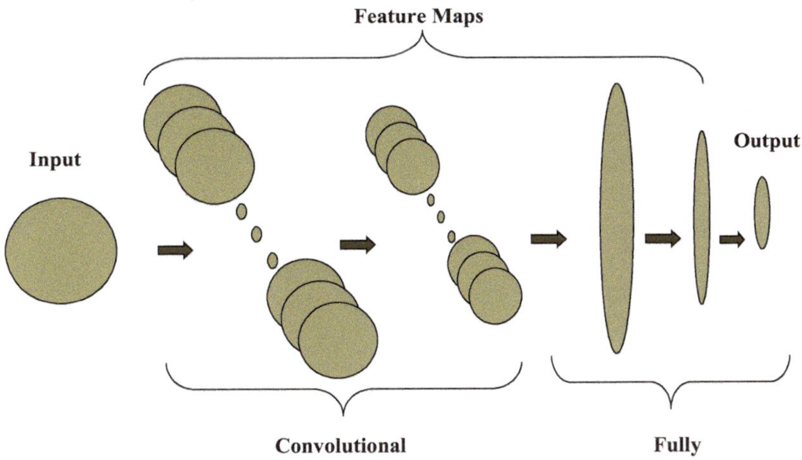

Fig. 3 Frequency domain feature extraction mechanism.

Time-frequency domain deep feature extraction network

Fig. 4 depicts how a CNN works in terms of time and frequency. The network is a three-dimensional matrix of $128 \times 23 \times 14$. The new network is based on the same concepts as the previous network for frequency-domain deep feature extraction. It is considerably easier to deal with the convolution kernel now that the three-dimensional convolution approach has been reimplemented. In two-dimensional convolution, the essential idea is the same.

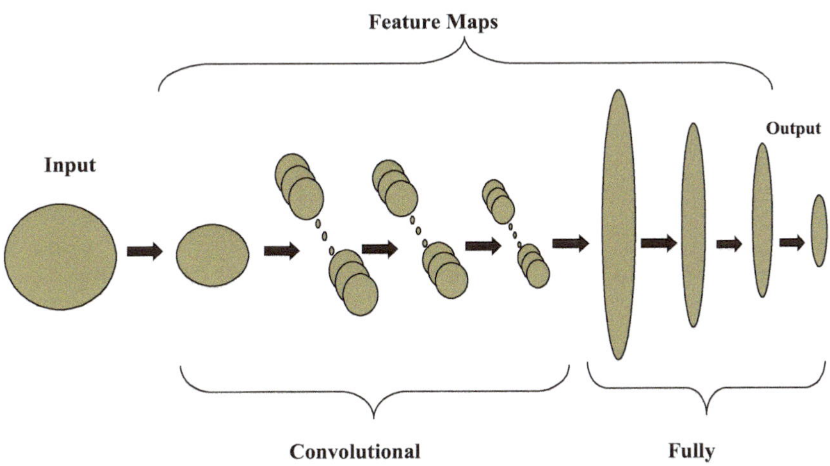

Fig. 4 Time-frequency domain feature extraction mechanism.

Multiview TSK blur system based on view weighting

Multiview features built using shallow and deep features provide good expressiveness and minimal information loss and knowing how to employ these features efficiently for epilepsy diagnosis is critical. A TSK fuzzy system is a rule-based intelligent model with great interpretability. This study builds an interpretable multiview classifier for epilepsy detection in EEG signals using the TSK fuzzy system.

Experimental study

Dataset

This study uses the CHB_MIT dataset for experimental research. The dataset collected EEG signals from 23 patients at Boston Children's Hospital. The records from 23 patients were divided into 24 groups (of which group 21 was the data added after the first patient was re-sampled several years later), with each group containing one patient's recordings of EEG signals for more than 10 consecutive hours. These continuous signals are sampled at a frequency of 256 Hz, that is, there are 256 sample points in one second of the signal. The EEG signal of each patient is collected from 18 points to form a single-channel data set, and the subsequent processing becomes multichannel data of 23 channels.

Validity analysis

Table 1 compares the four views' features using single-view and multiview TSK fuzzy systems. Acc, Sen, and Spec were found to be the most accurate, sensitive, and specific methods, as illustrated in Fig. 5.

This research employs the SVM, KNN, NB, and DT algorithms to build a multiview classifier based on view weighting to validate the efficiency of the multiview TSK fuzzy system. Table 2 and Fig. 6 show the experimental results.

Numerical analysis of deep feature extraction networks

In this study, we trained two CNNs for deep feature extraction. The number of iterations of the network was adjusted progressively to determine the impact, and it was discovered that the effects of the two CNNs achieved a stable state after a particular number of iterations. The number of iterations after the network achieves a stable state is set as the final number of iterations, allowing for a stronger classification result in less training time. Figs. 7 and 8

Table 1 Effect of combined features on classifiers.

Feature	Classifier	Accuracy (Acc)	Sensitivity (Sen)	Specificity (Spec)
FFT shallow feature	SVM	97.87	88.37	99.02
	*K*NN	97.38	82.07	99.23
	NB	87.84	89.6	87.61
	DT	95.59	79.98	97.48
	TSK-FLS	94.61	85.43	99.1
WPD shallow feature	SVM	97.87	88.37	99.02
	*K*NN	97.38	82.07	99.23
	NB	87.84	89.6	87.61
	DT	95.59	79.98	97.48
	TSK-FLS	94.61	85.43	99.1
FFT deep feature	SVM	97.87	88.37	99.02
	*K*NN	97.38	82.07	99.23
	NB	87.84	89.6	87.61
	DT	95.59	79.98	97.48
	TSK-FLS	94.61	85.43	99.1
WPD deep feature	SVM	97.87	88.37	99.02
	*K*NN	97.38	82.07	99.23
	NB	87.84	89.6	87.61
	DT	95.59	79.98	97.48
	TSK-FLS	94.61	85.43	99.1
Average		92.71	84.36	93.87
Proposed		**99.08**	**98.59**	**99.32**

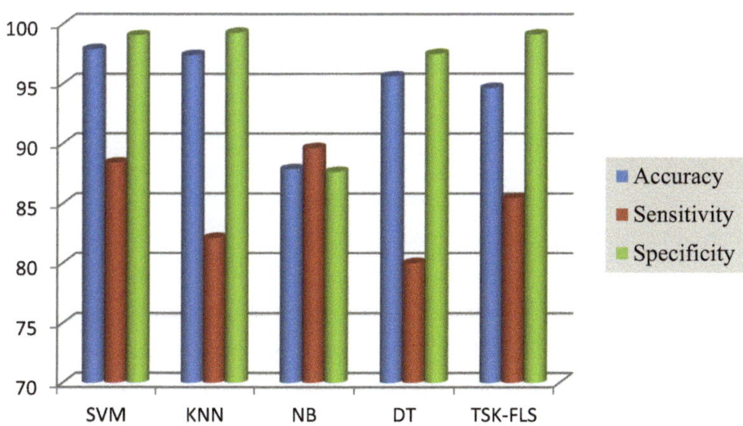

Fig. 5 WPD deep features on classifiers.

Table 2 Effect on multiview classifiers.

Classifier	Accuracy	Sensitivity	Specificity
Multiangle SVM	94.65	92.94	95.64
Multiangle KNN	95.32	84.21	97.54
Multiangle NB	72.18	64.56	74.38
Multiangle DT	92.37	91.12	94.68
Multiview TSK blur system	**99.25**	**98.27**	**99.43**

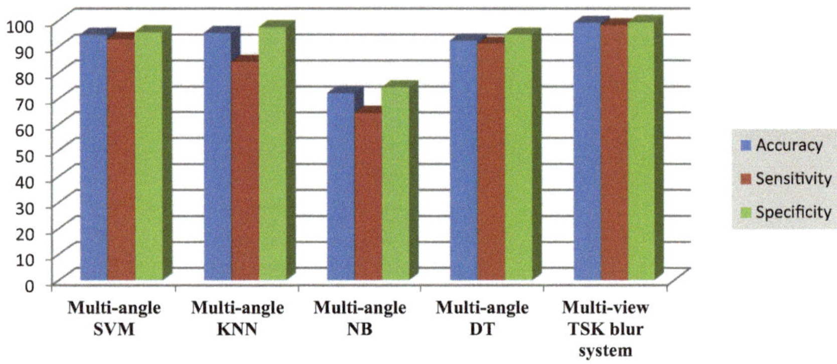

Fig. 6 Effect on multiview classifiers.

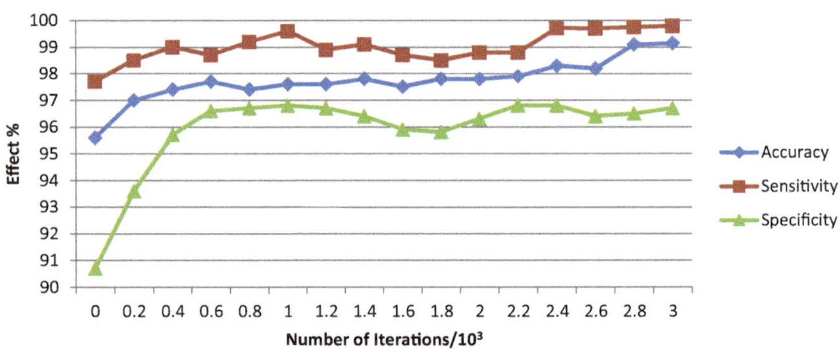

Fig. 7 Performance of frequency domain network.

only demonstrate the link between the number of network iterations and the final impact of the first patient data; the other groups are similar to the first.

Results show that the frequency domain features tend to be stable after 2000 iterations of training, thus the training times of the frequency domain network was set to 2000. Although the effect increases with the number of

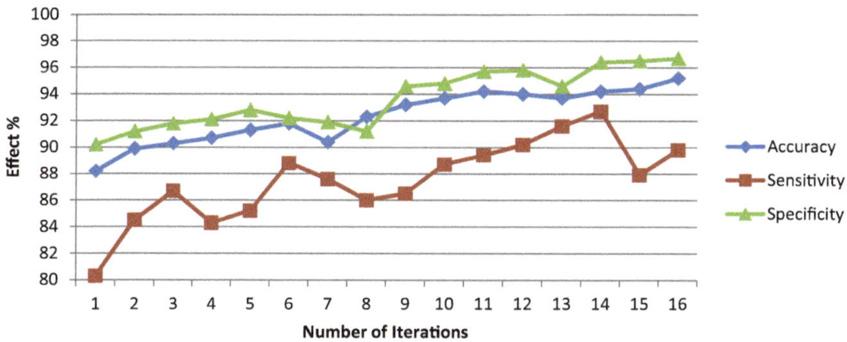

Fig. 8 Performance of time-frequency domain network.

trainings, the time-frequency domain experiences large fluctuation. Considering the effect and fluctuation of the feature, the training number of the time-frequency domain feature extraction network was set to 5000.

Conclusion

To detect epilepsy EEG signals, this chapter provided a multiview classification system that merges deep and shallow features. It was discovered through experimental data that the fusion features presented in this chapter outperform typical feature extraction methods. The TSK fuzzy system's multiview classifier learns the fusion features, which increases the algorithm's performance even further. Deep and shallow feature extraction methods, as well as the introduction of a multiview TSK blur system, have been found to be useful for detecting epilepsy EEG signals.

Even though the research in this field has become more important, it still has flaws that need to be fixed. Even though the multiview TSK blur system is better than the single-view system, multiview algorithms can always be better. For example, the multiview TSK blur system is better than the single-view TSK blur system, but there is still a lot of room for improvement. Thus, future research should look into these issues further.

References

Aarabi, A., Fazel-Rezal, R., & Aghakhani, Y. (2009). A fuzzy rule-based system for epileptic seizure detection in intracranial EEG. *Clinical Neurophysiology*, *120*(9), 1648–1657.

Anfinsen, C. B. (1973). Principles that govern the folding of protein chains. *Science*, *181*, 223–230.

Anfinsen, C. B., Haber, E., Sela, M., & White, F. H. (1961). The kinetics of formation of native ribonuclease during oxidation of the reduced polypeptide chain. *Proceedings of the National Academy of Sciences of the United States of America, 47*, 1309–1314.
Borujeny, G. T., Yazdi, M., Keshavarz-Haddad, A., & Borujeny, A. R. (2013). Detection of epileptic seizure using wireless sensor networks. *Journal of Medical Signals and Sensors, 3*(2), 63–68.
Burkitt, G., Silberstein, R., Cadush, P., & Wood, A. (2000). Steady-state visual evoked potentials and travelling waves. *Clinical Neurophysiology, 111*(2), 246–258.
Fleishman, S. J., Whitehead, T. A., Ekiert, D. C., Dreyfus, C., Corn, J. E., Strauch, E., et al. (2011). Computational design of proteins targeting the conserved stem region of influenza hemagglutinin. *Science, 332*, 816–821.
Harikumar, R., & Narayanan, B. S. (2003). Fuzzy techniques for classification of epilepsy risk level from EEG signals. In *1. Proceedings of Conference Convergent Technology Asia-Pacific Region* (pp. 209–213).
Haselsteiner, E., & Pfurtscheller, G. (2000). Using time dependent neural networks for eeg classification. *IEEE Transactions on Rehabilitation Engineering, 8*(4), 457–463.
Hsu, P. D., Lander, E. S., & Zhang, F. (2014). Development and applications of CRISPR-Cas9 for genome engineering. *Cell, 157*, 1262–1278.
Huang, R. S., Tsai, L. L., & Kuo, C. J. (2001). Selection of valid and reliable EEG features for predicting auditory and visual alertness levels. *Proceedings of the National Science Council, Republic of China, Part B, 25*, 17–25.
Jung, T. P., Makeig, S., Stensmo, M., & Sejnowski, T. J. (1997). Estimating alertness from the EEG power spectrum. *IEEE Transactions on Biomedical Engineering, 44*(1), 60–69.
Kostov, A., & Polak, M. (2000). Parallel man-machine training in development of EEG-based cursor control. *IEEE Transactions on Rehabilitation Engineering, 8*(2), 203–205.
Kuhlman, B., Dantas, G., Ireton, G. C., Varani, G., Stoddard, B. L., & Baker, D. (2003). Design of a Novel globular protein fold with atomic-level accuracy. *Science, 302*, 1364–1368.
LeCun, Y., Bottou, L., Bengio, Y., & Haffner, P. (1998). Gradient-based learning applied to document recognition. *Proceedings of the IEEE, 86*(11), 2278–2324.
Makeig, S., & Jung, T. P. (1996). Tonic, phasic, and transient EEG correlates of auditory awareness in drowsiness. *Brain Research. Cognitive Brain Research, 4*, 15–25.
Makeig, S., Jung, T. P., & Sejnowski, T. J. (2000). Awareness during drowsiness: Dynamics and electrophysiological correlates. *Canadian Journal of Experimental Psychology, 54*, 266–273.
Mark, P., & Aleksandar, K. (2003). Feature extraction in development of brain-computer interface: A case study. In *Proceedings of the 20th Annual International Conference of the IEEE Engineering in Medicine and Biology Society*.
Muller-Putz, G., Scherer, R., Brauneis, C., & Pfurtscheller, G. (2005). Steady state visual evoked potential (ssvep)-based communication: Impact of harmonic frequency components. *Journal of Neural Engineering, 2*(4), 123–130.
Ren, Y., & Wu, Y. (2014). Convolutional deep belief networks for feature extraction of EEG signal. In *Proceedings of the International Joint Conference on Neural Networks* (pp. 2850–2853). https://doi.org/10.1109/IJCNN.2014.6889383.
Sharbrough, G.-E., Chatrian, R., Lesser, H., Lders, M., Nuwer, T., & Picton, T. W. (1991). American electroencephalographic society guidelines for standard electrode position nomenclature. *Journal of Clinical Neurophysiology, 8*, 200–202.
Siegel, J. B., Zanghellini, A., Lovick, H. M., Kiss, G., Lambert, A. R., Clair, J. L. S., et al. (2010). Computational design of an enzyme catalyst for a stereoselective bimolecular Diels-Alder reaction. *Science, 329*, 309–313.

Van Orden, K. F., Jung, T. P., & Makeig, S. (2000). Combined eye activity measures accurately estimate changes in sustained visual task performance. *Biological Psychology*, *52*, 221–240.

Zheng, Z., Dong, X., Jian, Y., Leyuan, Z., Yang, D., & Aiguo, C. (2021). Identification of epileptic EEG signals through TSK transfer learning fuzzy system. *Frontiers in Neuroscience*, *15*. ISSN 1662-453X.

Zhong, S., & Joydeep, G. (2002). HMMs and coupled HMMs for multi-channel EEG classification. In *Vol. 2. Proceedings of IEEE international joint conference on neural networks* (pp. 1154–1159).

CHAPTER 7

Dementia diagnosis with EEG using machine learning

Neelam Sharma[a] and Maheshkumar H. Kolekar[b]
[a]Electrical and Instrumentation Engineering Department, Thapar Institute of Engineering and Technology, Patiala, Punjab, India
[b]Department of Electrical Engineering, Indian Institute of Technology Patna, Bihta, Patna, Bihar, India

Introduction

Dementia is a group of symptoms due to any chronic neuropsychiatric brain disorder, leading to impaired cognitive capability and memory loss (Scott & Barrett, 2007). The Diagnostic and Statistical Manual of Mental Disorders, Fifth Edition (DSM-5) defines significant cognitive impairment as a major neurocognitive disorder or dementia (American Psychiatric Association, 2013). It is also known as an age-related disorder. Its symptoms include impairment of brain functions, memory, thinking, orientation, comprehension, calculation, learning capacity, language, and judgment. Dementia patients do not lose consciousness, but they can become dependent on others because of neurological maladies and cognitive impairment. DSM-5 lists more than 12 etiological causes of dementia. Some of the most common types of dementia are Alzheimer's disease (AD), vascular dementia (VD), Lew body dementia, and Parkinson's disease (PD) (American Psychiatric Association, 2013). The major cause of dementia in Western countries is AD, and 60%–80% of dementia cases with AD occur in the United States (Bird, 2001) (see Table 1). AD is characterized by the presence of cortical amyloid plaques and neurofibrillary tangles in the brain (Hebert, Scherr, Bienias, Bennett, & Evans, 2003; Scott & Barrett, 2007). VD is caused by a disruption in blood supply to the brain, typically caused by stroke. It is the second most common cause of dementia worldwide. According to DSM-5 (American Psychiatric Association, 2013), AD and VD have the most complex symptomology as they have many similarities in diagnostic criteria and early symptoms. If dementia is caused by more than two neurodegenerative diseases, it is called as mixed dementia. The prevalence of frontotemporal lobar degeneration (FTLD), which is caused by frontal

Artificial Intelligence for Neurological Disorders
https://doi.org/10.1016/B978-0-323-90277-9.00007-9

Table 1 The prevalence of dementia type.

Dementia type	Symptoms	Cause	Prevalence
Alzheimer's disease (AD)	A slow cognitive decline that involves memory loss, apathy, and depression	Cortical amyloid plaques and neurofibrillary tangles	50%–75%
Vascular dementia (VD)	Similar symptoms to AD, but with less memory impairment; decline in executive functioning with emotional variation; proceeds in a step-by-step progression	Cerebrovascular disease, single infarcts in critical regions, or more diffuse multiinfarct disease	20%–30%
Franto-temporal dementia	Behavioral variant, fluent and progressive nonfluent aphasia, mood changes, disinhibition, and personality changes	Damage in frontal and temporal cortex	5%–10%
Dementia with Lewy bodies	Labeled variation in cognitive abilities, hallucinations, parietal/frontal involvement, and parkinsonism	Cortical Lewy bodies (alpha synuclein)	<5%

and temporal lobe atrophy, is 5%–10% in developed countries worldwide (Ferri, Sousa, Albanese, Ribeiro, & Honyashiki, 2009). Some rare cases of dementia are diagnosed by a surgical situation such as hypercalcemia, subdural hematoma, normal pressure hydrocephalus (NPH), and deficiencies of thyroid hormone, vitamin B12, and folic acid (World Health Organization, 2006). Due to lack of awareness in rural India, sometimes dementia is cosidered as normal aging and not treated as medical issue (Shaji, Smitha, Lal, & Prince, 2003).

To diagnose dementia, clinicians need to assess the impairment in cognitive functions, memory, and independent daily living skills. Thus, mild cognitive impairment (MCI) can be an early sign of different neurological conditions that finally leads to memory deficit or dementia. Cognition plays a critical role in many neurodegenerative diseases. MCI is defined as a mild reduction in one or two cognitive abilities, such as remembering things and recalling skills, focusing, managing bills and money, communicating, perceptual and motor skills, and recognizing people. It is also known as mild neurocognitive disorder (American Psychiatric Association, 2013).

Prevalence of dementia worldwide

Demographic aging is a natural phenomenon and is increasing in developing and low-income countries. India's population is experiencing a sharp demographic shift at present. More than 90 million people are 65 years and older, as per the 2019 census in India, which is only 6.57% of the whole population. However, in the future there will be a definite increase in the elderly population.

Along with a sharp rise in demographic shift, fast-paced social restructuring following economic growth is occurring. This is a challenging situation in regard to meeting the needs of the elderly population. Demographic aging also brings with it the problem of dementia, cases of which double in number every 5 years. Thus, India will have massive numbers of elderly people with dementia in the future. There are 50 million dementia patients globally and this number is expected to increase to 152 million by 2050. One case of dementia is diagnosed every 3 s. The present cost of dementia treatment in the United States is $1 trillion and this cost will double by 2030 (Alzheimer's Disease International, 2019). These calculations are relatively 12%–13% more than those projected by the World Alzheimer Report 2009 (Ferri et al., 2009). This report projected 115.4 million in 2050 compared to 41.5 million in 2015. It also estimated that more than 50% of dementia patients would be from low- and middle-income countries. More than 90% of dementia patients receive at-home care (Prince, 2015). Around 42,000 elder investigated at eight sites of India by Alzheimer's and Related Disorders Society of India (ARDSI) and observed wide variations in estimation of dementia prevalence (Shaji et al., 2010). The diagnostic gap is widening due to the absence of a sensitive and specific diagnostic tool for dementia. Such a tool must not have dialectal, educational, and cultural barriers. Dementia studies are limited in India, but one report estimated that Delhi, Bihar, and Jharkhand will experience a 200% increase in dementia cases in 2026 compared to 2006 (Shaji et al., 2010). Worldwide reports show that around 36 million dementia patients do not receive proper diagnosis (Prince, Bryce, & Ferri, 2011). Early dementia diagnosis will allow patients to access information, resources, and support in advance. Early diagnosis will also enhance patient quality of life and treatment.

Electroencephalogram

An electroencephalogram (EEG) is an image of recorded scalp electrical activity generated by the brain. EEG measures voltage fluctuations occurring from ionic current flows within the neurons of the brain. The high temporal resolution of EEG can measure brain signals at a rate of less than 1 ms, which is an advantage over other techniques of measuring brain activity. EEG recording is entirely noninvasive and can be used on adults and children without any limitations. Modern EEG has high spatial resolution where up to 256 electrode sites can record at the same time. The brain signals are produced in all mental states (normal and abnormal). EEG can reflect all brain activity, so it is a potent tool for neurological and psychological diseases. EEG patterns are commonly sinusoidal signals. EEG ranges from 0.5 to 100 μV in amplitude and comprises frequency components of up to 300 Hz. Various frequencies of sine waves are obtained from a raw EEG signal using Fourier or wavelet transform. Every cognitive state of the brain generates frequencies; sometimes certain frequencies are more dominant. Brain waves are categorized into the following:

- Delta waves (δ) (range 0–4 Hz): These waves have the highest amplitude. They are emitted by adults in slow-wave sleep and occur naturally in babies and children.
- Theta waves (θ) (range 4–8 Hz): These waves are usually observed in growing children. Theta waves occur during both drowsiness and arousal states in adolescents and adults. Theta power is prominent during meditation. Profuse theta waves with age depict abnormal activity.
- Alpha waves (α) (range 8–12 Hz): These waves occur during relaxed states of mind and usually when a person's eyes are closed. They attenuate with eye-opening or mental exertion.
- Sensory motor response (12–13 Hz): This response generates from the brain's sensory-motor cortex. It is responsible for human body movements.
- Beta waves (β) (range 12–30 Hz): Beta activity attenuates during active movements. Beta waves usually occur during working, involved or concerned thought, and intense attention.
- Gamma waves (γ) (range >30 Hz): These waves occur during different cognitive and motor functions such as cross-modal sensory processing and short-term memory. Gamma waves decline with cognitive deterioration.

Cognitive testing and EEG

Cognition is the process whereby individuals acquire knowledge from the environment. Mental states, such as arousal, workload, and working memory, reflect central nervous system and automatic nervous system activity. Cognitive neuroscience has become crucial for establishing the scientific foundation and brain function principles associated with human performance. Dementia refers to cognitive decline, which is why the conventional method of detection was neuropsychological test batteries; however, these tests fail to diagnose dementia alone due to lack of cross validation with physiology. Many studies have proven that EEG recordings with cognitive tasks achieve better diagnostic results than resting-state EEG signals. The EEG signal of a dementia patient demonstrates the functional changes in the cerebral cortex. The symptoms of dementia develop slowly in the early stages and are often confused with impairments due to normal aging with etiologies. The symptoms become more obvious as dementia progresses, as shown in Fig. 1, and thus cognition and functional ability begin to decline. A major challenge is differentiating the early stages of dementia from normal age-related changes in memory and thinking. Determining the date of onset of dementia and estimating its occurrence rate is difficult due to its progressive nature. Usually, dementia onset occurs with MCI. Hence, the likelihood of identifying early stages of dementia and MCI is very low. Also, all types of dementia symptoms are similar, but their expression differs from patient to patient. This factor diminishes the possibility of differentiating dementia from other types of cognitive decline (Phillips, Pond, & Goode, 2011).

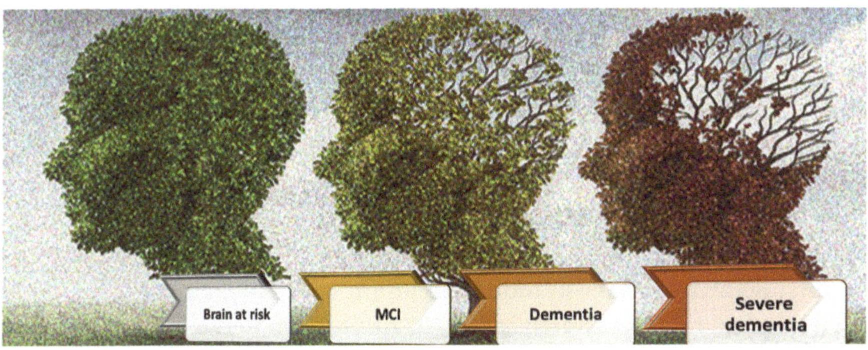

Fig. 1 Different stages of dementia.

One study of MCI patients found that coherence and power of EEG signal bands are greater in patients during a working memory task (Jiang, 2005). Previous studies have already explained age-related changes in EEG during cognitive tasks and showed significant differences between healthy young people and healthy older people (Widagdo, Pierson, & Helme, 1998). Similarly, EEG combined with a cognitive task can improve accuracy of detecting different types of dementia. Every type has different early signs of cognitive decline, as shown in Table 1. For example, memory impairment occurs first in AD, whereas decline in executive function occurs first in VD. Frontotemporal dementia shows a behavioral variant. Sometime patients also suffer from mixed-type dementia (Rossor, Fox, Mummery, Schott, & Warren, 2010). Much recent research has investigated the use of EEG to diagnose AD as well as other brain disorders. However, accuracy is still a challenge in diagnosing MCI, in which anatomical brain changes are minimal.

Moreover, it is not feasible to measure the cause of cognitive decline in MCI, which is confused with normal subjects. Even though dementia is possible with any neurodegenerative disease, everyone with MCI progresses to dementia. Some studies have tried to differentiate progressive from stable MCI (Missonnier et al., 2007). Therefore, a different kind of tool is necessary because every disease has different treatments, interventions, and cures. To design a proper tool for diagnosis, a combination of cognitive screening and EEG analysis with machine learning are required for specific disorders by considering all the challenged involved in the diagnosis as projected in Fig. 2. Cognitive screening tools are beneficial for the subject identification from the mass population in the beginning. This can save cost of further diagnosis (Sharma, Kolekar, & Chandra, 2015). Additionally, EEG analysis aids in exploring brain activity during a cognitive test, which is not feasible with imaging techniques. Lastly, ML techniques help to obtain the desired diagnostic accuracy.

Meanwhile, emerging technologies play an essential role in making neuropsychological tests more user-friendly and less time-consuming, such as those accessible via Android or iOS apps. These apps are cost-effective, transportable, and convenient for diagnosing large populations. Consequently, dementia diagnosis requires a fusion of cognitive tests, EEG signal processing, and ML, as shown in Fig. 2. EEG-based dementia diagnosis has improved over time. It is noninvasive, inexpensive, and portable. Recently, studies of EEG and ML techniques have shown good sensitivity and specificity in comparison to imaging studies in diagnosing AD, seizure, and

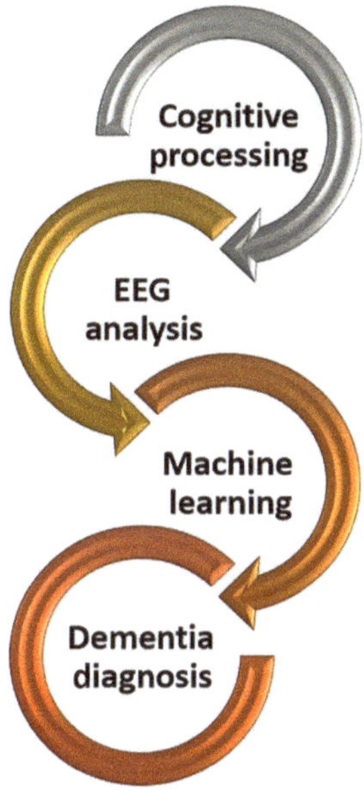

Fig. 2 EEG-based proposed model.

dementia (Bennys, Rondouin, Vergnes, & Touchon, 2001; Benvenuto, Jin, Casale, Lynch, & Granger, 2002; Jeong, 2004; Kolekar & Dash, 2015). Traditionally, EEG is acquired at a resting state for diagnosing diseases such as epilepsy, Creutzfeldt-Jakob disease, and other etiologies. EEG during cognitive processing is used for research purposes to investigate the physiology of the human brain more accurately. In the last few years, cognitive test performance and EEG information obtained during the test has proven to be a potential diagnostic tool for dementia (Durongbhan et al., 2019; Mcbride et al., 2014; Sharma et al., 2020, 2019). There are three important EEG parameters to consider when using this modality to diagnose dementia: slowing of EEG, reduction in complexity, perturbed synchronization (Dauwels, Vialatte, & Cichocki, 2010). Dementia diagnosis using EEG has four basic steps: (1) data acquisition, (2) signal preprocessing, (3) feature extraction, and (4) classification, as shown in Fig. 3.

Data acquisition

Traditionally, diagnostic EEG is done with the patient lying down with both eyes closed for a few minutes and then with both eyes open for a few minutes and later on photonic stimulus is given to the patients. The EEG pattern is analyzed by expert technicians and medical practitioners according to the

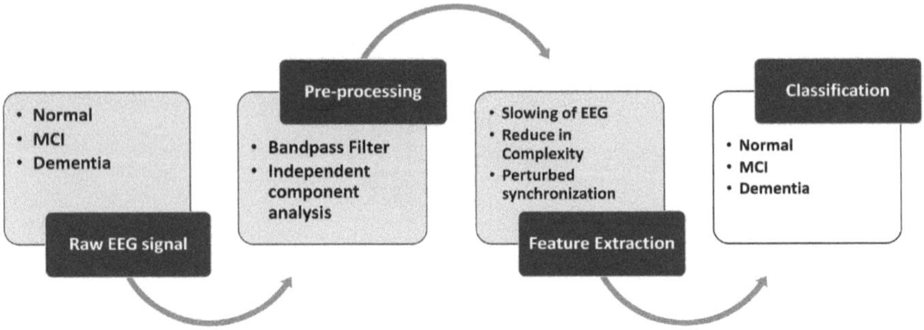

Fig. 3 EEG-based approach for classification of dementia.

patient's history and symptomatology. Severe dementia patients are hard to handle in a cognitive task, and they barely respond to any test question. Thus, the patient's EEG recording at resting state condition utilize as a reference to the milder stages of dementia. For early dementia diagnosis, cognitive task studies have analyzed EEG signal records along with patient performance on cognitive tasks (Sharma et al., 2019). Sometimes, event-related potentials (ERPs), such as visual and auditory stimuli potentials, are also recorded to investigate stimulus response in the brain.

Preprocessing of EEG signal

Preprocessing of the EEG signal is required because noise factors diminish the accuracy of the EEG record and most artifacts overlap with the frequencies of the signal. Physiological artifacts are due to muscle activities, pulse, and eye blinking. Nonphysiological artifacts are due to power line interference noise and sweat (Jung et al., 2000). Various signal preprocessing methods are used to eliminate the noise from a recorded EEG signal. Adaptive filtering and independent component analysis (ICA) are effective preprocessing methods (Kilicarslan, Grossman, & Contreras-Vidal, 2016; Roy & Shukla, 2015). The wavelet transform (WT) is an efficient denoising method to treat nonstationary EEG. A few recent studies have applied WT to remove visual artifacts and detect epileptic seizure (Krishnaveni, Jayaraman, Aravind, Hariharasudhan, & Ramadoss, 2006; Kumar & Kolekar, 2014; Zikov, Bibian, Dumont, Huzmezan, & Ries, 2002).

Feature extraction

EEG-based diagnosis of dementia can be broadly categorized into two categories of feature extraction: linear spectral analysis and nonlinear dynamics.

The linear and nonlinear methods of analysis measure the three essential parameters of dementia diagnosis: slowing of EEG, reduced complexity, and disturbed synchronization, as shown in Fig. 3. Table 2 shows the different features used in different studies.

Linear approach

EEG analysis in the frequency and time-frequency domains is considered a linear approach that involves Fourier and wavelet transform. The mother wavelet types utilized in EEG denoising are Daubechies, Coiflets, and Dyme. An early diagnostic method based on multiresolution wavelet analysis of ERP has been proposed. EEG signal is decomposed into coarse approximation by discrete wavelet transform (DWT) and a multiresolution analysis. In this diagnostic method, a neural network is trained to extract various features for diagnosis of dementia. Here, DWT was used to represent ERPs compactly (Polikar, Keinert, & Greer, 2001).

The linear relationship between electrodes in cases of dementia are studied by different types of feature such as EEG relative power, coherence, and Granger causality. Dementia slows down the EEG signals and causes a power increase in lower frequency bands (delta and theta bands) and a power decrease in higher frequency bands (alpha and beta bands), which is known as slowing of EEG. Nevertheless, when comparing AD patients with the healthy group, a power increase in the gamma band of EEG signals has been reported (Dauwels, Vialatte, & Cichocki, 2010; Jeong, 2004). A recent study (Rodriguez et al., 2014) proposed an EEG spectral analysis-based method for early diagnosis of AD. The research was carried out with three subject groups: probable AD, asymptomatic carrier, and healthy groups. Twenty-four artifact-free epochs of 2.56 ms were selected for the quantitative analysis, and the Fourier transform of each epoch was calculated. The broadband spectral parameters of mean frequency, relative power, and absolute power were calculated for delta, theta, alpha, and beta bands. Afterwards, 50 Hz notch filter is applied to remove electrical interference (Szava et al., 1994).

Spectral EEG analysis can quantify the severity of dementia and achieve a more reliable dementia diagnosis. Better features allow the classifier to identify signals more accurately. Spectral analysis also helps in differentiating between different types and causes of dementia. Neto, Allen, Aurlien, Nordby, and Eichele (2015) proposed a model that significantly differentiates AD and VD based on EEG features such as frequency amplitudes, the decay of amplitude from low to high frequencies, alpha frequency,

Table 2 Different cognitive tests for dementia and their techniques and features.

Study	Dataset	EEG/ERP	Technique and features
Sharma, Kolekar, and Jha (2020) with finger tapping test (FTT) and continuous performance test (CPT)	Dementia = 16; MCI/ ED = 16; healthy = 15	13 min	Power spectral density (PSD), variance, fractal dimension(FD), and Tsallis entropy (TE) using iterative filtering decomposition
Li, Xiao, Li, Li, and Yang (2020) cognitive reappraisal task	MCI = 15; healthy = 15	5 min	Variance, RMS amplitude, Hjorth parameters, zero crossings, skewness and kurtosis, five EEG band Shannon entropy, singular value decomposition, entropy, Fisher information, and spectral entropy
Sharma, Kolekar, Jha, and Kumar (2019) with FTT and CPT	Dementia = 15; MCI = 16; control = 13	13 min	PSD, SE, spectral features, and FD using wavelet packet decomposition
A visual oddball task (Wang, Xu, Zhao, & Lou, 2019)	VD =15; control= 21	2.64 s (ERP)	Directed transfer function of power
Memory recall task ()	VD = 5; MCI = 15; normal = 15	60 s	Relative powers, permutation entropy, and FD using fuzzy neighborhood preserving analysis with QR decomposition
Auditory oddball paradigm (Staudinger & Polikar, 2011)	AD = 79; control = 82	30 min (ERP)	Higuchi FD, spectral entropy (SE), spectral centroid (SC), spectral roll-off (SR), and zero-crossing rate (ZCR)
Boston naming test (Reyes-Coronel et al., 2016)	Rapid cognitive decline (RCD) = 15; non-RCD = 53	NA	SE, TE, Granger causality (GC), auto MI, conditional GC, canonical correlation (CC), dynamic CC, FD, PSD, partial and phase coherence
Counting backwards with finger tapping (Mcbride et al., 2014)	Control = 15; early MCI = 16; early stage AD = 17	10 min	PSD, spectral features, entropy first derivative

and power. A decline in coherence is an important linear feature of the EEG signal that describes the spectral correlation between two EEG signals recorded at an electrode pair. Coherence C can be mathematically calculated by

$$C = |f_{xy}|^2 / (f_{xx} \cdot f_{yy}) \tag{1}$$

where f_{xy} denotes a spectral estimation of a pair of electrodes recorded signals x and y for a given frequency band. Further, the mean phase coherence computed describes the phase synchronization between two signals, unlike mean square amplitude coherence (Che, Jung, Im, & Lee, 2007). Some recent research has introduced wavelet coherence, which is considered more sensitive to brainwave alterations during the cognitive process than traditional coherence (Vyšata, Vališ, Procházka, Rusina, & Pazdera, 2015).

Granger casualty is another linear measure used for extraction and the computation of directionality from EEG signals and is based on the bivariate autoregressive estimate of the EEG data. It includes partial directed coherence (PDC) and directed transfer function (DTF) extracted from multichannel data. DTF has an advantage over spectral coherence; it can determine the directionality in the coupling when the frequency spectra of the two brain regions have overlapping spectra (Sanei & Chambers, 2013). According to studies on Granger causality in cases of MCI, DTF significantly reduced in MCI and AD patients as did the direction of the information flow from parietal to frontal (Babiloni et al., 2009).

Nonlinear approach

An EEG signal has nonlinear behavior due to the brain's complex construction of synapses and neurons. The EEG signal presents the chaotic nature of the nervous system. Therefore, the nonlinear approach has the advantage to provide additional in-formation in comparison to the linear approach. Unlike the linear methods, nonlinear analysis methods measure the complexity and stability of the brain waves. Various nonlinear features such as entropy, fractal dimension (FD), Lyapunov exponent (LE), mutual information (MI), correlation dimension (CD), zero-crossing rate (ZCR), and weight permutation entropy (WPE) are used to detect MCI and dementia.

Initially, CD and LE are used as nonlinear features to detect the dynamical system's complexity (Grassberger & Procaccia, 2004; Wolf, Swift,Swinney, & Vastano, 1985). CD is the measure of FD, often referred to as dimensional complexity (Besthorn, Sattel, Geiger-Kabisch, Zerfass,

& Förstl, 1995). FD is a quotient providing a statistical degree of complexity. It measures the variation in pattern. There is a different type of FD, but few classical fractals, including CD, Kurts, and Higuchi dimensions (HD), are used in the case of AD and dementia diagnosis (Henderson, Ifeachor, Wimalaratna, Allen, & Hudson, 2000; Rodríguez-Bermúdez & García-Laencina, 2015; Sharma et al., 2020; Staudinger & Polikar, 2011). Some prior research has revealed that AD patients have fewer CD/FD values at parietal and temporal regions than healthy people do (Adeli, Ghosh-Dastidar, & Dadmehr, 2008; Das, Das, & Roy, 2002; Sanei & Chambers, 2013). However, a positive larger LE value signifies the chaotic system, while a negative LE shows fewer changes, and zero LE indicates no change (Adeli et al., 2008; Das et al., 2002; Sanei & Chambers, 2013). There are two well-known algorithms for LE used for EEG data: wolf's algorithm (Wolf et al., 1985), and Rosenstein's method (Rosenstein, Collins, & De Luca, 1993). These measures are limited to quantifying irregular behavior of brain patterns and are sensitive to parameters such as embedding dimension, time delay, and data length (Henderson et al., 2006). However, LE is related to the entropy of the chaotic or dynamic system. Entropy measures the uncertainty and greater uncertainty correlates to greater entropy and more chaotic nature of the brain. Entropy is defined as

$$x(n) = \int_{\min(x)}^{\max(x)} p_x \log(1/p_x) dx \tag{2}$$

where p_x is the probability density function of a signal $x(n)$. Entropy is very sensitive to noise. Recently, different types of entropy algorithms have been utilized for the analysis of EEG signals for detecting AD, MCI, or epilepsy (Kolekar & Dash, 2015). These algorithms include spectral entropy (Staudinger & Polikar, 2011), approximate entropy (Hornero, Abásolo, Escudero, & Gómez, 2009), sample entropy (Hornero et al., 2009; Timothy, Krishna, Menon, & Nair, 2014), Kolmogorov entropy (Grassberger & Procaccia, 1983), Tsallis entropy (Zhao et al., 2007), multiscale entropy (Hornero et al., 2009), permutation entropy (Timothy et al., 2014), and WPE (Deng et al., 2015). In addition to these features, some studies included more than one feature like ZCR, spectral centroid, spectral roll off, and spectral entropy with FD in order to increase the sensitivity and specificity of dementia diagnosis (Henderson et al., 2006; Staudinger & Polikar, 2011).

MI is the shared information between the two variables X and Y and is derived from the information theory. It estimates the probability density distribution (PDD) and joint PDD of two-time series and measures the statistical independence between two-time series by calculating various entropies.

$$I(X;Y) = \sum_{x,y} p_{XY}(x,y) \log \frac{p_{XY}(x,y)}{p_X(x)p_y(y)} = H(X) - H(X|Y) \tag{3}$$

Here $p_X(x)$ and $p_Y(y)$ are the marginal PDD, $H(X)$ and $H(Y)$ are the entropy and conditional properties, respectively. MI is divided into two types: cross MI, which measures linear and nonlinear dependence, and auto MI, which measures the dependencies between single-channel EEG and the same EEG with some specified time delay. A recent study showed decreased MI in the frontal and right anterior temporal regions indicated less corticocortical connections in AD patients (Jeong, Gore, & Peterson, 2001). One study revealed that MCI subjects found lower theta frequency of MI during a task than normal subjects and proved that the connection between the parietal and occipital lobes for age- and disease-related changes could be detected using CMI analysis (Liu et al., 2012). Like Granger casualty, permutation conditional mutual information (PCMI) is used to measure coupling strength and direction between two EEG signals. This is a better feature than Granger casualty for recognizing the coupling direction of unidirectional or bidirectional EEG signals (Li & Ouyang, 2010). Some recent research has shown the superiority of PCMI in analyzing MCI and epilepsy EEG data (Li, Liu, & Ouyang, 2013; Wen et al., 2016). A recent study proved that MI is a more important responsive factor in detecting MCI compared to coherence (Vyšata et al., 2015). In addition to these, there are some features used to analyze synchronization between two or more EEG signals, such as S estimator (Dauwels, Vialatte, Musha, & Cichocki, 2010), stochastic event synchronization (Dauwels, Vialatte, Musha, & Cichocki, 2010), and graph theory (Stam, Jones, Nolte, Breakspear, & Scheltens, 2007).

Table 2 demonstrates the latest work done in the area of dementia diagnosis using different cognitive tasks. There are many previous studies available that use the cognitive task in work but that did not acquire the related EEG signal for the task. Staudinger and Polikar (2011) studied a few EEG features methods, such as EEG spectral analysis, coherence, and complexity, along with corresponding cognitive functions, to diagnose AD. Their investigation was an ERP study with auditory stimulus. The decline in resting-state alpha waves shows early dementia. The alpha wave in cognitive task ensure that the patient is in the awake state and not drowsy

(Staudinger & Polikar, 2011). Mcbride et al. (2014) acquired EEG data via a 10-min finger tapping test (FTT) performed by 48 subjects, when subjects were counting backwards, and when they were in a resting state. Relative spectral power, additional spectral features, spectral power ratios, entropy/complexity features, and first derivative features were extracted in this study (Mcbride et al., 2014). Reyes-Coronel et al. (2016) used the Boston naming test on subjects with rapid cognitive decline (RCD) and extracted many EEG features such as Shannon and TE, Granger causality and listed in Table 2. This study had a small number of cognitive decline subjects (Reyes-Coronel et al., 2016). Al-Qazzaz, Ali, Ahmad, Islam, and Escudero (2017) worked on a memory recall task with 35 VD, MCI, and healthy subjects. Recording length was 1 min and relative power and complexity measures were extracted using fuzzy neighborhood-preserving analysis with QR decomposition. Wang et al. (2019) worked with visual stimulus ERP and extracted one feature: directed transfer function of power (Wang et al., 2019). In recent work, Li et al. (2020) diagnosed MCI using 5 min of EEG in a cognitive reappraisal task. This work extracted various EEG features such as variance, RMS amplitude, Hjorth parameters, zero crossings, skewness and kurtosis, five EEG band Shannon entropy, singular value decomposition entropy, Fisher information, and SE.

Classification of dementia

During the classification process, the main aim is to distinguish MCI and dementia with high accuracy, which is essential for a diagnostic test. Classification is a way to identify the category or class of a new observation based on a training set of data that contains the characteristics. ML and DL techniques can be applied in almost every research problem for classification. ML techniques are chosen often and perform well where supervised learning is required. There are a few classifiers that are known for detecting MCI or AD, such as linear discriminate analysis (LDA), support vector machine (SVM), K-nearest neighbor (KNN), and neural network (NN).

KNN is a well-known classical ML technique to classify unknown subjects (Durongbhan et al., 2019; Sharma et al., 2020). The default 10-fold cross-validation procedure is implemented to increase the performance and reduce the fluctuation in ML algorithms. The collected EEG data were separated into N folders and the $N-1$ folders were used for training the models. The rest of the data was used for testing, and this process continued for N times. Recently, KNN showed a better classification for dementia diagnosis with a maximum 92% accuracy in eyes-closed situation (Sharma et al., 2020). Decision tree (DT) is a decision-aided classifier with tree type models. DT is a flowchart type structure. DT has been implemented in a few studies of AD (Amezquita-Sanchez, Mammone, Morabito, Marino, & Adeli, 2019; Fiscon et al., 2018; Yin, Cao, Siuly, & Wang, 2019). EEG extracted features created the tree to categorize the subjects into two labels, and performance was measured by the Gini index (Yuvaraj, Acharya, & Hagiwara, 2018). C4.5 DT algorithm mostly applied as it can manage noisy datasets (Fiscon et al., 2018). The quality of the model was measured using a leave-one-out cross-validation method. DT was utilized for comparing with other classifiers in Yu, Lei, Song, Liu, and Wang (2020), Armañanzas, Iglesias, Morales, and Alonso-Nanclares (2016), and Yin et al. (2019).

SVM is a widely used classifier in EEG signal processing to classify neurodegenerative disorders and brain disorders like epilepsy (Al-Qazzaz et al., 2014; Kolekar & Dash, 2015; Sharma et al., 2019). It showed better accuracy (78%) than NN (60%–66%) for diagnosing AD in one study (Staudinger & Polikar, 2011). SVM is a supervised learning model widely used for regression and classification. It is often used to detect neurodegenerative disorders and is suitable for use when the data size is small or medium. The binary classifier selected to detect MCI at the early stage was chosen in Sharma et al. (2019), Khatun, Morshed, and Bidelman (2019), and Mcbride et al. (2014). SVM performance was also compared with a fuzzy logic-based method in Chiang and Pao (2016). Yin et al. (2019) developed a 3D evaluation method to select features to design an SVM classifier (Yin et al., 2019). The SVM considers the instances which will increase the margin to adjusting its hyperplane. Even the SVM could not solve the margin solidity problem, as the number of noisy instances in the EEG signal features was so large.

Table 3 lists the research work conducted on ML for dementia diagnosis. Most of the studies used SVM classifiers with 10-fold cross validation. One recent study used KNN with fivefold classification and achieved diagnostic accuracy of 98.57% and F-score of 98.57% (Li et al., 2020). Sharma et al. (2021) worked used FTT and obtained diagnostic accuracies of 97.12%, 94.86%, 94.74% for dementia, MCI, and healthy subjects, respectively. However, that was a binary classification for early dementia. Only four studies used multiclassification problem (Mcbride et al., 2014; Sharma et al., 2020, 2021). Al-Qazzaz et al. (2017) achieved 91.48% accuracy, 91.40% precision, and 91.48% sensitivity. These limited research findings and techniques motivate enhancement and provide an avenue for taking research in

Table 3 Literature review of ML work for dementia diagnosis.

State-of-the-art work	Classifier	Recording state	Group	Acc. (%)	Sen. (%)	Spe. (%)
Sharma, Kolekar, and Jha (2021)	SVM, 10-fold	EC	Healthy	90.23	84.92	92.67
			MCI	89.72	86.45	91.43
			Dementia	90.23	83.88	93.52
		EO	Healthy	94.66	89.38	97.11
			MCI	87.22	79.49	91.24
			Dementia	88.72	78.39	94.10
		FTT	Healthy	94.74	92.86	95.60
			MCI	94.86	93.77	95.43
			Dementia	97.12	93.41	99.05
Sharma et al. (2020)	*k*NN, 10-fold	EC	Healthy	89.58	85.14	91.61
			MCI	91.67	87.87	93.54
			Dementia	92.00	86.77	94.89
		EO	Healthy	90.36	83.33	93.53
			MCI	91.67	87.87	93.54
			Dementia	91.41	88.11	93.20
		FTT	Healthy	87.69	78.96	91.11
			MCI	87.69	78.87	92.51
			Dementia	88.62	87.37	89.33
		CPT	Healthy	91.87	90.46	92.52
			MCI	88.80	83.94	91.42
			Dementia	89.44	81.03	93.60
Li et al. (2020)	*k*NN, fivefold	Reappraisal task	Healthy vs. MCI	98.57	99.06	NA

Sharma et al. (2019)	SVM, 10-fold	EC	Healthy vs. MCI	79.50	73.00	85.00
			Healthy vs. dementia	83.70	90.00	78.00
			MCI vs. dementia	86.60	86.00	88.00
		EO	Healthy vs. MCI	84.10	86.00	81.00
			Healthy vs. dementia	82.00	82.00	82.00
			MCI vs. dementia	73.40	83.00	63.00
		FTT	Healthy vs. MCI	89.80	84.00	94.00
			Healthy vs. dementia	83.80	85.00	83.00
			MCI vs. dementia	84.00	89.00	78.00
		CPT	Healthy vs. MCI	73.90	63.00	82.00
			Healthy vs. dementia	87.90	88.00	88.00
			MCI vs. dementia	88.00	88.00	89.00
Durongbhan et al. (2019)	*k*NN, 10-fold	EC	AD vs. healthy	83.41	73.80	86.89
		EO	AD vs. healthy	83.32	72.57	87.52
Wang et al. (2019)	SVM, 10-fold	Visual oddball	VD vs. healthy	86.11	86.67	85.71
Al-Qazzaz et al. (2017)	SVM, 10-fold	EC	AD vs. healthy	84.61	100.00	80.00
		Memory recall	AD vs. healthy	91.48	91.48	NA
Reyes-Coronel et al. (2016)	SVM 10-fold	Boston naming test	RCD and non-RCD	80.9	80.00	81.10
Mcbride et al. (2014)	SVM, 10-fold	EC	Multiclass	79.20	NA	NA
		EO	Multiclass	83.30	NA	NA
		Counting with FTT	Multiclass	85.40	NA	NA
Staudinger and Polikar (2011)	SVM, sixfold	Auditory oddball	AD vs. healthy	77.61	NA	NA

this area to the next level. In one study, the NN accuracy was around 71% (Polikar et al., 2001). Dementia diagnosis using machine learning is currently an emerging research area (Buscema, Rossini, Babiloni, & Grossi, 2007). Many classifiers have been successfully used in other domains, such as a Bayesian classifier and hidden Markov model (Kolekar, 2011; Kolekar & Sengupta, 2004, 2015; Wu, Nagarajan, & Chen, 2016), which can be used for detecting dementia in future works.

Discussion

This chapter explored recent studies on using EEG signal data to detect dementia. Existing tools, like PET, MRI, and MEG, are expensive and time-consuming and thus there is increased attention on using inexpensive EEG data to assist in diagnosing neurological diseases. Among the many techniques used for detecting and classifying dementia, ML-based algorithms show the best accuracy. More specifically, SVM, DT, and KNN exhibited impressive performance for classifying EEG features. ML algorithms perform well when the dataset is not large. DL approaches were used in only a few studies but showed competitive performance. However, DL algorithms are computationally expensive and they perform better when the sample size is large. NNs did not perform well in any dementia and MCI classification. Future studies should focus on working collaboratively with hospitals and other researchers to create and share diverse and large datasets. Having a diverse and large dataset enables the model to be more accurate while classifying the subjects. ML algorithms also need the extracted features to be fed into the system as inputs. Raw data consists of artifacts and other noises, thus a good-quality noise-removing algorithm is important for achieving high accuracy. Conversely, DL algorithms are useful when the sample size is large and they do not need the extracted features to be fed into the system. DL algorithms extract features for themselves according to the requirements.

Conclusion

Early and automated diagnosis of dementia is a challenge. As such, future studies must investigate novel methods for detecting dementia with high accuracy. According to the studies presented in this chapter, ML algorithms with shallow architecture have performed well with small sample size, as it is always a challenge to find large datasets to use. Some studies were limited due to EEG data noise, and thus future works should aim to use larger datasets and implement noise-removing filters.

References

Adeli, H., Ghosh-Dastidar, S., & Dadmehr, N. (2008). A spatio-temporal wavelet-chaos methodology for EEG-based diagnosis of Alzheimer's disease. *Neuroscience Letters*, *444*(2), 190–194.

Al-Qazzaz, N. K., Ali, S. H. B., Ahmad, S. A., Chellappan, K., Islam, M. S., & Escudero, J. (2014). Role of EEG as biomarker in the early detection and classification of Dementia. *The Scientific World Journal*, *2014*(906038), 16.

Al-Qazzaz, N. K., Ali, S. H. B. M, Ahmad, S. A., Islam, M. S., & Escudero, J. (2017). Automatic artifact removal in EEG of normal and demented individuals using ICA–WT during working memory tasks. *Sensors*, *17*(6), 1326.

Alzheimer's Disease International. (2019). *World Alzheimer report 2019: Attitudes to dementia.*. London: Alzheimer's Disease Internationals.

American Psychiatric Association. (2013). *Diagnostic and statistical manual of mental disorders (DSM-5®)*. Washington, DC: American Psychiatric Publication.

Amezquita-Sanchez, J. P., Mammone, N., Morabito, F. C., Marino, S., & Adeli, H. (2019). A novel methodology for automated differential diagnosis of mild cognitive impairment and the Alzheimer's disease using EEG signals. *Journal of Neuroscience Methods*, *322*, 88–95.

Armañanzas, R., Iglesias, M., Morales, D. A., & Alonso-Nanclares, L. (2016). Voxel-based diagnosis of Alzheimer's disease using classifier ensembles. *IEEE Journal of Biomedical and Health Informatics*, *21*(3), 778–784.

Babiloni, C., Ferri, R., Binetti, G., Vecchio, F., Frisoni, G. B., Lanuzza, B., et al. (2009). Directionality of EEG synchronization in Alzheimer's disease subjects. *Neurobiology of Aging*, *30*(1), 93–102.

Bennys, K., Rondouin, G., Vergnes, C., & Touchon, J. (2001). Diagnostic value of quantitative EEG in Alzheimer's disease. *Neurophysiologie Clinique/Clinical Neurophysiology*, *31*(3), 153–160.

Benvenuto, J., Jin, Y., Casale, M., Lynch, G., & Granger, R. (2002). Identification of diagnostic evoked response potential segments in Alzheimer's disease. *Experimental Neurology*, *176*(2), 269–276.

Besthorn, C., Sattel, H., Geiger-Kabisch, C., Zerfass, R., & Förstl, H. (1995). Parameters of EEG dimensional complexity in Alzheimer's disease. *Electroencephalography and Clinical Neurophysiology*, *95*(2), 84–89.

Bird, T. D. (2001). Alzheimer's disease and other primary dementias. In E. Braunwald, A. Fauci, S. Hauser, J. Jameson, & D. Kapser (Eds.), *Vol. 2. Harrisons principles of internal medicine* (pp. 2391–2398). New York: McGraw Hill (Eds.).

Buscema, M., Rossini, P., Babiloni, C., & Grossi, E. (2007). The IFAST model, a novel parallel nonlinear EEG analysis technique, distinguishes mild cognitive impairment and Alzheimer's disease patients with high degree of accuracy. *Artificial Intelligence in Medicine*, *40*(2), 127–141.

Che, H., Jung, Y.-J., Im, C.-H., & Lee, S. (2007). Extraction of qEEG variables to diagnose early dementia. In *29th Annual international conference of the IEEE, engineering in medicine and biology society, EMBS 2007* (pp. 4115–4118). IEEE.

Chiang, H.-S., & Pao, S.-C. (2016). An EEG-based fuzzy probability model for early diagnosis of Alzheimer's disease. *Journal of Medical Systems*, *40*(5), 125.

Das, A., Das, P., & Roy, A. B. (2002). Applicability of Lyapunov exponent in EEG data analysis. *Complexity International*, *9*, 1–8.

Dauwels, J., Vialatte, F., & Cichocki, A. (2010). Diagnosis of Alzheimer's disease from EEG signals: Where are we standing? *Current Alzheimer Research*, 7(6), 487–505.

Dauwels, J., Vialatte, F., Musha, T., & Cichocki, A. (2010). A comparative study of synchrony measures for the early diagnosis of Alzheimer's disease based on EEG. *NeuroImage*, *49*(1), 668–693.

Deng, B., Liang, L., Li, S., Wang, R., Yu, H., Wang, J., & Wei, X. (2015). Complexity extraction of electroencephalograms in Alzheimer's disease with weighted-permutation entropy. *Chaos: An Interdisciplinary Journal of Nonlinear Science*, *25*(4), 043105.

Durongbhan, P., Zhao, Y., Chen, L., Zis, P., De Marco, M., Unwin, Z. C., et al. (2019). A dementia classification framework using frequency and time-frequency features based on EEG signals. *IEEE Transactions on Neural Systems and Rehabilitation Engineering*, *27*(5), 826–835.

Ferri, C. P., Sousa, R., Albanese, E., Ribeiro, W. S., & Honyashiki, M. (2009). World alzheimer report 2009-executive summary. In M. Prince, & J. Jackson (Eds.) (pp. 1–22). London: Alzheimer's Disease International (Eds.).

Fiscon, G., Weitschek, E., Cialini, A., Felici, G., Bertolazzi, P., De Salvo, S., et al. (2018). Combining EEG signal processing with supervised methods for Alzheimer's patients classification. *BMC Medical Informatics and Decision Making*, *18*(1), 1–10.

Grassberger, P., & Procaccia, I. (1983). Estimation of the Kolmogorov entropy from a chaotic signal. *Physical Review A*, *28*(4), 2591.

Grassberger, P., & Procaccia, I. (2004). Measuring the strangeness of strange attractors. In *The theory of chaotic attractors* (pp. 170–189). New York: Springer.

Hebert, L. E., Scherr, P. A., Bienias, J. L., Bennett, D. A., & Evans, D. A. (2003). Alzheimer disease in the US population: Prevalence estimates using the 2000 census. *Archives of Neurology*, *60*(8), 1119–1122.

Henderson, G., Ifeachor, E., Hudson, N., Goh, C., Outram, N., Wimalaratna, S., et al. (2006). Development and assessment of methods for detecting dementia using the human electroencephalogram. *IEEE Transactions on Biomedical Engineering*, *53*(8), 1557–1568.

Henderson, G. T., Ifeachor, E. C., Wimalaratna, H., Allen, T., & Hudson, R. (2000). Prospects for routine detection of dementia using the fractal dimension of the human electroencephalogram. *IEEE Proceedings-Science, Measurement and Technology*, *147*(6), 321–326.

Hornero, R., Abásolo, D., Escudero, J., & Gómez, C. (2009). Nonlinear analysis of electroencephalogram and magnetoencephalogram recordings in patients with Alzheimer's disease. *Philosophical Transactions of the Royal Society of London A: Mathematical, Physical and Engineering Sciences*, *367*(1887), 317–336.

Jeong, J. (2004). EEG dynamics in patients with Alzheimer's disease. *Clinical Neurophysiology*, *115*(7), 1490–1505.

Jeong, J., Gore, J. C., & Peterson, B. S. (2001). Mutual information analysis of the EEG in patients with Alzheimer's disease. *Clinical Neurophysiology*, *112*(5), 827–835.

Jiang, Z.-Y. (2005). Study on EEG power and coherence in patients with mild cognitive impairment during working memory task. *Journal of Zhejiang University Science B*, *6*(12), 1213–1219.

Jung, T.-P., Makeig, S., Westerfield, M., Townsend, J., Courchesne, E., & Sejnowski, T. J. (2000). Removal of eye activity artifacts from visual event-related potentials in normal and clinical subjects. *Clinical Neurophysiology*, *111*(10), 1745–1758.

Khatun, S., Morshed, B. I., & Bidelman, G. M. (2019). A single-channel EEG-based approach to detect mild cognitive impairment via speech-evoked brain responses. *IEEE Transactions on Neural Systems and Rehabilitation Engineering*, *27*(5), 1063–1070.

Kilicarslan, A., Grossman, R. G., & Contreras-Vidal, J. L. (2016). A robust adaptive denoising framework for real-time artifact removal in scalp EEG measurements. *Journal of Neural Engineering*, *13*(2), 026013.

Kolekar, M. H. (2011). Bayesian belief network based broadcast sports video indexing. *Multimedia Tools and Applications*, *54*(1), 27–54.

Kolekar, M. H., & Dash, D. P. (2015). A nonlinear feature based epileptic seizure detection using least square support vector machine classifier. In *TENCON 2015 IEEE region 10 conference, Macao* (pp. 1–6). IEEE.

Kolekar, M. H., & Sengupta, S. (2004). Hidden Markov model based video indexing with discrete cosine transform as a likelihood function. In *Proceedings of the IEEE INDICON India annual conference* (pp. 157–159). IEEE.

Kolekar, M. H., & Sengupta, S. (2015). Bayesian network-based customized highlight generation for broadcast soccer videos. *IEEE Transactions on Broadcasting, 61*(2), 195–209.

Krishnaveni, V., Jayaraman, S., Aravind, S., Hariharasudhan, V., & Ramadoss, K. (2006). Automatic identification and removal of ocular artifacts from EEG using wavelet transform. *Measurement Science Review, 6*(4), 45–57.

Kumar, A., & Kolekar, M. H. (2014). Machine learning approach for epileptic seizure detection using wavelet analysis of EEG signals. In *International conference on medical imaging, m-health and emerging communication systems (MedCom) Greater Noida* (pp. 412–416). IEEE.

Li, J., Liu, X., & Ouyang, G. (2013). Permutation conditional mutual information and its application to epileptic EEG. In *International conference on computer sciences and applications (CSA)* (pp. 733–736). IEEE.

Li, X., & Ouyang, G. (2010). Estimating coupling direction between neuronal populations with permutation conditional mutual information. *NeuroImage, 52*(2), 497–507.

Li, Y., Xiao, S., Li, Y., Li, Y., & Yang, B. (2020). Classification of mild cognitive impairment from multi-domain features of resting-state EEG. In *2020 42nd annual international conference of the IEEE engineering in medicine & biology society (EMBC)* (pp. 256–259). IEEE.

Liu, C.-J., Huang, C.-F., Chou, C.-Y., Kuo, W.-J., Lin, Y.-T., Hung, C.-M., et al. (2012). Age- and disease-related features of task-related brain oscillations by using mutual information. *Brain and Behavior, 2*(6), 754–762.

Mcbride, J. C., Zhao, X., Munro, N. B., Smith, C. D., Jicha, G. A., Hively, L., et al. (2014). Spectral and complexity analysis of scalp EEG characteristics for mild cognitive impairment and early Alzheimer's disease. *Computer Methods and Programs in Biomedicine, 114*(2), 153–163.

Missonnier, P., Deiber, M.-P., Gold, G., Herrmann, F. R., Millet, P., Michon, A., et al. (2007). Working memory load–related electroencephalographic parameters can differentiate progressive from stable mild cognitive impairment. *Neuroscience, 150*(2), 346–356.

Neto, E., Allen, E. A., Aurlien, H., Nordby, H., & Eichele, T. (2015). EEG spectral features discriminate between Alzheimer's and vascular dementia. *Frontiers in Neurology, 6*, 25.

Phillips, J., Pond, D., & Goode, S. (2011). *Timely diagnosis of dementia: Can we do better.* Canberra: Alzheimer's Australia.

Polikar, R., Keinert, F., & Greer, M. H. (2001). Wavelet analysis of event related potentials for early diagnosis of Alzheimer's disease. In *Wavelets in signal and image analysis* (pp. 453–478). Netherlands: Springer.

Prince, M., Bryce, R., & Ferri, C. (2011). *World Alzheimer Report 2011: The benefits of early diagnosis and intervention.* London: Alzheimer's Disease International.

Prince, M. J. (2015). *World Alzheimer Report 2015: The global impact of dementia: An analysis of prevalence, incidence, cost and trends.* Alzheimer's Disease International.

Reyes-Coronel, C., Waser, M., Garn, H., Deistler, M., Dal-Bianco, P., Benke, T., et al. (2016). Predicting rapid cognitive decline in Alzheimer's disease patients using quantitative EEG markers and neuropsychological test scores. In *2016 38th Annual international conference of the IEEE engineering in medicine and biology society (EMBC)* (pp. 6078–6081). IEEE.

Rodriguez, R., Lopera, F., Alvarez, A., Fernandez, Y., Galan, L., Quiroz, Y., & Bobes, M. A. (2014). Spectral analysis of EEG in familial Alzheimer's disease with E280A presenilin-1 mutation gene. *International Journal of Alzheimer's Disease, 2014*(180741), 10.

Rodríguez-Bermúdez, G., & García-Laencina, P. J. (2015). Analysis of EEG signals using nonlinear dynamics and chaos: A review. *Applied Mathematics & Information Sciences*, *9*(5), 2309.

Rosenstein, M. T., Collins, J. J., & De Luca, C. J. (1993). A practical method for calculating largest Lyapunov exponents from small data sets. *Physica D: Nonlinear Phenomena*, *65*(1), 117–134.

Rossor, M. N., Fox, N. C., Mummery, C. J., Schott, J. M., & Warren, J. D. (2010). The diagnosis of young-onset dementia. *The Lancet Neurology*, *9*(8), 793–806.

Roy, V., & Shukla, S. (2015). A NLMS based approach for artifacts removal in multichannel EEG signals with ICA and double density wavelet transform. In *Fifth international conference on communication systems and network technologies (CSNT)* (pp. 461–466). IEEE.

Sanei, S., & Chambers, J. A. (2013). *EEG signal processing*. West Sussex, England: John Wiley & Sons.

Scott, K. R., & Barrett, A. M. (2007). Dementia syndromes: Evaluation and treatment. *Expert Review of Neurotherapeutics*, 7(4), 407–422.

Shaji, K. S., Jotheeswaran, A. T., Girish, N., Bharath, S., Dias, A., Pattabiraman, M., & Varghese, M. (2010). *The Dementia India report: Prevalence, impact, costs and services for dementia*. Alzheimer's and Related Disorders Society of India.

Shaji, K. S., Smitha, K., Lal, K. P., & Prince, M. J. (2003). Caregivers of people with Alzheimer's disease: A qualitative study from the Indian 10/66 dementia research network. *International Journal of Geriatric Psychiatry*, *18*(1), 1–6.

Sharma, N., Kolekar, M. H., & Chandra, S. (2015). Role of the cognitive screening tool in early dementia diagnosis. *International Journal of Advances in Science, Engineering and Technology*, *3*(4), 101–104.

Sharma, N., Kolekar, M. H., & Jha, K. (2020). Iterative filtering decomposition based early dementia diagnosis using EEG with cognitive tests. *IEEE Transactions on Neural Systems and Rehabilitation Engineering*, *28*(8), 1–9.

Sharma, N., Kolekar, M. H., & Jha, K. (2021). EEG based dementia diagnosis using multiclass support vector machine with motor speed cognitive test. *Biomedical Signal Processing and Control*, *63*, 102102.

Sharma, N., Kolekar, M. H., Jha, K., & Kumar, Y. (2019). EEG and cognitive biomarkers based mild cognitive impairment diagnosis. *IRBM*, *40*(2), 113–121.

Stam, C. J., Jones, B. F., Nolte, G., Breakspear, M., & Scheltens, P. (2007). Small-world networks and functional connectivity in Alzheimer's disease. *Cerebral Cortex*, *17*(1), 92–99.

Staudinger, T., & Polikar, R. (2011). Analysis of complexity based EEG features for the diagnosis of Alzheimer's disease. In *Annual international conference of the IEEE, engineering in medicine and biology society, EMBC-2011* (pp. 2033–2036). IEEE.

Szava, S., Valdes, P., Biscay, R., Galan, L., Bosch, J., Clark, I., & Jimenez, J. C. (1994). High resolution quantitative EEG analysis. *Brain Topography*, *6*(3), 211–219.

Timothy, L. T., Krishna, B. M., Menon, M. K., & Nair, U. (2014). Permutation entropy analysis of EEG of mild cognitive impairment patients during memory activation task. In *Fractals, wavelets, and their applications: Vol. 92* (pp. 395–406). Springer International Publishing.

Vyšata, O., Vališ, M., Procházka, A., Rusina, R., & Pazdera, L. (2015). Linear and nonlinear EEG synchronization in Alzheimer's disease. *Neurophysiology*, *47*(1), 46–52.

Wang, C., Xu, J., Zhao, S., & Lou, W. (2019). Identification of early vascular dementia patients with EEG signal. *IEEE Access*, 7, 68618–68627.

Wen, D., Bian, Z., Li, Q., Wang, L., Lu, C., & Li, X. (2016). Resting-state EEG coupling analysis of amnestic mild cognitive impairment with type 2 diabetes mellitus by using permutation conditional mutual information. *Clinical Neurophysiology*, *127*(1), 335–348.

Widagdo, M. M., Pierson, J. M., & Helme, R. D. (1998). Age-related changes in qEEG during cognitive tasks. *International Journal of Neuroscience*, *95*(1–2), 63–75.

Wolf, A., Swift, J. B., Swinney, H. L., & Vastano, J. A. (1985). Determining Lyapunov exponents from a time series. *Physica D: Nonlinear Phenomena*, *16*(3), 285–317.

World Health Organization. (2006). *Neurological disorders: Public health challenges*. World Health Organization.

Wu, W., Nagarajan, S., & Chen, Z. (2016). Bayesian machine learning: EEG/MEG signal processing measurements. *IEEE Signal Processing Magazine*, *33*(1), 14–36.

Yin, J., Cao, J., Siuly, S., & Wang, H. (2019). An integrated MCI detection framework based on spectral-temporal analysis. *International Journal of Automation and Computing*, *16*(6), 786–799.

Yu, H., Lei, X., Song, Z., Liu, C., & Wang, J. (2020). Supervised network-based Fuzzy learning of EEG signals for Alzheimer's disease identification. *IEEE Transactions on Fuzzy Systems*, *28*(1), 60–71.

Yuvaraj, R., Acharya, U. R., & Hagiwara, Y. (2018). A novel Parkinson's disease diagnosis index using higher-order spectra features in EEG signals. *Neural Computing and Applications*, *30*(4), 1225–1235.

Zhao, P., Van-Eetvelt, P., Goh, C., Hudson, N., Wimalaratna, S., & Ifeachor, E. (2007). Characterization of EEG in Alzheimer's disease using information theoretic methods. In *29th Annual international conference of the IEEE, engineering in medicine and biology society, EMBS 2007* (pp. 5127–5131). IEEE.

Zikov, T., Bibian, S., Dumont, G. A., Huzmezan, M., & Ries, C. R. (2002). A wavelet based de-noising technique for ocular artifact correction of the electroencephalogram. In *24th annual conference and the annual fall meeting of the biomedical engineering society EMBS/BMES conference: Vol. 1* (pp. 98–105). IEEE.

CHAPTER 8

Computational methods for translational brain-behavior analysis

Olugbemi T. Olaniyan[a], Charles O. Adetunji[b], Olorunsola Adeyomoye[c], Ayobami Dare[d], Mayowa J. Adeniyi[e], and Alex Enoch[f]
[a]Department of Physiology, Laboratory for Reproductive Biology and Developmental Programming, Rhema University, Aba, Nigeria
[b]Department of Microbiology, Applied Microbiology, Biotechnology and Nanotechnology Laboratory, Edo State University Uzairue, Iyamho, Nigeria
[c]Department of Physiology, University of Medical Sciences, Ondo, Nigeria
[d]Department of Physiology, School of Laboratory Medicine and Medical Sciences, College of Health Sciences, Westville Campus, University of KwaZulu-Natal, Durban, South Africa
[e]Department of Physiology, Edo State University Uzairue, Iyamho, Nigeria
[f]Department of Human Physiology, Ahmadu Bello University Zaria, Zaria, Nigeria

Introduction

Translational clinical research has been identified as patient-oriented clinical investigation focused on the clinical subspecialty combined with the knowledge and techniques of biomedical sciences (Nathan, 2005). This chapter focuses on techniques that help in recognizing measurable entities in disease progression in seriously injured trauma patients. These features normally give direction to numerous underlying disrupted biological modes of actions that may be applied in the management of diseases (Karvunidis, Mares, Thongboonkerd, & Matejovic, 2009; Neugebauer & Tjardes, 2004; Polpitiya, McDunn, Burykin, Ghosh, & Cobb, 2009; Vodovotz, Csete, Bartels, Chang, & An, 2008). The application of computational and biochemical data models could provide healthcare providers with proper understanding of disease progression to help them properly diagnose diseases. The application of system biology could also help in determining the nature of diseases. Furthermore, it has been shown that certain treatments that depend on an individual's underlying physiology are spurring rapid growth of personalized therapies. There are steps involved in translational techniques which necessitate the scientists to recognized "gold standard" data patterns that that entails quantifiable biochemical data related to treatment, diagnosis, and prognosis.

Artificial Intelligence for Neurological Disorders
https://doi.org/10.1016/B978-0-323-90277-9.00025-0

Moreover, parameters and protocols that can accommodate these data must be developed for diverse clinical applications (Adetunji, Adetunji, & Olaniyan, 2021; Adetunji, Ajayi, et al., 2021; Adetunji et al., 2021a, 2021b; Adetunji, Inobeme, et al., 2021; Adetunji & Varma, 2020; Akram et al., 2020; Egbuna et al., 2020; Roli, Adetunji, Mishra, Adetunji, & Mishra, 2020; Suddhasuchi et al., 2020). These parameters and protocols require computational techniques for proper identification of disease associations and biochemical patterns as well as to produce patient data-driven reports that enable the clinician to adequately diagnose the nature of diseases and administer proper care (Butte, 2008; Salluh & Bozza, 2008). Moreover, several scientists have identified linguistic markers or patterns that show a relationship with mental disorders such as suicide, schizophrenia, and depression. These markers play a critical role that helps in assessing patient metal state and identifying mental health disorders using online digital records (Abdullah et al., 2016; Aliannejadi & Crestani, 2018; Al-Mosaiwi & Johnstone, 2018). Therefore, this chapter focuses on several computational approaches for the acquisition of data, processing of data, and device control in translational brain behavioral analysis.

Computational physiology

Physiology is an integrative discipline that studies how the structure and activities of cells mediate complex processes in living systems (such as cognition, cellular respiration, and movement). The physical underpinning of physiology, from biophysics to integrative control, was demonstrated in the twentieth century,. The last five decades have been devoted to molecular physiology, explaining mechanisms of physiology at the genetic or proteomic level, while recent years have recorded significant process in the development of experimental techniques such as polymerase chain reaction (PCR) assays and DNA analysis for high through-put measurement, in addition to comprehensive databases for quantitative science. However, the most significant consequence of this milestone in experimental techniques is the generation of large amounts of data at the cellular level that are difficult to comprehend in terms of body physiology (Hunter & Borg, 2003). As a result of nature's complexity and an explosion of data, a mathematical model of natural concepts is required. As such, it is critical that physiology incorporate the concept of "computational physiology," in which mathematicians and bioengineers collaborate with physiologists and molecular biologists to integrate physiology at various levels of organization with the growing genomic and proteomic databases. Furthermore, understanding the physiological implications of diseases requires linking quantitative physiology models to signal cascades, metabolic pathways, and other cellular activities (Crampin et al., 2003).

Understanding the pathogenesis of disease and its management requires the generation of models that cut across various levels of organization and this has resulted in some challenges to the analysis of various disease conditions and their effective management. Thus, computational physiology, a fast-growing field in medicine, has become highly relevant in providing assistance for adequate analysis of medical conditions. The major aim of computational physiology is to produce mechanistic models of physiological systems, personalize these methods using patient data, and apply these models to enhance disease diagnosis and management (Winslow, Trayanova, Geman, & Miller, 2012). This is achieved by integrating information from a molecular level to the level of cells, tissues, organs, and systems. This multiscale modeling technique has been applied in cancer, diabetes, kidney, and brain disorders, and is receiving attention in other systems. Computational physiology models are classically produced from data collected from experimental models of disease using animals, which are later specialized and translated using human datasets. As much as possible, the experiments and modeling are carried out in a modular pattern where diverse physiological concepts are scrutinized and later organized into an algorithm. These algorithms are then grouped together, and their performances are evaluated to ensure reproducibility to determine if they accurately predict the actual experimental data (Hunter, 2020). A major organ where computational physiology has been applied is the heart. The first model developed on cardiac myocyte action potential published more than 50 years ago. The creation of whole-heart image-based models combining unique anatomical and biophysical aspects such as cardiac electromechanics, electrophysiology, and fluid dynamics has resulted from this cellular modeling (Hodgkin & Huxley, 1952). Scientists are now interested in applying similar techniques to analyze brain diseases, reproductive problems, diabetes, and other physiological alterations.

Medical and data scientists

A medical scientist is a scientist trained in physiology or other related disciplines who tries to better understand how the human body works and develop sophisticated diagnostic tools or new therapeutic techniques to cure or treat disease. A medical scientist works in universities, pharmaceutical firms, biomedical corporations, research centers, and industries and his or her job entails a combination of research and analysis, as well as teaching and writing comprehensive articles (Nwogu, 1997). Among the major responsibilities of a medical scientist in brain behavior are:

- Conducting research to develop methodology, instrumentation, and procedures for medical use, as well as analyzing data and communicating conclusions.
- Studying toxicity, bacteria, or microbes, or researching cell structure in the brain and preparing and analyzing organs, tissues, and cell samples.
- Designing and directing research investigations into human or animal disease, prevention strategies, and treatments for brain disease.
- Studying the origins, progression, life cycle, and method of transmission of brain illnesses.
- Assessing the impact of medications, gases, insecticides, parasites, and microbes on various levels in the brain.

Big Data wranglers, also known as data scientists, collect and analyze massive amounts of structured and unstructured data. A data scientist's job entails a mix of computer science, statistics, and math. They collect, process, and model data before interpreting the outcomes to develop actionable plans for analyzing brain function. Data scientists are analytic professionals that use their knowledge of technology and social science to identify patterns and handle data (Knopp, Beaghley, Frank, Orrie, & Watson, 2016).

The numerous roles of a data scientist include:

- Asking the right questions to begin data recovery process in the brain
- Acquiring, processing, and cleaning data relating to brain function
- Combining and saving the information acquired
- Conducting preliminary data analysis and exploratory data analysis to design algorithms
- Choosing one or more prospective models and algorithms to work with
- Using data science approaches such as machine learning, statistical modeling, and artificial intelligence to analyze brain behavior
- Measuring and improving results based on the input received
- Using the same steps to address a new problem

Translational brain behavioral pattern

Brain functions and behavioral patterns in disease conditions are analyzed and validated through cognition, motor control, and affective skills disorders.

Many research activities in neuroscience are directed toward improving and developing powerful advanced behavioral tools to investigate neural function, networks, patterns, and dysfunctions. Computational physiology involving in vivo electrophysiological monitoring techniques can analyze deep behavioral and biochemical assays to generate highly quantitative datasets. Various applications of translational brain behavioral patterns involve characterizing detailed behavioral phenotypes and genotypes, early diagnosis of disease, and precision analysis of new therapeutic agents. Stress and other environmental factors like pollution and heavy metals have been shown to cause serious barriers to and modulation of positive mental health behavior, causing negative emotions, sleep deprivation, neurodegeneration, reduction in white and grey matter integrity in the axonal tracts, and decreased functional connectivity. Thus, in vivo electrophysiological monitoring techniques can be utilized to elucidate many of these processes.

Cognitive mapping and neural coding

The brain has the potential to map out the spatial environment (Bellmund, Gärdenfors, Moser, & Doeller, 2018). This is necessary for memory and navigation. Much research in this area utilizes animal models for investigations. Evidence has shown that grid, border, and head direction cells are responsible for spatial mapping and these cells are localized within the hippocampus and associated structures (Epstein, Patai, Julian, & Spiers, 2017). The ability to imagine ourselves in the past, future, or at different locations further supports reports that the brain can perform cognitive mapping. A system capable of tracking dozens of bats simultaneously at high accuracy and resolution was used to collect data from these animals in a study designed to investigate spatial navigation in bats. The tracking analysis revealed that bats rarely engage in random search, preferring to forage in a long and straightforward manner with frequent shortcuts (Toledo et al., 2020). Several years back, the Morris water maze was adopted by neuroscientists to study behavioral responses and cognitive mapping processes (Vorhees & Williams, 2006). The Morris water maze was developed for animals to escape from water and it helps to assess spatial learning and memory and therefore is useful in assessing the effects of aging, medications, and lesions (Terry Jr., 2009). This procedure involves using visual cues to investigate and locate a hidden platform to escape from water. In a study conducted by Rendeiro et al. (2009), rats were made to navigate space in an opaque swimming pool to locate a platform. The study was designed to investigate if the rats could develop a novel trajectory pattern to navigate or if they could navigate based on spatial learning from repeated exposure. The rats navigated using cognitive experience based on repeated exposure to locate the platform and this further corroborates the observation that the brain has the potential for spatial navigation.

Neural coding involves the relationship between stimuli and ensembles or network of neurons. Evidence have shown that in animals and humans, there must be interactions with the environment before their development is complete (Panzeri, Harvey, Piasini, Latham, & Fellin, 2017). This suggests that they build neural codes that are necessary for turning sensory input into motor output as their neurons develop. The biophysical nature of neurons gives them the capability to generate and propagate action potentials over a long distance within a short duration (Kress & Mennerick, 2009). Therefore, it has been proposed that signal processing and coding could even be achieved by subthreshold oscillation and resonance of neuronal representations that are deposited in the frequency domain (Villacorta & Panetsos, 2005).

Neuroelectrophysiology modeling

Mathematical and statistical models are significant in neuroscience and have made the description of electrical activity of individual neurons possible. Even now, large networks of neurons are described using different models (D'Angelo et al., 2013). With advancement in the field of neuroscience, more computational models are still needed to understand neurologic functions. Although many investigations in neuroscience involve the use of new modern technologies, the use of mathematical and statistical ideas has provided new insight into analyzing and interpreting large collections of data. Several decades ago, logic was used to explain neuronal activity and it was postulated that the human mind could be explained using logic (Friedrich & Friederici, 2009). Now, efforts are being made to develop models that can be used in neural computation. Furthermore, computer programs have been developed and used to perform functions that are peculiar to the human mind (Kriegeskorte & Douglas, 2018). Many scientists have developed computational structures that can be used to investigate the mechanisms of neuronal activities. For example, in 1949 Hebb developed a model that stated that when a neuron A is near and fires a neuron B, growth and metabolic changes take place that increase the strength of both neurons. As the saying goes, "Neurons that fire together wire together" (Keysers & Gazzola, 2014). In the past, mathematical models were proposed in neurobiology studies which has to do with the electrical properties of neuronal impulse transmission, it was stated that neurons are electrically excitable in a similar manner to circuits which involves resistor and capacitor which are in parallel and its differential equations generated from data (Dabrowski, Castaño, & Tartar, 2013). This mathematical model is still in use today. A rate coding mathematical concept was also proposed in which the intensity of firing neurons does not affect the voltage of the action potential of neurons. Neuroelectrophysiology still refers to this as the all-or-none principle (Weinberg, 2015).

Clinical translation of cognitive mapping and neural coding

The brain is endowed with superb neural structures and connectivity that function as a compass, enabling us to know where we are. Cognitive mapping is a multidisciplinary concept that was first used to describe spatial learning in rats (Tolman, 1948). It is a mental imagery that helps in acquisition, coding, storing, and decoding of spatial information and information regarding relative positions, locations, and characteristics of occurrences in the environment. It is the tendency to comprehend the spatial disposition of a given environment (Ambrosini & Bowman, 2002; Fiol & Huff, 1992). With cognitive maps, it is possible to design and accumulate spatial information that will make visualization of an image possible with attendant enhancement of cognitive abilities. Studies have discovered a link between hippocampal place cell firing and local field oscillation theta with regard to theta phase precession, which has been identified as a temporal coding process leading to behavioral sequence memory acquisition and asymmetric Hebbian plasticity (Wagatsuma & Yamaguchi, 2007).

In fact, the hippocampus's place cells and the grid cells in the medial entorhinal cortex's dorsolateral aspect are cognitive map hubs (O'Keefe & Nadel, 1978). Due to the peculiarity of hippocampal connectivity compared with other brain areas, it serves as an integration point for spatial and non-spatial information, bringing together information regarding the relative locations of an object and its other attributes (Manns & Eichenbaum, 2009). Head direction cells (HD cells), which were first discovered in the postsubiculum (Taube, Muller, & Ranck, 1990) and that discharge when the head of an animal tilts to a particular direction in relation to an allocentric reference frame, are other neural structures that help in spatial locations in the environment. These cells can also be found in many different parts of the brain (Taube, 2007).

In their navigation model, Gerstner and Abbott (1997) proved that synaptic remodeling happens during random exploration of an environment in such a way that place-cell brain activity encodes a shift in location from any position to a set destination while avoiding walls and other obstacles. Wagatsuma and Yamaguchi (2007) designed a theta phase coding in their review because there is a relationship between place-cell discharge and local field oscillation theta with respect to theta phase precession. Memory-encoding functions include one-time experience instantaneous memory development, as well as the tendency for memory integration of behavioral sequences into a cognitive map network. When it comes to memory recall, theta phase coding allows the hippocampus to encode the spatial position in the current behavioral context with nonspecific sensory input when many sequences have been coded. Reversion of theta rhythm-dependent activities into information in the behavioral time scale is involved in the hippocampus's utilization and retrieval of temporal sequences.

Computational models of spatial cognition employing robots have been studied. Milford and Wyeth (2008), for example, demonstrated that pose cells constructed in the RatSLAM paradigm can encode positions in a large-scale environment for robotic long-term navigation activities. To build grid cells' firing style, Jauffret, Cuperlier, and Gaussier (2015) used a modulo projection considering path integration. Zeng and Si (2017) developed a model that improves knowledge of neural mechanisms of spatial cognition and aids in the development of more accurate brain models by including continuous attractor networks of head direction cells and conjunctive grid cells to combine velocity information by conjunctive encodings of space and movement.

Systems biology in translational and computational biology

System biology is an interdisciplinary field that deals with the analysis of biological systems using mathematical and computational models. Due to innovations in biology, much information has been gained from molecular studies of metabolomics, proteomics, genomics, and transcriptomics.

Application of system biology in translational brain tumor research

Application of system biology to cancer studies helps in the integration of many fields such as genetics, epigenetics, signaling networks, cellular behavior, histology, clinical medicine, engineering, physics, mathematics, computer science, epidemiology, and others (Huang, Kuo, & Huang, 2011; Kreeger & Lauffenburger, 2010) to advance the quantum of scientific information generated in cancer research. Among the successes derived from this integration are design of mathematical modeling of cancer cell population growth (Tyson, Garbett, Frick, & Quaranta, 2012), cellular responses to therapeutic intervention (Traina et al., 2010) and cancer metabolism (Lewis & Abdel-Haleem, 2013), creation of cancer genome atlases (a list of genetic mutations responsible for cancer) (Collins & Barker, 2007), design of algorithms that make it possible to identify players of cancer progression using existing datasets (Pe'er & Hacohen, 2011), development of statistical and mechanistic modeling of signaling networks (Tyson et al., 2011), and construction of quantitative modeling of cancer evolutionary processes (Anderson & Quaranta, 2008).

Summary

The understanding of brain behavioral patterns requires mathematical modeling to analyze neuronal activity, metabolic pathways, and various signaling cascades. This is done through the utilization of data, multiscale modeling techniques, and algorithms to predict the translational patterns of events in any condition. The electrical activity generated by a neuron plays a significant role in neuroelectrophysiological modeling and analyzing and interpreting large collections of data. Currently, computer programs and algorithms are used in mechanistic investigation of brain behavior patterns, neural structures, and connectivity.

Conclusion

Computational physiology approaches help to tackle different challenges involved in brain-behavioral computer interface pattern analysis through the utilization of physiological data and biomarkers to understand the dynamics involve in the central nervous system behavioral pattern. Computational models can also enable manipulation and measurement of neuronal activity, neurotransmitter functions, and metabolic processes through multidimensional biophysiological datasets. Thus, this chapter focused on several computational physiology approaches such as principal component analysis, canonical correlation analysis, single value decomposition, and others for data acquisition, processing, and device control in translational brain behavioral analysis.

References

Abdullah, S., Matthews, M., Frank, E., Doherty, G. J., Gay, G., & Choudhury, T. (2016). Automatic detection of social rhythms in bipolar disorder. *Journal of the American Medical Informatics Association, 23*(3), 538–543.

Adetunji, J. B., Adetunji, C. O., & Olaniyan, O. T. (2021). African walnuts: A natural depository of nutritional and bioactive compounds essential for food and nutritional security in Africa. In O. O. Babalola (Ed.), *Food security and safety*. Cham: Springer.

Adetunji, C. O., Ajayi, O. O., Akram, M., Olaniyan, O. T., Chishti, M. A., Inobeme, A., et al. (2021). Medicinal plants used in the treatment of influenza a virus infections. In K. Dua, S. Nammi, D. Chang, D. K. Chellappan, G. Gupta, & T. Collet (Eds.), *Medicinal plants for lung diseases*. Singapore: Springer.

Adetunji, C. O., Akram, M., Olaniyan, O. T., Ajayi, O. O., Inobeme, A., Olaniyan, S., et al. (2021a). Medicinal plants used in the treatment of pulmonary hypertension. In K. Dua, S. Nammi, D. Chang, D. K. Chellappan, G. Gupta, & T. Collet (Eds.), *Medicinal plants for lung diseases*. Singapore: Springer. https://doi.org/10.1007/978-981-33-6850-7_14.

Adetunji, C. O., Akram, M., Olaniyan, O. T., Ajayi, O. O., Inobeme, A., Olaniyan, S., et al. (2021b). Targeting SARS-CoV-2 novel Corona (COVID-19) virus infection using medicinal plants. In K. Dua, S. Nammi, D. Chang, D. K. Chellappan, G. Gupta, & T. Collet (Eds.), *Medicinal plants for lung diseases*. Singapore: Springer.

Adetunji, C. O., Inobeme, A., Olaniyan, O. T., Ajayi, O. O., Olaniyan, S., & Adetunji, J. B. (2021). Application of nanodrugs derived from active metabolites of medicinal plants for the treatment of inflammatory and lung diseases: Recent advances. In K. Dua, S. Nammi, D. Chang, D. K. Chellappan, G. Gupta, & T. Collet (Eds.), *Medicinal plants for lung diseases*. Singapore: Springer.

Adetunji, C. O., & Varma, A. (2020). Biotechnological application of Trichoderma: A powerful fungal isolate with diverse potentials for the attainment of food safety, management of pest and diseases, healthy planet, and sustainable agriculture. In C. Manoharachary, H. B. Singh, & A. Varma (Eds.), *Soil biology: Vol. 61. Trichoderma: Agricultural applications and beyond*. Cham: Springer.

Akram, M., Jabeen, F., Daniyal, M., Zainab, R., Haq, U. U., Adetunji, C. O., et al. (2020). Genetic engineering of novel products of health significance: Recombinant DNA technology. In C. Egbuna, & T. G. Dable (Eds.), *Functional foods and nutraceuticals*. Cham: Springer. https://doi.org/10.1007/978-3-030-42319-3_26.

Aliannejadi, M., & Crestani, F. (2018). Personalized context-aware point of interest recommendation. *ACM Transactions on Information Systems, 36*, 4.

Al-Mosaiwi, M., & Johnstone, T. (2018). In an absolute state: Elevated use of absolutist words is a marker specific to anxiety, depression, and suicidal ideation. *Clinical Psychological Science, 6*(4), 529–542.

Ambrosini, V., & Bowman, C. (2002). Mapping successful organizational routines. In A. S. Huff, & M. Jenkins (Eds.), *Mapping strategic knowledge* (pp. 19–45). London; Thousand Oaks, CA: Sage Publications, ISBN:0761969497.

Anderson, A. R., & Quaranta. (2008). Integrative mathematical oncology. *Nature Reviews. Cancer, 8*(3), 227–234.

Bellmund, J., Gärdenfors, P., Moser, E. I., & Doeller, C. F. (2018). Navigating cognition: Spatial codes for human thinking. *Science (New York, N.Y.), 362*(6415), eaat6766.

Butte, A. J. (2008). Translational bioinformatics: Coming of age. *Journal of the American Medical Informatics Association, 15*(6), 709–714.

Collins, F. S., & Barker, A. D. (2007). Mapping the cancer genome. Pinpointing the genes involved in cancer will help chart a new course across the complex landscape of human malignancies. *Scientific American, 296*(3), 50–57.

Crampin, E. J., Halstead, M., Hunter, P., Nielsen, P., Noble, D., Smith, N., et al. (2003). Computational physiology and the physiome project. *Experimental Physiology*, *89*(1), 1–26.
Dabrowski, K. M., Castaño, D. J., & Tartar, J. L. (2013). Basic neuron model electrical equivalent circuit: An undergraduate laboratory exercise. *Journal of Undergraduate Neuroscience Education*, *12*(1), A49–A52.
D'Angelo, E., Solinas, S., Garrido, J., Casellato, C., Pedrocchi, A., Mapelli, J., et al. (2013). Realistic modeling of neurons and networks: Towards brain simulation. *Functional Neurology*, *28*(3), 153–166.
Egbuna, C., Gupta, E., Ezzat, S. M., Jeevanandam, J., Mishra, N., Muhammad Akram, N., et al. (2020). Aloe species as valuable sources of functional bioactives. In C. Egbuna, & G. Dable Tupas (Eds.), *Functional foods and nutraceuticals*. Cham: Springer.
Epstein, R. A., Patai, E. Z., Julian, J. B., & Spiers, H. J. (2017). The cognitive map in humans: Spatial navigation and beyond. *Nature Neuroscience*, *20*(11), 1504–1513.
Fiol, C. M., & Huff, A. S. (1992). "maps for managers: Where are we? Where do we go from here?" (PDF). *Journal of Management Studies*, *29*(3), 267–285.
Friedrich, R., & Friederici, A. D. (2009). Mathematical logic in the human brain: Syntax. *PLoS One*, *4*(5), e5599.
Gerstner, W., & Abbott, L. (1997). Learning navigational maps through potentiation and modulation of hippocampal place cells. *Journal of Computational Neuroscience*, *4*, 79–94.
Hodgkin, A. L., & Huxley, A. F. (1952). A quantitative description of membrane current and its application to conduction and excitation in nerve. *The Journal of Physiology*, *117*, 500–544.
Huang, Y. W., Kuo, S., & Huang, W. (2011). An overview of epigenetics and chemoprevention. *FEBS Letters*, *585*(13), 2129–2136.
Hunter, P. (2020). Modeling framework for computational physiology. In H. Altenbach, & A. Öchsner (Eds.), *Encyclopedia of continuum mechanics*. Berlin Heidelberg: Springer.
Hunter, P. J., & Borg, T. K. (2003). Integration from proteins to organs: The physiome project. *Nature Reviews. Molecular Cell Biology*, *4*, 237–243.
Jauffret, A., Cuperlier, N., & Gaussier, P. (2015). From grid cells and visual place cells to multimodal place cell: A new robotic architecture. *Frontiers in Neurorobotics*, *9*, 1.
Karvunidis, T., Mares, J., Thongboonkerd, V., & Matejovic, M. (2009). Recent progress of proteomics in critical illness. *Shock*, *31*(6), 545–552.
Keysers, C., & Gazzola, V. (2014). Hebbian learning and predictive mirror neurons for actions, sensations and emotions. *Philosophical Transactions of the Royal Society of London. Series B, Biological Sciences*, *369*(1644), 20130175.
Knopp, B. M., Beaghley, S., Frank, A., Orrie, R., & Watson, M. (2016). *Defining the roles, responsibilities, and functions for data science within the defense intelligence agency*. Library of Congress Cataloging-in-Publication Data is available for this publication, ISBN:978-0-8330-9658-6.
Kreeger, P. K., & Lauffenburger. (2010). Cancer systems biology: A network modeling perspective. *Carcinogenesis*, *31*(1), 2–8.
Kress, G. J., & Mennerick, S. (2009). Action potential initiation and propagation: Upstream influences on neurotransmission. *Neuroscience*, *158*(1), 211–222.
Kriegeskorte, N., & Douglas, P. K. (2018). Cognitive computational neuroscience. *Nature Neuroscience*, *21*(9), 1148–1160.
Lewis, N. E., & Abdel-Haleem, A. M. (2013). The evolution of genome-scale models of cancer metabolism. *Frontiers in Physiology*, *4*, 237.
Manns, J. R., & Eichenbaum, H. (2009). A cognitive map for object memory in the hippocampus. *Learning & Memory*, *16*(10), 616–624.
Milford, M. J., & Wyeth, G. F. (2008). Mapping a suburb with a single camera using a biologically inspired SLAM system. *IEEE Transactions on Robotics*, *24*, 1038–1053.
Nathan, D. G. (2005). The several Cs of translational clinical research. *The Journal of Clinical Investigation*, *115*(4), 795–797.

Neugebauer, E. A., & Tjardes, T. (2004). New approaches to shock and trauma research: Learning from multidisciplinary exchange. *The Journal of Trauma, 56*(5), 1156–1165.

Nwogu, K. N. (1997). The medical research paper: Structure and functions. *English for Specific Purposes, 16*(2), 119–138.

O'Keefe, J., & Nadel, L. (1978). *The hippocampus as a cognitive map*. Oxford; New York: Clarendon Press; Oxford University Press, ISBN:0198572069.

Panzeri, S., Harvey, C. D., Piasini, E., Latham, P. E., & Fellin, T. (2017). Cracking the neural code for sensory perception by combining statistics, intervention, and behavior. *Neuron, 93*(3), 491–507.

Pe'er, D., & Hacohen, N. (2011). Principles and strategies for developing network models in cancer. *Cell, 144*(6), 864–873.

Polpitiya, A. D., McDunn, J. E., Burykin, A., Ghosh, B. K., & Cobb, J. P. (2009). Using systems biology to simplify complex disease: Immune cartography. *Critical Care Medicine, 37*(1 Suppl), S16–S21.

Rendeiro, C., Spencer, J. P., Vauzour, D., Butler, L. T., Ellis, J. A., & Williams, C. M. (2009). The impact of flavonoids on spatial memory in rodents: From behaviour to underlying hippocampal mechanisms. *Genes & Nutrition, 4*(4), 251–270.

Roli, O. I., Adetunji, C. O., Mishra, P., Adetunji, J. B., & Mishra, R. R. (2020). Caffeine: Nutraceutical and health benefit of caffeine-containing commodities and products. In P. Mishra, R. R. Mishra, & C. O. Adetunji (Eds.), *Innovations in food technology*. Singapore: Springer.

Salluh, J. I., & Bozza, P. T. (2008). Biomarkers of sepsis: Lost in translation? *Critical Care Medicine, 36*(7), 2192–2194.

Suddhasuchi, D., Sharangi, A. B., Egbuna, C., Jeevanandam, J., Ezzat, S. M., Adetunji, C. O., et al. (2020). Health benefits of isoflavones found exclusively of plants of the Fabaceae Family. In C. Egbuna, & T. G. Dable (Eds.), *Functional foods and nutraceuticals*. Cham: Springer. https://doi.org/10.1007/978-3-030-42319-3_22.

Taube, J. S. (2007). The head direction signal: Origins and sensory-motor integration. *Annual Review of Neuroscience, 30*, 181–207.

Taube, J. S., Muller, R. U., & Ranck, J. B. (1990). Head-direction cells recorded from the postsubiculum in freely moving rats. I Description and quantitative analysis. *The Journal of Neuroscience, 10*, 420–435.

Terry, A. V., Jr. (2009). Spatial navigation (water maze) tasks. In J. J. Buccafusco (Ed.), *Methods of behavior analysis in neuroscience* (2nd ed.). Boca Raton, FL: CRC Press/Taylor & Francis. (chapter 13). Available from: https://www.ncbi.nlm.nih.gov/books/NBK5217/.

Toledo, S., Shohami, D., Schiffner, I., Lourie, E., Orchan, Y., Bartan, Y., et al. (2020). Cognitive map-based navigation in wild bats revealed by a new high-throughput tracking system. *Science (New York, N.Y.), 369*(6500), 188–193.

Tolman, E. C. (1948). Cognitive maps in rats and men. *Psychological Review, 55*(4), 189–208.

Traina, T. A., Dugan, U., Higgins, B., Kolinsky, K., Theodoulou, M., Hudis, C. A., et al. (2010). Optimizing chemotherapy dose and schedule by Norton-Simon mathematical modeling. *Breast Disease, 31*(1), 7–18.

Tyson, J. J., Baumann, W. T., Chen, C., Verdugo, A., Tavassoly, I., Wang, Y., et al. (2011). Dynamic modelling of oestrogen signalling and cell fate in breast cancer cells. *Nature Reviews. Cancer, 11*(7), 523–532.

Tyson, D. R., Garbett, S. P., Frick, P. L., & Quaranta, V. (2012). Fractional proliferation: A method to deconvolve cell population dynamics from single-cell data. *Nature Methods, 9*(9), 923–928.

Villacorta, J. A., & Panetsos, F. (2005). Information coding by ensembles of resonant neurons. *Biological Cybernetics, 92*(5), 339–347.

Vodovotz, Y., Csete, M., Bartels, J., Chang, S., & An, G. (2008). Translational systems biology of inflammation. *PLoS Computational Biology, 4*(4), e1000014.

Vorhees, C. V., & Williams, M. T. (2006). Morris water maze: Procedures for assessing spatial and related forms of learning and memory. *Nature Protocols*, *1*(2), 848–858.
Wagatsuma, H., & Yamaguchi, Y. (2007). Neural dynamics of the cognitive map in the hippocampus. *Cognitive Neurodynamics*, *1*(2), 119–141.
Weinberg, S. H. (2015). Membrane capacitive memory alters spiking in neurons described by the fractional-order Hodgkin-Huxley model. *PLoS One*, *10*(5), e0126629.
Winslow, R. L., Trayanova, N., Geman, D., & Miller, M. I. (2012). Computational medicine: Translating models to clinical care. *Science Translational Medicine*, *4*(158), 158rv11.
Zeng, T., & Si, B. (2017). Cognitive mapping based on conjunctive representations of space and movement. *Frontiers in Neurorobotics*, *11*, 61.

CHAPTER 9

Clinical applications of deep learning in neurology and its enhancements with future directions

Olugbemi T. Olaniyan[a], Charles O. Adetunji[b], Olorunsola Adeyomoye[c], Ayobami Dare[d], Mayowa J. Adeniyi[e], and Alex Enoch[f]

[a]Department of Physiology, Laboratory for Reproductive Biology and Developmental Programming, Rhema University, Aba, Nigeria
[b]Department of Microbiology, Applied Microbiology, Biotechnology and Nanotechnology Laboratory, Edo State University Uzairue, Iyamho, Nigeria
[c]Department of Physiology, University of Medical Sciences, Ondo, Nigeria
[d]Department of Physiology, School of Laboratory Medicine and Medical Sciences, College of Health Sciences, Westville Campus, University of KwaZulu-Natal, Durban, South Africa
[e]Department of Physiology, Edo State University Uzairue, Iyamho, Nigeria
[f]Department of Human Physiology, Ahmadu Bello University Zaria, Zaria, Nigeria

Introduction

Artificial intelligence (AI) is a section of computer sciences that involves machine learning, deep learning, and representation learning (Jiang et al., 2017). Deep learning and machine learning have great clinical relevance, for example, in radiology where they are used in prognosis, risk assessment, prediction of therapy responses, classification, and diagnosis (Choy et al., 2018; Fazal, Patel, Tye, & Gupta, 2018; Giger, 2018; Hosny, Parmar, Quackenbush, Schwartz, & Aerts, 2018; Liew, 2018; Mayo & Leung, 2018; McBee et al., 2018; Nichols, Herbert Chan, & Baker, 2018; Savadjiev et al., 2018). These techniques have also been applied in the evaluation of brain imaging and classification of strokes, demyelinating diseases, neurodegenerative disorders, epilepsy, and psychiatric disorders (Davatzikos, 2018; Feng, Badgeley, Mocco, & Oermann, 2018; Kamal, Lopez, & Sheth, 2018; Mateos-Pérez et al., 2018; Middlebrooks, Ver Hoef, & Szaflarski, 2017; Plis et al., 2014; Zaharchuk, Gong, Wintermark, Rubin, & Langlotz, 2018).

Furthermore, it has been established that the process involved in the initial learning applications in neuroradiology play more attention on "downstream" region of the brain by applying computer vision techniques for identification, detection of lesions, and segmentation of anatomical

Artificial Intelligence for Neurological Disorders
https://doi.org/10.1016/B978-0-323-90277-9.00026-2

structures such as white matter hyper intensities, hemorrhage, metastases, primary brain tumors, stroke, aneurysms, lacunes, and microbleeds, (Chartrand et al., 2017; Tang et al., 2018). Also, in the upstream region of the brain, the application of AI is important for medical imaging, image processing, enhancement and generation. Moreover, the application of deep learning has several advantages in image enhancement and image generation, including shortening the duration of imaging studies and improving image quality by removing image artifacts, lowering radiation and contrast dose, and normalizing/harmonizing images (Zaharchuk et al., 2018).

Deep learning represents a type of AI that can roughly model the structure of neurons in the brain, thus playing a vital role in developing solutions to numerous challenges in computer vision, robotics, and natural language processing (LeCun, Bengio, & Hinton, 2015). Deep learning constituted a great part of machine learning which might be linked to convergence of theoretic advances that are presented in the hardware and computer software in the presence of adequate computational power. The application of deep learning can help in the application of newly generated data that can be used in a wide variety of tasks. One example of machine learning presentation is the ImageNet Challenge (Krizhevsky, Sutskever, & Hinton, 2018).

The advantage of deep learning has been documented in the rapid development of industry and the medical sector, especially medical imaging as well as diagnosis and management of neurologic diseases. Furthermore, there are many advances that have been recorded in the application of machine learning most especially in the industry entailing medical imaging (Erickson, Korfiatis, Akkus, & Kline, 2017). Machine learning is a technique that has robust application in the innovations of data science that enable computers to perform like humans and to identify and classify patterns. Moreover, machines have the capacity to perform more functions than humans do (Wang & Summers, 2012). Also, they have a greater possibility as a sustainable technology that could be applied for diagnostic imaging, diagnostic radiology imaging interpretation and some other vital application in medicine (Jordan & Mitchell, 2015).

Medical data and artificial intelligence in neurology

Medical data is an aspect of Big Data, which refers to overly large volumes of data. These data are stored in databases and can be analyzed, processed, and interpreted to obtain valuable information that can help in the diagnosis, treatment, and prognosis of neurologic disease (Yang et al., 2020). With the massive increase in the amount of data generated, the medical database has moved beyond the traditional models of storage and analysis. Both structured and unstructured medical data are now processed at high velocity (Ristevski & Chen, 2018). These medical data are easy to collect, analyze, and interpret and require minimal effort from medical professionals. AI simulates human behavior (Amisha, Malik, Pathania, & Rathaur, 2019). The use of this intelligent model to process and interpret large volumes of data in the 21st century is gaining momentum. Furthermore, with machine learning algorithms computers can now use data to learn make predictions or solve problems (Nichols, Herbert Chan, & Baker, 2019). AI can be used in biomedicine to classify images and predict the risk of death from disease. In neurology, AI can be used to predict areas of brain lesions based on results obtained from computer tomography (CT) scans and functional magnetic resonance imaging (fMRI) (Hatt, Tixier, Visvikis, & Cheze Le Rest, 2017; Segato, Marzullo, Calimeri, & De Momi, 2020). The use of model-based and data-driven methodologies to study the function of AI in monitoring and regulating neuro-critical care indices such as intracranial pressure, seizure, blood pressure, and ventilation has been reviewed (Bishop, 2012). These two technologies have proven effective at examining a large amount of patient data. As such, they have the potential to reduce healthcare expenses as well as medical errors and delays (Bishop, 2012). In another study, machine learning models including deep neural networks, random forests, and logistic regression were used to predict long-term outcomes in persons who suffered acute ischemic stroke (Kamal et al., 2018). The results showed that these machine learning techniques, notably deep neural networks, enhanced the prediction of long-term outcomes in patients with ischemic stroke.

Neurology-centered medical system

The health care system today is faced with many challenges, some of which include poor accommodation for patients, slow adoption of information and communication technology, shortage of workforce, and more (Roncarolo, Boivin, Denis, Hébert, & Lehoux, 2017). Thus, there is a need for specialized healthcare systems to facilitate efficient and effective service delivery. Neurology is a specialized medical discipline and research in this area has gained serious attention over the years. A neurologist is a principal component of the neurology-centered medical system. The role of a neurologist cannot be underestimated because of the need to understand the pathogenesis of many neurologic diseases. Furthermore, with the increased therapies for neurologic disease, more specialists are needed to assist in utilizing, analyzing, and interpreting data obtained using modern technology (Maulden, 2003). A neurologist can provide care for patients with seizures, Alzheimer's disease (AD), Parkinson's disease (PD), movement disorders, and many other neurologic disorders (Narayanaswami, 2015). Many neurologic diseases require urgent attention and therefore most patients present to emergency departments or units. Cerebrovascular accidents, vertigo, and seizures are some neurologic events that require urgent care (Shiber, Macindoe, Flower, Knight Iv, & Bösel, 2012). Emergency rooms are the ideal setting for patients who require advance imaging such as fMRI or consultation with neuro-specialists. Many neurological instruments have been developed to assist in assessing neurological functions. Many of these tools rely on neuroimaging techniques, such as computer tomography, magnetic resonance imaging, and positron emission tomography, among others (Xue, Chen, Lu, & Dong, 2010). These neurological procedures have helped with neurological disease diagnosis, management, and prognosis.

Clinical applications of artificial intelligence and deep learning

AI is an area of research concerned with the creation of computer models capable of performing certain functions that require human intellect, such as cognition, problem-solving, and thinking (Beam & Kohane, 2016). However, the amount of data created in the healthcare system is growing. Although AI has been around for more than five decades, its use in the healthcare system has been minimal. However, recent improvements in computing efficiency and connectivity have expanded the horizon of AI research and potential application in various fields. Thus, AI is expected to influence several fields in the healthcare system including, radiology, dermatology, and pathology due to the generation of large volume of data (in the form of images) and the need for proper analysis and interpretation to improve healthcare delivery (Fenech, Strukelj, & Buston, 2018). The growing amount of data collected in the healthcare system suggests that AI applications in this field are essential, since this data may be used in machine and deep learning to design algorithms that imitate how the human brain functions to execute complex tasks, thereby enhancing healthcare delivery (Ching et al., 2018). In neurology, AI utilizes the provided data (e.g., images) and, via different stages of calculations, develops on its own specific indices that may be used to determine specific outcomes such as diagnosis of disease conditions (Chartrand et al., 2017).

Deep learning is a sort of AI that consists of several stages of learning between inputs and outputs, allowing the computer to understand multiple data formats (Jiang et al., 2017). In deep learning, two methods are often used to achieve the major goal. The first is a supervised method, which uses series of labeled data to predict a known result accurately (e.g., confirming the result of diagnosis). The second is an unsupervised method, which uses a set of data to explain trends or progression of a disease condition (Ching et al., 2018). A major type of deep learning in the healthcare system is the convolutional neural network (Mieloszyk & Bhargava, 2018).

In recent years, neurological diagnosis and treatment have improved due to the clinical application of AI in the medical field. Several AI algorithms has been developed using neuroimaging to determine the pathogenesis and progression of neurological disorders (Dallora, Eivazzadeh, Mendes, Berglund, & Anderberg, 2017). This has resulted in the creation of an automated disease detection system that translates neuroimages into quantitative data and analyzes it to help diagnosis, supporting physicians in their decision-making (Gillies, Kinahan, & Hricak, 2016). The application of AI in neurology has significantly reduced medical errors because this technique can process large numbers of data, without any interference of lapse of memory, psychological response, or stress. Also, AI can coordinate and integrate large amounts of information, identify patients that require screening tests, prioritize patients that need urgent attention, and standardize reports (Loria, 2018). Aside from neuroimaging, deep learning's therapeutic application can improve large-scale hypothesis generation and provide detailed explanations of the physiology, anatomy, and signaling cascade implicated in brain function (Vu et al., 2018). Deep learning algorithms are expected to have widespread clinical application because they can automatically extract and classify features (such as segmentation, multimodal prognostication, ischemic changes, thrombosis, plagues, edema) on neuroimages obtained from radiological scans such as CT and MRI (Chen et al., 2016; Rehme et al., 2015; Thornhill et al., 2014). AI algorithms have been applied in the prediction of stroke because they can detect early signs of the disease by differentiating between resting and stroke-related paralysis (Villar, Gonzalez, Sedano, Chira, & Trejo-Gabriel-Galan, 2015). Deep learning may also be used to predict future stroke features by assessing the association between various physiological indicators in stroke patients, which could aid in early therapeutic intervention of patients (Zhang, Xie, Ye, & Pang, 2013).

An AI algorithm has been used in spinal cord injury for assistive rehabilitation to help patients resume walking after injury. In addition, research has shown that a deep learning algorithm can predict the severity of PD by generating a score card that can be used in conjunction with standard PD measures, allowing for more frequent and objective evaluation of treatment efficacy (Zhan et al., 2018).

Artificial intelligence for medical imaging and precision medicine

The emergence of new technologies has had a considerable impact on the practice of medical imaging in recent years, and this has transformed the "modus operandi" in this sector. Digital imaging, as well as picture archiving and communication systems (PACS), are examples of new technologies that have drastically altered the reporting system for medical images and enhanced image preservation. This reporting system generates a vast amount of data for machine and deep learning. The recent rise of AI, which makes use of sophisticated computers and vast amounts of data, has resulted in substantial advancements in radiology and medical imaging.

In AI, both qualitative and quantitative data obtained from medical images are used to develop algorithms that allow for proper identification of features that are used to predict outcomes in similar cases, thereby assisting radiologists during decision-making and improving diagnostic outcomes (Leite, 2019). In medical imaging, AI is widely applied in patient screening, automatic clinical decision support, image acquisition, automatic detection of features, automatic interpretation of results, image management and archiving, postprocessing of images, analysis of image quality, automatic dose estimation, reporting and analysis of radiological images, and automatic integration of medical imaging data (Choy et al., 2018).

Patient screening: AI can optimize the scheduling procedure for patients attending radiological appointments, thereby reducing the chance of them missing their appointments. Also, AI can improve safety during screening or enhance safety reports, which has been proven to improve radiological services (Marella, Sparnon, & Finley, 2017).

Image acquisition: AI can make an imaging system more intelligent by improving positioning and characterization of findings via proper data processing. For instance, AI may detect a lesion and provide further options for modifying the sequence to improve the image's characteristics (Kohli, Prevedello, Filice, & Geis, 2017).

Automatic detection of findings: AI can promote the extraction of incidental and critical findings (e.g., fractures, pneumothorax, thyroid nodules, and organ laceration) on medical images, which are usually overlooked during diagnosis by the radiologist (Burns, Yao, Muñoz, & Summers, 2016; Zeng, Ye, Yang, et al., 2015).

Automatic interpretation of findings: AI can improve interpretation of findings, which could assist the radiologist during diagnosis. AI could interpret some radiological features (such as contour, margins, texture) of the incidental findings to improve precision during diagnosis (Liu, Balagurunathan, Atwater, et al., 2017).

Postprocessing of images: This includes segmentation, registration, and image quantification, all of which can be improved with the application of AI (Akkus, Galimzianova, Hoogi, Rubin, & Erickson, 2017).

Image quality analysis: AI algorithms can help reduce artifacts from radiological scans and improve image quality (Wolterink, Leiner, Viergever, & Isgum, 2017).

Automated radiation dose estimation: AI can be used for specific classification and dose estimation from CT data sets with greater than 96% accuracy (Coates, Souhami, & El Naqa, 2016).

Radiological reporting analysis: AI techniques are widely applicable in natural language processing (NLP) and extraction of data from medical images using NLP has been shown to improve quality, performance evaluation, and testing of clinical decision support. AI can also assist in tracking radiologist reports and decrease the incidence of disconnection during follow-up recommendations (Cai, Giannopoulos, Yu, et al., 2016).

Automatic data integration: AI can use different algorithms to analyze multiple data generated from different sources such as pathological reports, medical images, and radiological reports. In addition, AI can effectively integrate these data by modifying their interaction with other data sources (Li, Wu, & Ngom, 2018).

Precision medicine is a technique to ensure that clinicians and researchers increase their prediction accuracy about the type of therapeutic measure that will be effective in a set of people for a particular ailment (Adetunji, Adetunji, & Olaniyan, 2021; Adetunji, Ajayi, et al., 2021; Adetunji et al., 2021a, 2021b; Adetunji, Inobeme, et al., 2021; Adetunji & Varma, 2020; Akram et al., 2020; Egbuna et al., 2020; Roli, Adetunji, Mishra, Adetunji, & Mishra, 2020; Suddhasuchi et al., 2020). Precision in medicine can be greatly improved with the application of AI, which may help to reduce false-positive results (Setio, Ciompi, Litjens, et al., 2016). Application of

AI in medical images has been used make predictions with a success rate of about 92%, while independent review of the same radiological image by a pathologist recorded a success rate of 96.6%. As a result, by combining AI predictions with a physician's expertise, the precision success rate may be enhanced to 99.5%, with an 85% reduction in the rate of human error.

Examples of neurology AI

AI centers on how tasks normally executed by human beings are achieved by machines. In neurology, a machine learning algorithm was trained using 500 MRI scans of brain tumor patients (Kickingereder et al., 2019). Using data from these images, the algorithm was able to assess and diagnose the tumors. A support vector machine (SVM) developed with the aid of MRI-dependent blood volume distribution by Emblem et al. was able to predict preoperative glioma survival. Furthermore, AI has been developed in neuroimaging, recording of speech, cognitive evaluation scores, stroke diagnosis, measurement of AD biomarkers, automation of segmentation of hemorrhagic infarcts (Scherer et al., 2016), detection of early warning signs, and much more. Researchers developed an algorithm capable of predicting 30-day mortality for patients with traumatic brain injury (Raj, Luostarinen, Pursiainen, Posti, & Takala, 2019). Cerebral perfusion pressure, intracranial pressure, and mean arterial pressure are all areas in which AI algorithms have been trained.

Challenges of deep learning applied to neuroimaging techniques

Deep learning and AI face numerous obstacles in general as well as in neurology (Valliani, Ranti, & Oermann, 2019). Such challenges include:

Sufficiency of quality data: The more quality data available, the better the deep learning. Alteration of available data elicits changes in a deep learning system.

Cost of training: AI is very expensive to train due to large complex models.

Expectation discrepancy: Disparity between what people expect of AI and what it can actually do is another challenge.

Security challenges: Although, there are cybersecurity-related applications as far as deep learning is concerned, network-related malicious attacks cannot be ruled out completely.

AI for assessing response to targeted neurological therapies

Spieler et al. (2021) reported that the use of AI in health care is reshaping medical practice, particularly in the treatment of cancer through radiological techniques like prognostic imaging findings, locoregional therapy, and radiomics. There are huge volumes of medical data being generated in clinical settings that can be utilized in clinical investigation, diagnosis, and treatment using AI through programming language interfaces and machine learning-derived image analysis and interpretation. AI has a significant role to play in the future healthcare delivery system and precision medicine through image analysis, speech therapy, movement recognition, and disease prediction. Davenport and Kalakota (2019) reported that for wide adoption and utilization of AI in healthcare systems, there must be proper standardization, regulation, and integration. The authors noted that AI will augment clinical practice as a result of tremendous development and growth in computational physiology and biophysics, algorithms, and AI techniques, thus enhancing diagnostics, monitoring, therapeutics, and precision medicine. Rudie, Rauschecker, Nick Bryan, Davatzikos, and Mohan (2019) reported that many advanced applications of AI have been deployed in pharmacological studies such as novel identification of noncoding variant domains of genomes, functionality, mechanistic interactions between drugs, optimization, dosing, and patient selection and classification. Tachmazidis, Chen, Adamou, and Antoniou (2021) reported that attention-deficit hyperactivity disorder as a neurodegenerative disorder is considered an important public health issue that requires early and advanced diagnostic approaches due to the huge financial burden the disorder places on the affected individual. Advancement in the development of AI and its adoption in the management of several neurodegenerative diseases has been proven to be a useful strategy in combating ADHD through early diagnosis, treatment, and adaptability. Patel et al. (2019) reported that AI is increasingly influencing many aspects of healthcare practice and delivery systems, particularly when it comes to neurological diseases. Improvement in patient outcomes through precision neurology can be achieved via effective adoption of AI algorithms and smart devices for monitoring, improved prognosis, smart decision-making, and image analysis.

Conclusion and future perspectives

This chapter highlighted the importance of AI in the healthcare industry for managing medical data, thus improving healthcare's data-driven environment, optimization, clinical pathways, preclinical research, patient care, and population-level applications. It described several algorithms and techniques used to assist clinicians in making prompt and smart decisions, particularly for neurological disease treatment, monitoring, prediction, and analysis. Many areas of neurology that can be strengthened via the application of AI include radiology, pathology, dermatology, and physiology. Machine learning and deep learning will significantly enhance automated medical imaging, disease detection, data conversion and interpretation, decision support systems, rehabilitation, prediction of neurological outcomes, digitization of histology slides to generate high-resolution images, early stroke prediction, neuroimaging, modification of genomics-informed precision medicine, response to diverse neurochemicals, drugs, and external factors to enhance tailored treatment.

References

Adetunji, J. B., Adetunji, C. O., & Olaniyan, O. T. (2021). African walnuts: A natural depository of nutritional and bioactive compounds essential for food and nutritional security in Africa. In O. O. Babalola (Ed.), *Food security and safety*. Cham: Springer.

Adetunji, C. O., Ajayi, O. O., Akram, M., Olaniyan, O. T., Chishti, M. A., Inobeme, A., et al. (2021). Medicinal plants used in the treatment of influenza a virus infections. In K. Dua, S. Nammi, D. Chang, D. K. Chellappan, G. Gupta, & T. Collet (Eds.), *Medicinal plants for lung diseases*. Singapore: Springer.

Adetunji, C. O., Akram, M., Olaniyan, O. T., Ajayi, O. O., Inobeme, A., Olaniyan, S., et al. (2021a). Medicinal plants used in the treatment of pulmonary hypertension. In K. Dua, S. Nammi, D. Chang, D. K. Chellappan, G. Gupta, & T. Collet (Eds.), *Medicinal plants for lung diseases*. Singapore: Springer. https://doi.org/10.1007/978-981-33-6850-7_14.

Adetunji, C. O., Akram, M., Olaniyan, O. T., Ajayi, O. O., Inobeme, A., Olaniyan, S., et al. (2021b). Targeting SARS-CoV-2 novel Corona (COVID-19) virus infection using medicinal plants. In K. Dua, S. Nammi, D. Chang, D. K. Chellappan, G. Gupta, & T. Collet (Eds.), *Medicinal plants for lung diseases*. Singapore: Springer.

Adetunji, C. O., Inobeme, A., Olaniyan, O. T., Ajayi, O. O., Olaniyan, S., & Adetunji, J. B. (2021). Application of nanodrugs derived from active metabolites of medicinal plants for the treatment of inflammatory and lung diseases: Recent advances. In K. Dua, S. Nammi, D. Chang, D. K. Chellappan, G. Gupta, & T. Collet (Eds.), *Medicinal plants for lung diseases*. Singapore: Springer.

Adetunji, C. O., & Varma, A. (2020). Biotechnological application of Trichoderma: A powerful fungal isolate with diverse potentials for the attainment of food safety, management of pest and diseases, healthy planet, and sustainable agriculture. In C. Manoharachary, H. B. Singh, & A. Varma (Eds.), *Vol. 61. Trichoderma: Agricultural applications and beyond. Soil biology*. Cham: Springer.

Akkus, Z., Galimzianova, A., Hoogi, A., Rubin, D. L., & Erickson, B. J. (2017). Deep learning for brain MRI segmentation: State of the art and future directions. *Journal of Digital Imaging*, *30*(4), 449–459.

Akram, M., Jabeen, F., Daniyal, M., Zainab, R., Haq, U. U., Adetunji, C. O., et al. (2020). Genetic engineering of novel products of health significance: Recombinant DNA

technology. In C. Egbuna, & T. G. Dable (Eds.), *Functional foods and nutraceuticals*. Cham: Springer. https://doi.org/10.1007/978-3-030-42319-3_26.
Amisha, Malik, P., Pathania, M., & Rathaur, V. K. (2019). Overview of artificial intelligence in medicine. *Journal of Family Medicine and Primary Care*, *8*(7), 2328–2331.
Beam, A. L., & Kohane, I. S. (2016). Translating artificial intelligence into clinical care. *Journal of the American Medical Association*, *316*(22), 2368–2369.
Bishop, C. M. (2012). Model-based machine learning. *Philosophical Transactions. Series A, Mathematical, Physical, and Engineering Sciences*, *371*(1984), 20120222.
Burns, J. E., Yao, J., Muñoz, H., & Summers, R. M. (2016). Automated detection, localization, and classification of traumatic vertebral body fractures in the thoracic and lumbar spine at CT. *Radiology*, *278*(1), 64–73.
Cai, T., Giannopoulos, A. A., Yu, S., et al. (2016). Natural language processing technologies in radiology research and clinical applications. *Radiographics*, *36*(1), 176–191.
Chartrand, G., Cheng, P. M., Vorontsov, E., Drozdzal, M., Turcotte, S., Pal, C. J., et al. (2017). Deep learning: A primer for radiologists. *Radiographics*, *37*(7), 2113–2131. https://doi.org/10.1148/rg.2017170077.
Chen, Y., Dhar, R., Heitsch, L., Ford, A., Fernandez-Cadenas, I., Carrera, C., et al. (2016). Automated quantification of cerebral edema following hemispheric infarction: Application of a machine-learning algorithm to evaluate CSF shifts on serial head CTs. *Neuroimage Clinical*, *12*, 673–680.
Ching, T., Himmelstein, D. S., Beaulieu-Jones, B. K., Kalinin, A. A., Do, B. T., Way, G. P., et al. (2018). Opportunities and obstacles for deep learning in biology and medicine. *bioRxiv*.
Choy, G., Khalilzadeh, O., Michalski, M., Do, S., Samir, A. E., Pianykh, O. S., et al. (2018). Current applications and future impact of machine learning in radiology. *Radiology*, *288*(2), 318–328. https://doi.org/10.1148/radiol.2018171820.
Coates, J., Souhami, L., & El Naqa, I. (2016). Big data analytics for prostate radiotherapy. *Frontiers in Oncology*, *6*, 149.
Dallora, A. L., Eivazzadeh, S., Mendes, E., Berglund, J., & Anderberg, P. (2017). Machine learning and microsimulation techniques on the prognosis of dementia: A systematic literature review. *PLoS One*, *12*(6), e0179804.
Davatzikos, C. (2018). Machine learning in neuroimaging: Progress and challenges. *NeuroImage*, *197*, 652–656. https://doi.org/10.1016/j.neuroimage.2018.10.003.
Davenport, T., & Kalakota, R. (2019). The potential for artificial intelligence in healthcare. *Future Healthcare Journal*, *6*(2), 94–98.
Egbuna, C., Gupta, E., Ezzat, S. M., Jeevanandam, J., Mishra, N., Muhammad Akram, N., et al. (2020). Aloe species as valuable sources of functional bioactives. In C. Egbuna, & G. Dable Tupas (Eds.), *Functional foods and nutraceuticals*. Cham: Springer.
Erickson, B. J., Korfiatis, P., Akkus, Z., & Kline, T. L. (2017). Machine learning for medical imaging. *Radiographics*, *37*(2), 505–515.
Fazal, M. I., Patel, M. E., Tye, J., & Gupta, Y. (2018). The past, present and future role of artificial intelligence in imaging. *European Journal of Radiology*, *105*, 246–250. https://doi.org/10.1016/j.ejrad.2018.06.020.
Fenech, M., Strukelj, N., & Buston, O. (2018). *Ethical, social, and political challenges of artificial intelligence in health*. London (GB): Future Advocacy. https://wellcome.ac.uk/sites/default/files/ai-in-health-ethical-socialpolitical-challenges.pdf. (Accessed 10 September 2018).
Feng, R., Badgeley, M., Mocco, J., & Oermann, E. K. (2018). Deep learning guided stroke management: A review of clinical applications. *Journal of Neurointerventinal Surgery*, *10*, 358–362. https://doi.org/10.1136/neurintsurg-2017-013355.
Giger, M. L. (2018). Machine learning in medical imaging. *Journal of the American College of Radiology*, *15*, 512–520. https://doi.org/10.1016/j.jacr.2017.12.028.

Gillies, R. J., Kinahan, P. E., & Hricak, H. (2016). Radiomics: Images are more than pictures, they are data. *Radiology*, *278*(2), 563–577.
Hatt, M., Tixier, F., Visvikis, D., & Cheze Le Rest, C. (2017). Radiomics in PET/CT: More than meets the eye? *Journal of Nuclear Medicine*, *58*(3), 365–366.
Hosny, A., Parmar, C., Quackenbush, J., Schwartz, L. H., & Aerts, H. J. W. L. (2018). Artificial intelligence in radiology. *Nature Reviews. Cancer*, *18*, 500–510. https://doi.org/10.1038/s41568-018-0016-5.
Jiang, F., Jiang, Y., Zhi, H., Dong, Y., Li, H., Ma, S., et al. (2017). Artificial intelligence in healthcare: Past, present and future. *Stroke and Vascular Neurology*, *2*(4), 230–243. https://doi.org/10.1136/svn-2017-000101.
Jordan, M. I., & Mitchell, T. M. (2015). Machine learning: Trends, perspectives, and prospects. *Science*, *349*(6245), 255–260.
Kamal, H., Lopez, V., & Sheth, S. A. (2018). Machine learning in acute ischemic stroke neuroimaging. *Frontiers in Neurology*, *9*, 945. https://doi.org/10.3389/fneur.2018.00945.
Kickingereder, P., et al. (2019). Articles automated quantitative tumour response assessment of MRI in neuro-oncology with artificial neural networks: A multicentre, retrospective study. *The Lancet*, 1–13.
Kohli, M., Prevedello, L. M., Filice, R. W., & Geis, J. R. (2017). Implementing machine learning in radiology practice and research. *AJR. American Journal of Roentgenology*, *208*(4), 754–760.
Krizhevsky, A., Sutskever, I., & Hinton, G. (2018). *ImageNet classification with deep convolutional neural networks*. http://www.cs.toronto.edu/~fritz/absps/imagenet.pdf. (Accessed 23 January 2018).
LeCun, Y., Bengio, Y., & Hinton, G. (2015). Deep learning. *Nature*, *521*, 436–444. https://doi.org/10.1038/nature14539.
Leite, C. (2019). Artificial intelligence, radiology, precision medicine, and personalized medicine. *Radiologia Brasileira*, *52*(6), VII–VIII.
Li, Y., Wu, F. X., & Ngom, A. (2018). A review on machine learning principles for multi-view biological data integration. *Briefings in Bioinformatics*, *19*(2), 325–340. https://doi.org/10.1093/bib/bbw113 (Epub ahead of print).
Liew, C. (2018). The future of radiology augmented with artificial intelligence: A strategy for success. *European Journal of Radiology*, *102*, 152–156. https://doi.org/10.1016/j.ejrad.2018.03.019.
Liu, Y., Balagurunathan, Y., Atwater, T., et al. (2017). Radiological image traits predictive of cancer status in pulmonary nodules. *Clinical Cancer Research*, *23*(6), 1442–1449.
Loria, K. (2018). Putting the AI in radiology. *Radiology Today*, *19*(1). Davenport TH, Dreyer KJ. AI will change radiology, but it won't replace radiologists. *Harv Bus Rev*. 2018.
Marella, W. M., Sparnon, E., & Finley, E. (2017). Screening electronic health record-related patient safety reports using machine learning. *Journal of Patient Safety*, *13*(1), 31–36.
Mateos-Pérez, J. M., Dadar, M., Lacalle-Aurioles, M., Iturria-Medina, Y., Zeighami, Y., & Evans, A. C. (2018). Structural neuroimaging as clinical predictor: A review of machine learning applications. *Neuroimage Clinical*, *20*, 506–522. https://doi.org/10.1016/j.nicl.2018.08.019.
Maulden, S. A. (2003). Information technology, the internet, and the future of neurology. *The Neurologist*, *9*(3), 149–159.
Mayo, R. C., & Leung, J. (2018). Artificial intelligence and deep learning—Radiology's next frontier? *Clinical Imaging*, *49*, 87–88. https://doi.org/10.1016/j.clinimag.2017.11.007.
McBee, M. P., Awan, O. A., Colucci, A. T., Ghobadi, C. W., Kadom, N., Kansagra, A. P., et al. (2018). Deep learning in radiology. *Academic Radiology*, *25*, 1472–1480. https://doi.org/10.1016/j.acra.2018.02.018.
Middlebrooks, E. H., Ver Hoef, L., & Szaflarski, J. P. (2017). Neuroimaging in epilepsy. *Current Neurology and Neuroscience Reports*, *17*, 32. https://doi.org/10.1007/s11910-017-0746-x.

Mieloszyk, R. J., & Bhargava, P. (2018). Convolutional neural networks: The possibilities are almost endless. *Current Problems in Diagnostic Radiology*, *47*(3), 129–130.
Narayanaswami, P. (2015). What is the value of a neurologist? *Neurology*, *84*(2), 209–210.
Nichols, J. A., Herbert Chan, H. W., & Baker, M. A. B. (2018). Machine learning: Applications of artificial intelligence to imaging and diagnosis. *Biophysical Reviews*, *11*, 111–118. https://doi.org/10.1007/s12551-018-0449-9.
Nichols, J. A., Herbert Chan, H. W., & Baker, M. (2019). Machine learning: Applications of artificial intelligence to imaging and diagnosis. *Biophysical Reviews*, *11*(1), 111–118.
Patel, U. K., Anwar, A., Saleem, S., Malik, P., Rasul, B., Patel, K., et al. (2019). Artifcial intelligence as an emerging technology in the current care of neurological disorders. *Journal of Neurology*, 1–20. https://doi.org/10.1007/s00415-019-09518-3.
Plis, S. M., Hjelm, D. R., Salakhutdinov, R., Allen, E. A., Bockholt, H. J., Long, J. D., et al. (2014). Deep learning for neuroimaging: A validation study. *Frontiers in Neuroscience*, *8*, 229. https://doi.org/10.3389/fnins.2014.00229.
Raj, R., Luostarinen, T., Pursiainen, E., Posti, J. P., & Takala, R. S. K. (2019). Machine learning-based dynamic mortality prediction after traumatic brain injury. *Scientific Reports*, 1–13.
Rehme, A. K., Volz, L. J., Feis, D. L., Bomilcar-Focke, I., Liebig, T., Eickhoff, S. B., et al. (2015). Identifying neuroimaging markers of motor disability in acute stroke by machine learning techniques. *Cerebral Cortex*, *25*(9), 3046–3056.
Ristevski, B., & Chen, M. (2018). Big data analytics in medicine and healthcare. *Journal of Integrative Bioinformatics*, *15*(3), 20170030.
Roli, O. I., Adetunji, C. O., Mishra, P., Adetunji, J. B., & Mishra, R. R. (2020). Caffeine: Nutraceutical and health benefit of caffeine-containing commodities and products. In P. Mishra, R. R. Mishra, & C. O. Adetunji (Eds.), *Innovations in food technology*. Singapore: Springer.
Roncarolo, F., Boivin, A., Denis, J. L., Hébert, R., & Lehoux, P. (2017). What do we know about the needs and challenges of health systems? A scoping review of the international literature. *BMC Health Services Research*, *17*(1), 636.
Rudie, J. D., Rauschecker, A. M., Nick Bryan, R., Davatzikos, C., & Mohan, S. (2019). Emerging applications of artificial intelligence in neuro-oncology. *Radiology*, *290*(3), 607–618. https://doi.org/10.1148/radiol.2018181928.
Savadjiev, P., Chong, J., Dohan, A., Vakalopoulou, M., Reinhold, C., Paragios, N., et al. (2018). Demystification of AI-driven medical image interpretation: Past, present and future. *European Radiology*, *29*, 1616–1624. https://doi.org/10.1007/s00330-018-5674-x.
Scherer, M., Cordes, J., Younsi, A., Sahin, Y. A., Götz, M., Möhlenbruch, M., ... Orakcioglu, B. (2016). Development and validation of an automatic segmentation algorithm for quantification of intracerebral hemorrhage. *Stroke*, *47*(11), 2776–2782. https://doi.org/10.1161/STROKEAHA.116.013779.
Segato, A., Marzullo, A., Calimeri, F., & De Momi, E. (2020). Artificial intelligence for brain diseases: A systematic review. *APL Bioengineering*, *4*(4), 041503.
Setio, A. A., Ciompi, F., Litjens, G., et al. (2016). Pulmonary nodule detection in CT images: False positive reduction using multi-view convolutional networks. *IEEE Transactions on Medical Imaging*, *35*(5), 1160–1169.
Shiber, J. R., Macindoe, C., Flower, O., Knight Iv, W. A., & Bösel, J. (2012). Neurological emergencies. *Emergency Medicine International*, *2012*, 208193.
Spieler, B., Sabottke, C., Moawad, A. W., Gabr, A. M., Bashir, M. R., Kinh, R., et al. (2021). Artificial intelligence in assessment of hepatocellular carcinoma treatment response. *Abdominal Radiology*. https://doi.org/10.1007/s00261-021-03056-1.
Suddhasuchi, D., Sharangi, A. B., Egbuna, C., Jeevanandam, J., Ezzat, S. M., Adetunji, C. O., et al. (2020). Health benefits of Isoflavones found exclusively of plants

of the Fabaceae Family. In C. Egbuna, & T. G. Dable (Eds.), *Functional foods and nutraceuticals*. Cham: Springer. https://doi.org/10.1007/978-3-030-42319-3_22.

Tachmazidis, I., Chen, T., Adamou, M., & Antoniou, G. (2021). A hybrid AI approach forsupporting clinical diagnosis ofattention defcit hyperactivity disorder (ADHD) in adults. *Health Information Science and System*, *9*(1), 1–8. https://doi.org/10.1007/s13755-020-00123-7.

Tang, A., Tam, R., Cadrin-Chênevert, A., Guest, W., Chong, J., Barfett, J., et al. (2018). Canadian association of radiologists white paper on artificial intelligence in radiology. *Canadian Association of Radiologists Journal*, *69*, 120–135. https://doi.org/10.1016/j.carj.2018.02.002.

Thornhill, R. E., Lum, C., Jaberi, A., Stefanski, P., Torres, C. H., Momoli, F., et al. (2014). Can shape analysis differentiate free-floating internal carotid artery thrombus from atherosclerotic plaque in patients evaluated with CTA for stroke or transient ischemic attack? *Academic Radiology*, *21*(3), 345–354.

Valliani, A. A. A., Ranti, D., & Oermann, E. K. (2019). Deep learning and neurology: A systematic review. *Neurology and Therapy*, *8*, 351–365.

Villar, J. R., Gonzalez, S., Sedano, J., Chira, C., & Trejo-Gabriel-Galan, J. M. (2015). Improving human activity recognition and its application in early stroke diagnosis. *International Journal of Neural Systems*, *24*(4).

Vu, M. T., Adali, T., Ba, D., Buzsaki, G., Carlson, D., Heller, K., et al. (2018). A shared vision for machine learning in neuroscience. *The Journal of Neuroscience*, *38*(7), 1601–1607.

Wang, S., & Summers, R. M. (2012). Machine learning and radiology. *Medical Image Analysis*, *16*(5), 933–951.

Wolterink, J. M., Leiner, T., Viergever, M. A., & Isgum, I. (2017). Generative adversarial networks for noise reduction in low-dose CT. *IEEE Transactions on Medical Imaging*, *36*(12), 2536–2545.

Xue, G., Chen, C., Lu, Z. L., & Dong, Q. (2010). Brain imaging techniques and their applications in decision-making research. *Xin li xue bao. Acta Psychologica Sinica*, *42*(1), 120–137.

Yang, J., Li, Y., Liu, Q., Li, L., Feng, A., Wang, T., et al. (2020). Brief introduction of medical database and data mining technology in big data era. *Journal of Evidence-Based Medicine*, *13*(1), 57–69.

Zaharchuk, G., Gong, E., Wintermark, M., Rubin, D., & Langlotz, C. P. (2018). Deep learning in neuroradiology. *American Journal of Neuroradiology*, *39*, 1776–1784. https://doi.org/10.3174/ajnr.A5543.

Zeng, J. Y., Ye, H. H., Yang, S. X., et al. (2015). Clinical application of a novel computer-aided detection system based on three-dimensional CT images on pulmonary nodule. *International Journal of Clinical and Experimental Medicine*, *8*(9), 16077–16082.

Zhan, A., Mohan, S., Tarolli, C., Schneider, R., Adams, J., Sharma, S., et al. (2018). Using smartphones and machine learning to quantify Parkinson disease severity: The mobile Parkinson disease score. *JAMA Neurology*, *75*(7), 876–880. https://doi.org/10.1001/jamaneurol.2018.0809.

Zhang, Q., Xie, Y., Ye, P., & Pang, C. (2013). Acute ischaemic stroke prediction from physiological time series patterns. *The Australasian Medical Journal*, *6*(5), 280–286.

CHAPTER 10

Ensemble sparse intelligent mining techniques for cognitive disease

Olugbemi T. Olaniyan[a], Charles O. Adetunji[b], Olorunsola Adeyomoye[c], Ayobami Dare[d], Mayowa J. Adeniyi[e], and Alex Enoch[f]
[a]Department of Physiology, Laboratory for Reproductive Biology and Developmental Programming, Rhema University, Aba, Nigeria
[b]Department of Microbiology, Applied Microbiology, Biotechnology and Nanotechnology Laboratory, Edo State University Uzairue, Iyamho, Nigeria
[c]Department of Physiology, University of Medical Sciences, Ondo, Nigeria
[d]Department of Physiology, School of Laboratory Medicine and Medical Sciences, College of Health Sciences, Westville Campus, University of KwaZulu-Natal, Durban, South Africa
[e]Department of Physiology, Edo State University Uzairue, Iyamho, Nigeria
[f]Department of Human Physiology, Ahmadu Bello University Zaria, Zaria, Nigeria

Introduction

The application of machine learning approaches and numerous pattern recognition and identification of neurodegeneration related diseases through the application of neuroimaging data is attracting serious attention (Hinrichs et al., 2009). Machine learning is used in neurology to develop effective techniques for identifying samples of the human brain (e.g., subcortical and cortical structures). For example, automated magnetic resonance imaging (MRI) measures were used in one study as a noninvasive diagnostic approach to identify mild cognitive impairment in different populations (Desikan et al., 2009). Recent advances in statistical learning technologies have driven the development of several high-dimensional classification algorithms proficient at dealing with neuroimaging data. Machine learning has been used to design classifiers for separating and optimizing datasets (Adetunji, Adetunji, and Olaniyan, 2021e; Adetunji, Ajayi, et al., 2021a; Adetunji, Akram, et al., 2021b; Adetunji, Akram, et al., 2021c; Adetunji, Inobeme, et al., 2021d; Adetunji & Varma, 2020; Akram et al., 2020; Egbuna et al., 2020; Roli, Adetunji, Mishra, Adetunji, & Mishra, 2020; Suddhasuchi et al., 2020). The most widely applied classifier is the support vector machine, which exhibits excellent performance in classifying high-dimensional data (Zhang, Wang, Zhou, Yuan, & Shen, 2011).

Artificial Intelligence for Neurological Disorders
https://doi.org/10.1016/B978-0-323-90277-9.00027-4

Alzheimer's disease is an irreversible neurodegenerative disorder that results in advanced loss of memory and cognition function. The application of powerful imaging tools such as fluorodeoxyglucose positron emission tomography (FDG-PET) and MRI of functional and structural brain images have helped in understanding the neural changes related to this neurodegenerative disorder (Davatzikos, Bhatt, Shaw, Batmanghelich, & Trojanowski, 2010; Fan, Resnick, Wu, & Davatzikos, 2008). Therefore, in this chapter, we review recent advances in machine learning research, algorithms, and their applications in forecasting cognitive diseases, particularly Alzheimer's disease, Parkinson's disease, vascular dementia, and mild cognitive impairment. We also discuss the challenges of implementing machine learning tools into therapy planning and diagnostic practice.

Cognitive disease

Cognitive diseases, also known as neurocognitive diseases, are conditions that deteriorate an individual's cognitive functions so significantly that normal functions are impaired without treatment (Lopez, 2013). The most frequent cause of dementia, Alzheimer's disease, is an example of a neurocognitive disorder. Other diseases causing dementia include Huntington's disease, supranuclear palsy, cortical basal degeneration, multisystem atrophy, and Creutzfeldt-Jakob disease (Emmady & Tadi, 2021). Although the exact cause of Alzheimer's dementia is unclear, studies have shown that amyloid plaques and tau protein deposition in neurons play a significant role (Swerdlow, 2007). These protein deposits can be seen in histological sections of Alzheimer's disease patients' brains (DeTure & Dickson, 2019). Individuals with Alzheimer's-type dementia present with symptoms such as memory loss, mood swings, aggression, self-neglect, personality changes, forgetfulness, communication problems, social withdrawal, and many others (Bature, Guinn, Pang, & Pappas, 2017). In certain cases, the comorbidity of dementia with other underlying diseases may occur, and the treatment of these diseases is necessary for effective treatment of dementia (Bunn et al., 2014). Functional imaging of the brain may help in early diagnosis and monitoring disease progression. Delirium is a type of cognitive impairment that causes problems with attention, consciousness, and cognition. It's a state of bewilderment that can also include other neurological issues like sleep–wake cycle problems, emotional disturbances, ADHD, and more (Fong, Tulebaev, & Inouye, 2009). Dementia is a significant risk factor for delirium. Medication, infection, trauma, surgery, and Parkinson's disease are also risk factors (Vasilevskis, Han,

Hughes, & Ely, 2012). Delirium can result in permanent cognitive decline, motor impairment, and mortality. Management of delirium can be pharmacological, in which the medical precipitating factors are treated, or nonpharmacological, which involves providing sensory, emotional, and environmental support (Grover & Avasthi, 2018). There are many life-threating conditions that can trigger delirium, therefore early diagnosis and treatment are critical. Evidence has shown that delirium can occur in children with serious medical illness, which is associated with adverse outcomes including increases in length of hospital stay, hospital costs, and mortality (Grover, Malhotra, Bharadwaj, Bn, & Kumar, 2009).

Machine learning and deep ensemble sparse regression network

Several feature identification and machine learning algorithms have been extensively employed for analyzing neuroimaging data to discover relevant traits and improve diagnosis of cognitive problems (Hinrichs et al., 2009; Magnin et al., 2009). Importantly, based on the information collected from the training data, machine learning algorithms can be used to detect a pattern for a new test sample. For example, structural MRI can be used to assess anatomical information about the cortex and hypothalamus in the human brain and find neuroanatomical differences across various samples (Desikan et al., 2009). Using data from neuroimages such as MRI scans, numerous classification techniques have been utilized to diagnose persons with cognitive illness (Cuingnet et al., 2011). However, most of these existing diagnostic procedures share the extraction and selection of significant features from neuroimaging data, but the methodologies employed to extract these features differ and are distinct to each technique.

In machine learning, features are extracted from neuroimaging data (MRI and PET scans) and used to train a global classifier (i.e., design an algorithm) that is then utilized for classification and diagnosis. However, due to several interfering factors and a short sample size, developing a general classifier that is comprehensive and detailed enough to provide adequate classification and better prediction during diagnosis is difficult. Hence, machine learning may not give an accurate diagnostic result. Thus, there is a need for a better technique such as a deep ensemble sparse intelligent network that can provide detailed analysis and improve diagnosis. In this technique, the neuroimaging data (brain image) is segmented into numerous local patches and several features are indiscriminately selected from different segments of local patches to develop a weak classifier, which is specifically used to achieve consistency in brain imaging. Following that, several weak classifiers are combined and used to arrive at a final diagnosis. Thus, a deep ensemble sparse intelligent network develops several individual algorithms from numerous local patches and synergistically uses these algorithms for a detailed analysis and more precise decision. In comparison to other techniques, this method more accurately detects brain diseases such as Alzheimer's disease and moderate cognitive impairment.

Feature selection is another significant technique with high precision to simultaneously select the most distinguishing features and remove those that are redundant (Fan, Shen, & Davatzikos, 2005). This objective can be accurately achieved using an ensemble sparse network, which can carefully divide the brain image into numerous discriminant segments to select important features from multiple components. These obtained features can then be combined for adequate classification (Yoon et al., 2007). Furthermore, feature selection by machine learning algorithms depends on the training dataset, thus, this technique develops a single classifier with suboptimal training required for the test dataset, which may probably result from the problem of data overfitting. However, instead of developing a single classifier with a training dataset, ensemble sparse intelligence uses meta-learning approaches to select features from different segments of the brain image and combines the predictions of these segments to improve the generalization ability and robustness of each classifier. The problem of data overfitting can be reduced using this strategy, which is known as ensemble methodology. Among the ensemble techniques that have been used in neuroimaging are Bagging and Boosting as well as random space ensemble techniques that generate several weak classifiers by resampling numerous sample subsets (Breiman, 1996; Fan et al., 2008; Kuncheva, Rodriguez, Plumpton, Linden, & Johnston, 2010).

Intelligent medical diagnostics with ensemble sparse intelligent mining techniques

In medical diagnosis, notably for cognitive illnesses like Alzheimer's disease, ensembles sparse intelligent mining techniques have been applied. This methodology aids in the identification of Alzheimer's disease by merging numerous weak classifiers derived from several local patches of neuroimaging data for precise diagnosis. Using this method, each individual weak classifier is designed using a sparse representation-based classifier (SRC), which has demonstrated great performance for strong classification of imaging data. A subset of local patches is randomly resampled to generate each individual weak classifier to capture the local spatial consistency of features (Kim, Scalzo, Telesca, & Hu, 2015). Because each subset of patches defines a subspace of the entire brain feature space, each weak classifier may be trained more readily in a smaller subspace, resulting in a greater dimensionality-to-subject ratio. By substituting a single classifier with an ensemble of numerous classifiers, the accuracy of the final classification can be enhanced even further. The image characteristics from the sampled patches are put together (by utilizing simple majority vote) into a feature vector for representation of each training sample after random sampling of a portion of patches from the patch pool (Liu et al., 2012).

High-dimensional data science in cognitive diseases

Data science involves using scientific techniques and algorithms to extract knowledge from datasets. High-dimensional data set is related to data mining, machine learning, and big data (Blei & Smyth, 2017). It aids in the incorporation of standardized data into electronic health records to improve the quality of care for the underserved. Furthermore, it helps in public health disease surveillance and guides intervention development. The use of data science for cognitive evaluation is gaining momentum. Studies have shown that data science is useful for early-stage prediction, due to the complexity of cognitive disease (Ristevski & Chen, 2018). Cognitive impairment is a major threat to public and private health and can further develop into Alzheimer's-type dementia, which can be expensive and difficult to treat. Data science has made it possible to use big or large data collections in identifying risk factors for complementary and effective therapeutic approaches (Pastorino et al., 2019). It can be used to build complex computational models that can perform cognitive tasks. These models can be used to simulate brain activity in an unprecedented way for information processing. Data science is a rapidly growing field that can provide insight into neural network connectivity as well as help to understand the etiology and pathogenesis of cognitive disease (Golriz Khatami, Mubeen, & Hofmann-Apitius, 2020). A study was designed to develop physical exercise programs using data science to assess cognitive function in Parkinson's disease patients. In Parkinson's disease patients, these programs improve cognition, neuronal processing speed, sustained attention, and mental flexibility (Fayyaz, Jaffery, Anwer, Zil-E-Ali, & Anjum, 2018). Similarly in another study, a virtual reality computing technology was developed and used for training older adults with mild cognitive impairment. This technology was observed to have the potential to rehabilitate brain health and neural efficiency (Kim, Pang, & Kim, 2019). Data science is currently being used to develop a Find-MyApp program, a cost-effective and community-based intervention that has the potential to facilitate personal management and engagement in social activities to promote health and enhance cognition due to the large number of people living with dementia (Neal et al., 2021). Data science helps not only in diagnosis and management of cognitive disease, but it can also be used to develop computer-based programs that can enhance or promote mental health in cognitive disease patients.

Diagnostic challenges with artificial intelligence

Diagnosis of diseases involves identification of health disorders or challenges that may be either very difficult or easy to carry out based on the available diagnostic tools. The manual method of disease diagnosis is prone to numerous errors, thus artificial intelligence such as machine learning, fuzzy logic, and deep learning have been adopted as a smart diagnostic and predictive tools with minimal or no error. Despite the progress recorded, there are still some limitations in terms of the technology deployed in clinical practice. These include implementation difficulties in terms of logistics, barriers in adoption and utilization, applicability in clinical settings, user friendliness, accidental fitting of confounders, dataset shift, unintended discriminatory bias, outcome noise, unintended negative factors of novel algorithms on medical or clinical outcomes, and errors of generalization into novel populations. More research effort is needed to identify unfair and biased algorithms, reduce fragility and generalization, and improve interpretation methods in artificial intelligence/human interfaces.

Summary

Cognitive diseases are incurable disorders that hamper cognitive functions and abilities (Simpson, 2014). In addition to aging, several predisposing environmental factors can precipitate disease onset. Therefore, there are incontestable needs to develop technology-driven techniques for improving early diagnosis and prognosis and preventing misdiagnosis. All studies presented in this chapter focused on advances in machine learning research algorithms and their applications for cognitive disease diagnosis, prognosis, and forecasting.

Machine learning algorithms are useful in the detection, classification, and prediction of early pathological cognitive decline in older adults. Algorithms like Ewer et al.'s (2012) logistic regression claim to be useful in distinguishing Alzheimer's disease from neurocognitive disease, with Cox regression simulating the conversion period from mild cognitive disorder to Alzheimer's disease. Payan and Montana (2015) created sparse autoencoders with a 3D cognitive neural network (CNN) that can predict Alzheimer's disease versus mild cognitive sickness with accuracy of 95.39%. Liu et al. developed an ensemble-based technique for predicting neurocognitive illness using a sparse representation-based classifier. Gomar et al. (2011) examined how cognitive markers and biomarkers could predict the progression of moderate cognitive impairment to Alzheimer's disease. Zhang, Shen, Initiative, et al. (2012) proposed a multimodal, multitask learning system that can perform both regression and classification at the same time. Moradi et al. (Moradi, Gaser, Huttunen, & Tohka, 2014) were able to use the Alzheimer's disease neuroimaging project to train a classifier with 77% accuracy in classifying patients into mild cognitive impairment, Alzheimer's disease, and healthy categories.

Tierney et al. (1996) investigated the possibility of developing a neuropsychological-based exam to predict the start of Alzheimer's disease. They were able to assess group differences in neurocognitive disease and Alzheimer's disease using analysis of covariance. Using a dynamic Bayesian network, a causal relationship from functional MRI was extracted and utilized to develop a neural anatomical network. With this, Burge, Clark, Lane, Link, and Qiu (2004) were able to classify patients into healthy and dementia categories with an accuracy of 73%. With the help of a support vector machine model, Challis et al. (2015) compared a Bayesian Gaussian process logistic model based on performance. They observed that by using functional MRI as input data the models were able to comparably predict mild cognitive illness and neurocognitive illness with an accuracy of 80%.

Cognitive decline such as loneliness, agitation (Cheng, 2017), and social isolation (Meeks & Jeste, 2009), which are difficult to understand (Biddle et al., 2019), have been elucidated more clearly by artificial intelligence. Furthermore, with novel attributes such as handwriting (text), sensor (digital), speech (audio), and wearable device monitors, new indicators for detecting cognitive decline can be identified (Kourtis, Regele, Wright, & Jones, 2019). Wearable cognitive devices exhibit potential for continuous monitoring and tracking of changes in cognitive function.

Fusion of data from MRI and PET using deep machine methods was accomplished by Suk, Lee, Shen, and Initiative (2014) with a method that involved integration of selected image-derived voxel patches into a deep Boltzmann machine. The data fusion resulted in improvement in classification as well as confirmation of brain regions associated with Alzheimer's disease. The use of intermediate data integration (early and intermediate) and novel multimodal and multitask learning to combine biomarkers from cerebrospinal fluid, PET, and MRI resulted in better classification of Alzheimer's disease compared to neurocognitive disease.

Conclusion and future perspectives

Neurocognitive diseases are incurable disorders that affect many people. As such, there is a need to develop techniques geared toward early diagnosis, prognosis, and forecasting of these diseases. This chapter outlined advances in machine learning research algorithms and their applications for cognitive disease diagnosis, prognosis, and forecasting. It also addressed issues related to classification of patients with neurocognitive diseases using machine learning. The chapter also presented investigative works that studied the roles of data fusion. Although there is immense data available on using machine learning algorithms for the classification of neurocognitive diseases, more research is needed to examine their applicability to other conditions like delirium and dementia.

References

Adetunji, J. B., Adetunji, C. O., & Olaniyan, O. T. (2021). African walnuts: a natural depository of nutritional and bioactive compounds essential for food and nutritional security in Africa. In O. O. Babalola (Ed.), *Food security and safety*. Cham: Springer.

Adetunji, C. O., Ajayi, O. O., Akram, M., Olaniyan, O. T., Chishti, M. A., Inobeme, A., Olaniyan, S., Adetunji, J. B., Olaniyan, M., & Awotunde, S. O. (2021). Medicinal plants used in the treatment of influenza a virus infections. In K. Dua, S. Nammi, D. Chang, D. K. Chellappan, G. Gupta, & T. Collet (Eds.), *Medicinal plants for lung diseases*. Singapore: Springer.

Adetunji, C. O., Akram, M., Olaniyan, O. T., Ajayi, O. O., Inobeme, A., Olaniyan, S., Hameed, L., & Adetunji, J. B. (2021b). Medicinal plants used in the treatment of pulmonary hypertension. In K. Dua, S. Nammi, D. Chang, D. K. Chellappan, G. Gupta, & T. Collet (Eds.), *Medicinal plants for lung diseases*. Singapore: Springer. https://doi.org/10.1007/978-981-33-6850-7_14.

Adetunji, C. O., Akram, M., Olaniyan, O. T., Ajayi, O. O., Inobeme, A., Olaniyan, S., Hameed, L., & Adetunji, J. B. (2021c). Targeting SARS-CoV-2 novel corona (COVID-19) virus infection using medicinal plants. In K. Dua, S. Nammi, D. Chang, D. K. Chellappan, G. Gupta, & T. Collet (Eds.), *Medicinal plants for lung diseases*. Singapore: Springer.

Adetunji, C. O., Inobeme, A., Olaniyan, O. T., Ajayi, O. O., Olaniyan, S., & Adetunji, J. B. (2021). Application of Nanodrugs derived from active metabolites of medicinal plants for the treatment of inflammatory and lung diseases: recent advances. In K. Dua, S. Nammi, D. Chang, D. K. Chellappan, G. Gupta, & T. Collet (Eds.), *Medicinal plants for lung diseases*. Singapore: Springer.

Adetunji, C. O., & Varma, A. (2020). Biotechnological application of trichoderma: a powerful fungal isolate with diverse potentials for the attainment of food safety, management of pest and diseases, healthy planet, and sustainable agriculture. In C. Manoharachary, H. B. Singh, & A. Varma (Eds.), *Vol. 61. Trichoderma: Agricultural applications and beyond. Soil biology*. Cham: Springer.

Akram, M., Jabeen, F., Daniyal, M., Zainab, R., Haq, U.u., Adetunji, C. O., Egbuna, C., Ephraim-Emmanuel, B. C., Patrick-Iwuanyanwu, K. C., & Ogbo, A. B. (2020). Genetic engineering of novel products of health significance: recombinant DNA technology. In C. Egbuna, & T. G. Dable (Eds.), *Functional foods and nutraceuticals*. Cham: Springer. https://doi.org/10.1007/978-3-030-42319-3_26.

Bature, F., Guinn, B. A., Pang, D., & Pappas, Y. (2017). Signs and symptoms preceding the diagnosis of Alzheimer's disease: A systematic scoping review of literature from 1937 to 2016. *BMJ Open*, 7(8), e015746.

Biddle, K. D., Uquillas, O., Jacobs, H. I. L., Zide, B., Kirn, D. R., Rentz, D. M., Psy, D., Johnson, K. A., Sperling, R. A., & Donovan, N. J. (2019). Social engagement and amyloid-B-related cognitive decline incognitively normal older adults. *The American Journal of Geriatric Psychiatry*, *27*, 1247–1256.

Blei, D. M., & Smyth, P. (2017). Science and data science. *Proceedings of the National Academy of Sciences of the United States of America*, *114*(33), 8689–8692.

Breiman, l. (1996). Bagging predictors. *Machine Learning*, *24*, 123–140.

Bunn, F., Burn, A. M., Goodman, C., Rait, G., Norton, S., Robinson, L., Schoeman, J., & Brayne, C. (2014). Comorbidity and dementia: A scoping review of the literature. *BMC Medicine*, *12*, 192.

Burge, J., Clark, V. P., Lane, T., Link, H., & Qiu, S. (2004). Bayesian classification of FMRI data: Evidence for altered neural networks in dementia. In *University of New Mexico, Tech. Rep. TR-CS-2004-28*.

Challis, E., Hurley, P., Serra, L., Bozzali, M., Oliver, S., & Cercignani, M. (2015). Gaussian process classification of Alzheimer's disease and mild cognitive impairment from resting-state FMRI. *NeuroImage*, *112*, 232–243.

Cheng, S. T. (2017). Dementia care giver burden: a research update and critical analysis. *Current Psychiatry Reports*, *19*(9), 64. https://doi.org/10.1007/s11920-017-0818-2. PMCID: PMC5550537.

Cuingnet, R., Gerardin, E., Tessieras, J., Auzias, G., Lehéricy, S., Habert, M. O., Chupin, M., Benali, H., & Colliot, O. (2011). Automatic classification of patients with Alzheimer's disease from structural mri: A comparison of ten methods using the adni database. *NeuroImage*, *56*, 766–781.

Davatzikos, C., Bhatt, P., Shaw, L. M., Batmanghelich, K. N., & Trojanowski, J. Q. (2010). Prediction of MCI to AD conversion, via MRI, CSF biomarkers, and pattern classification. *Neurobiology of Aging, 2010*.

Desikan, R. S., Cabral, H. J., Hess, C. P., Dillon, W. P., Glastonbury, C. M., Weiner, M. W., Schmansky, N. J., Greve, D. N., Salat, D. H., Buckner, R. L., & Fischl, B. (2009). Automated MRI measures identify individuals with mild cognitive impairment and Alzheimer's disease. *Brain*, *132*, 2048–2057.

DeTure, M. A., & Dickson, D. W. (2019). The neuropathological diagnosis of Alzheimer's disease. *Molecular Neurodegeneration*, *14*(1), 32.

Egbuna, C., Gupta, E., Ezzat, S. M., Jeevanandam, J., Mishra, N., Akram, M., Sudharani, N., Adetunji, C. O., Singh, P., Ifemeje, J. C., Deepak, M., Bhavana, A., Walag, A. M. P., Ansari, R., Adetunji, J. B., Laila, U., Olisah, M. C., & Onyekere, P. F. (2020). Aloe species as valuable sources of functional bioactives. In C. Egbuna, & T. G. Dable (Eds.), *Functional foods and nutraceuticals*. Cham: Springer.

Emmady, P. D., & Tadi, P. (2021). Dementia. [Updated 2021 Mar 11]. In *StatPearls [Internet]*. Treasure Island (FL): StatPearls Publishing. Available from https://www.ncbi.nlm.nih.gov/books/NBK557444/.

Fan, Y., Resnick, S. M., Wu, X., & Davatzikos, C. (2008). Structural and functional biomarkers of prodromal alzheimer's disease: A high-dimensional pattern classification study. *NeuroImage*, *41*, 277–285.

Fan, Y., Shen, D., & Davatzikos, C. (2005). Classification of structural images via high-dimensional image warping, robust feature extraction, and SVM. *Medical Image Computing and Computer-Assisted Intervention*, *8*, 1–8.

Fayyaz, M., Jaffery, S. S., Anwer, F., Zil-E-Ali, A., & Anjum, I. (2018). The effect of physical activity in Parkinson's disease: A mini-review. *Cureus*, *10*(7), e2995.

Fong, T. G., Tulebaev, S. R., & Inouye, S. K. (2009). Delirium in elderly adults: Diagnosis, prevention and treatment. *Nature Reviews Neurology*, *5*(4), 210–220.

Golriz Khatami, S., Mubeen, S., & Hofmann-Apitius, M. (2020). Data science in neurodegenerative disease: Its capabilities, limitations, and perspectives. *Current Opinion in Neurology*, *33*(2), 249–254.

Gomar, J. J., Bobes-Bascaran, M. T., Conejero-Goldberg, C., Davies, P., Goldberg, T. E., Initiative, A. D. N., et al. (2011). Utility of combinations of biomarkers, cognitive markers, and risk factors to predict conversion from mild cognitive impairment to alzheimer diseasein patients in the Alzheimer's disease neuroimaging initiative. *Archives of General Psychiatry*, *68*(9), 961–969.

Grover, S., & Avasthi, A. (2018). Clinical practice guidelines for management of delirium in elderly. *Indian Journal of Psychiatry*, *60*(Suppl 3), S329–S340.

Grover, S., Malhotra, S., Bharadwaj, R., Bn, S., & Kumar, S. (2009). Delirium in children and adolescents. *International Journal of Psychiatry in Medicine*, *39*(2), 179–187.

Hinrichs, C., Singh, V., Mukherjee, L., Xu, G., Chung, M. K., & Johnson, S. C. (2009). Spatially augmented LPboosting for AD classification with evaluations on the ADNI dataset. *NeuroImage*, *48*, 138–149.

Kim, O., Pang, Y., & Kim, J. H. (2019). The effectiveness of virtual reality for people with mild cognitive impairment or dementia: A meta-analysis. *BMC Psychiatry*, *19*(1), 219.

Kim, S., Scalzo, F., Telesca, D., & Hu, X. (2015). Ensemble of sparse classifiers for high-dimensional biological data. *International Journal of Data Mining and Bioinformatics*, *12*, 167–183.

Kourtis, L. C., Regele, O. B., Wright, J. M., & Jones, G. B. (2019). Digital biomarkers for Alzheimer's disease: Themobile/wearable devices opportunity. *npj Digital Medicine*, 1–9.

Kuncheva, I. I., Rodriguez, J. J., Plumpton, C. O., Linden, D. E., & Johnston, S. J. (2010). Random subspace ensembles for fmri classification. *IEEE Transactions on Medical Imaging*, *29*, 531–542.

Liu, M., Zhang, D., Shen, D., & The Alzheimer's Disease Neuroimaging Initiative. (2012). Ensemble sparse classification of Alzheimer's disease. *NeuroImage*, *60*(2), 1106–1116.

Lopez, O. L. (2013). Mild cognitive impairment. *Continuum (Minneapolis, Minn.)*, *19*(2 dementia), 411–424.

Magnin, B., Mesrob, I., Kinkingnéhun, S., Pélégrini-issac, M., Colliot, O., Sarazin, M., Dubois, B., Lehéricy, S., & Benali, H. (2009). Support vector machine-based classification of Alzheimer's disease from whole-brain anatomical MRI. *Neuroradiology*, *51*, 73–83.

Meeks, T. W., & Jeste, D. V. (2009). Neurobiology of wisdom. *Archives of General Psychiatry*, *66*, 355–365.

Moradi, E., Gaser, C., Huttunen, H., & Tohka, J. (2014). *Alzheimer's: MRI based dementia classification using semi-supervised learning and domain adaption* (pp. 65–73).

Neal, D. P., Kerkhof, Y., Ettema, T. P., Muller, M., Bosmans, J., Finnema, E., Graff, M., Dijkstra, K., Stek, M. L., & Dröes, R. M. (2021). Evaluation of FindMyApps: Protocol for a randomized controlled trial of the effectiveness and cost-effectiveness of a tablet-based intervention to improve self-management and social participation of community-dwelling people with mild dementia, compared to usual tablet use. *BMC Geriatrics*, *21*(1), 138.

Pastorino, R., De Vito, C., Migliara, G., Glocker, K., Binenbaum, I., Ricciardi, W., & Boccia, S. (2019). Benefits and challenges of big data in healthcare: An overview of the European initiatives. *European Journal of Public Health*, *29*(Supplement_3), 23–27.

Payan, A., & Montana, G. (2015). *Predicting Alzheimer's disease: A neuroimaging study with 3dconvolutional neural networks*. arXiv preprint arXiv:1502.02506.

Ristevski, B., & Chen, M. (2018). Big data analytics in medicine and healthcare. *Journal of Integrative Bioinformatics*, *15*(3), 20170030.

Roli, O. I., Adetunji, C. O., Mishra, P., Adetunji, J. B., & Mishra, R. R. (2020). Caffeine: Nutraceutical and health benefit of caffeine-containing commodities and products. In P. Mishra, R. R. Mishra, & C. O. Adetunji (Eds.), *Innovations in food technology*. Singapore: Springer.

Simpson, J. R. (2014). DSM-5 and neurocognitive disorders. *The Journal of the American Academy of Psychiatry and the Law*, *42*(2), 159–164.

Suddhasuchi, D., Sharangi, A. B., Egbuna, C., Jeevanandam, J., Ezzat, S. M., Adetunji, C. O., Tijjani, H., Olisah, M. C., Patrick-Iwuanyanwu, K. C., Adetunji, J. B., Ifemeje, J. C., Akram, M., Moboladji, B. M., & Onyeike, P. C. (2020). Health benefits of isoflavones found exclusively of plants of the Fabaceae family. In C. Egbuna, & T. G. Dable (Eds.), *Functional foods and nutraceuticals*. Cham: Springer. https://doi.org/10.1007/978-3-030-42319-3_22.

Suk, H. I., Lee, S. W., Shen, D., & Initiative, A. D. N. (2014). Hierarchical feature representation and multimodal fusion with deep learning for AD/MCI diagnosis. *NeuroImage*, *101*, 569–582.

Swerdlow, R. H. (2007). Pathogenesis of Alzheimer's disease. *Clinical Interventions in Aging*, *2*(3), 347–359.

Tierney, M., Szalai, J., Snow, W., Fisher, R., Nores, A., Nadon, G., Dunn, E., & George-Hyslop, P. S. (1996). Prediction of probable Alzheimer's disease in memory-impaired patients aprospective longitudinal study. *Neurology*, *46*(3), 661–665.

Vasilevskis, E. E., Han, J. H., Hughes, C. G., & Ely, E. W. (2012). Epidemiology and risk factors for delirium across hospital settings. *Best Practice & Research. Clinical Anaesthesiology*, *26*(3), 277–287.

Yoon, U., Lee, J. M., Im, K., Shin, Y. W., Cho, B. H., Kim, I. Y., Kwon, J. S., & kim, S. I. (2007). Pattern classification using principal components of cortical thickness and its discriminative pattern in schizophrenia. *NeuroImage*, *34*, 1405–1415.

Zhang, D., Shen, D., Initiative, A. D. N., et al. (2012). Multi-modal multi-task learning for joint prediction of multiple regression and classification variables in Alzheimer's disease. *NeuroImage*, *59*(2), 895–907.

Zhang, D., Wang, Y., Zhou, L., Yuan, H., & Shen, D. (2011). Multimodal classification of Alzheimer's disease and mild cognitive impairment. *NeuroImage*, *55*, 856–867.

CHAPTER 11

Cognitive therapy for brain diseases using deep learning models

Olugbemi T. Olaniyan[a], Charles O. Adetunji[b], Olorunsola Adeyomoye[c], Ayobami Dare[d], Mayowa J. Adeniyi[e], and Alex Enoch[f]
[a]Department of Physiology, Laboratory for Reproductive Biology and Developmental Programming, Rhema University, Aba, Nigeria
[b]Department of Microbiology, Applied Microbiology, Biotechnology and Nanotechnology Laboratory, Edo State University Uzairue, Iyamho, Nigeria
[c]Department of Physiology, University of Medical Sciences, Ondo, Nigeria
[d]Department of Physiology, School of Laboratory Medicine and Medical Sciences, College of Health Sciences, Westville Campus, University of KwaZulu-Natal, Durban, South Africa
[e]Department of Physiology, Edo State University Uzairue, Iyamho, Nigeria
[f]Department of Human Physiology, Ahmadu Bello University Zaria, Zaria, Nigeria

Introduction

Over the last three decades, hospitals and healthcare systems have produced a vast quantity of unstructured data such as medical imaging data, genomic information, and free text and data streams from monitoring devices. Different healthcare systems and hospitals generate different types of unstructured data (Senders et al., 2018). The evaluation of such data has led to development of techniques used by scientists for proper investigation, recognition, and management of brain pathologies and recognition of reactions and risk. Medical imaging has led to a great revolution in the field of medicine because of its wide applications in image processing, early and quick detection of disease, and the study, management, and treatment of diseases (Adetunji, Adetunji, and Olaniyan, 2021e; Adetunji, Ajayi, et al., 2021a; Adetunji, Akram, et al., 2021b; Adetunji, Akram, et al., 2021c; Adetunji, Inobeme, et al., 2021d; Adetunji & Varma, 2020; Akram et al., 2020; Egbuna et al., 2020; Roli, Adetunji, Mishra, Adetunji, & Mishra, 2020; Suddhasuchi et al., 2020). Some of the merits of medical imaging are its low cost and low risk (Sailer, van Zwam, Wildberger, & Grutters, 2015; Yasaka & Abe, 2018).

There are several techniques that have led to drastic improvement in the investigation of the human brain, especially during the performance of

Artificial Intelligence for Neurological Disorders
https://doi.org/10.1016/B978-0-323-90277-9.00017-1

noninvasive evaluations of brain structure and during the diagnosis of brain abnormalities caused by various diseases. These imaging techniques include positron emission tomography (PET), magnetic resonance imaging (MRI), and computed tomography (CT) (Fu, Patel, Wu, Li, & Gao, 2020; Liu et al., 2019). Machine learning is a type of artificial intelligence algorithm that automatically "learns" to recognize or predict unknown and future conditions using data. Machine learning has several applications for accurate qualitative and quantitative investigation of pathological and normal brain structures (Akkus, Galimzianova, Hoogi, Rubin, & Erickson, 2017; Sakai & Yamada, 2019; Vieira, Pinaya, & Mechelli, 2017).

The processing of data using machine learning techniques in the recognition of brain conditions entails and degenerative diseases, Alzheimer's disease, cancer, schizophrenia, infectious, dementia, as well as pathological tissues (Balafar, Ramli, Saripan, & Mashohor, 2010). Furthermore, deep learning and artificial intelligence can be used to execute several tasks, including neurosurgical tasks (Gudigar et al., 2020; McBee et al., 2018; Valliani, Ranti, & Oermann, 2019). Machine learning and deep learning involve the use of a Graphics Processing Unit (GPU) to obtain information from 2D and 3D images typical of the medical domain (Gudigar et al., 2020). This chapter addresses demographic, environmental, imaging, clinical, and genetic factors linked to the understanding of complex algorithms in cognitive therapy for brain diseases utilizing deep learning models for precision medicine.

Brain diseases affecting cognitive functions

There are many diseases that affect cognitive functions. One group is known as neurodegenerative disease. Neurodegenerative diseases are disorders characterized by a progressive deterioration in the normal structure and function of neurons over time. Neural degeneration can exist at any level of organization as far as the nervous system is concerned; it can occur at subcellular, cellular, tissue, and system levels. Unfortunately, neurodegenerative diseases, like other degenerative diseases, are incurable (Priller et al., 2006), with age being one of the risk factors. Diseases identified as neurodegenerative include Alzheimer's disease, Parkinson's disease, Huntington's disease, multiple sclerosis, and many others. In these diseases, deterioration occurs within neurotransmitter productions and subcellular structures. Degenerative diseases tend to exhibit different pathophysiological features and mechanisms than other diseases. Research activities are under greater pressure to identify the nexus among neurodegenerative diseases with respect to presentations, disease progression, and pathophysiological mechanisms to improve management.

Alzheimer's disease is the most prevalent neurodegenerative disease. In Alzheimer's, deterioration of neurons occurs in the cerebrum leading to attenuation of the cingulate gyrus, hippocampus, parietal lobe, temporal lobe, and frontal lobe of the brain. Alzheimer's disease enunciates the dangers associated with attenuation of the amyloid precursor protein, a protein involved in the repair, growth, and survival of neurons (Priller et al., 2006). Common pathophysiological features of the disease include the presence of beta-amyloid. Beta-amyloid is produced when amyloid precursor protein is broken down in the presence of beta-secretase and alpha-secretase (Hooper, 2005). Beta-amyloid protein condenses, forming what is known as amyloid plaques (Ohnishi & Takano, 2004; Tiraboschi, Hansen, Thal, & Corey-Bloom, 2004).

Parkinson's disease elucidates the function of a neurotransmitter secreted by the substantia nigra known as dopamine and disruption of the nigrostriatal dopaminergic pathway. As the second most common neurodegenerative disease, Parkinson's is characterized by hypokinetic and hyperkinetic features such as rigidity, tremor, and so on. It is unclear what specific mechanisms are responsible for the deterioration of dopaminergic neurons of the pars compacta in the substantia nigra. It is possible that dysfunction of a transport protein called RAB1 results in the accumulation of alpha-synuclein-ubiquitin (Varkey et al., 2010).

In Huntington's disease, there is a malfunction of GABA-secreting neurons of the pars reticulata in the substantia nigra (Purves et al., 2001). Although it is a rare neurodegenerative disorder, it is characterized by hyperkinetic features such as chorea (Crossman, 2000). Huntington's disease is also associated with neurogenesis (Barnat et al., 2020). Multiple sclerosis presents a scenario in which the body's immune proteins launch an attack against the covering of the neuronal axons causing injury due to inflammation (Kaufman & Milstein, 2013; Stys & Tsutsui, 2019) reduction in conduction of information from the central nervous system. Due to the absence of myelin sheath in the central nervous system, the brain and spinal cord are the soft hubs for the attack.

Multimodal information

Connectome mapping

The human brain is made of billions of neurons (Herculano-Houzel, 2009) connected with one another. The meeting point of neurons is called a synapse, which plays a vital role in cognition, memory, learning, language, and more. The complete mapping of neural connections in the brain is known as the connectome. Connectomics is the study of connectomes. It is a medium through which to achieve understanding of communications between neural cells in the functional regions of the brain. The human connectome project represents the commitment of scientists toward understanding the structural and functional connectivity network within healthy individuals that will create substantive data to stimulate groundbreaking research in brain diseases such as autism, Alzheimer's disease, dyslexia (Alagiakrishnan, Sankaralingam, Ghosh, Mereu, & Senior, 2013), and others.

Post-operative seizure

Post-operative seizure is an ictal phenomenon that occurs following resective surgery. Predictors of postoperative seizure may include a prolonged history of seizure, history of secondary generalized convulsive seizures, general imaging abnormalities, extra-temporal seizure onset, old age, and ictal and interictal findings (Kumar & Sharma, 2021). Animal studies on the possible underlying mechanisms of seizure identified production of free radicals due to neurosurgical insult and deranged electrolyte balance across the neuronal membrane (Manaka, Ishijima, & Mayanagi, 2003). Electroencephalographic findings during seizure show generalized spikes or sharp waves, 3-Hz or faster spike–wave complexes.

Gene signature

A stimulus is a chemical in the cell environment. Usually, when a stimulus is presented, activation of the signal transduction pathway results in alteration in gene expression levels. This alteration in gene expression levels is known as gene signature. A gene signature is a collection of genes within a cell exhibiting a characteristic pattern of a gene expression group, which arises because of unaltered or altered biological phenomenon or disease condition. Gene signature is clinically applied in cancer diagnosis, prognosis, and forecast (Nguyen, Welty, & Cooperberg, 2015; Warner, 2016). Transcriptome studies typically generate many signature genes associated with a disease. Gene signature may be useful in selecting treatments for patients (Hassane et al., 2008).

Overview of deep learning techniques

Deep learning is a branch of machine learning that utilizes a neural network to process, analyze, and store big data. With recent advances in technology, a large volume of biomedical data is now being processed, analyzed, and interpreted to assess health and disease status in humans (Cao et al., 2018). Deep learning algorithms are used to classify these complex data in images, text, and sound. These algorithms have a high level of accuracy and may even surpass human intelligence (Shen, Wu, & Suk, 2017). In deep learning, computers are trained to understand, learn, process, and perform tasks that humans typically carry out. Examples of deep learning applications include visual assistants, facial cognition, speech translation, and so on (Ching et al., 2018). Deep learning can extract features directly from raw data, and a single trained network can yield great results across applications. In healthcare research, deep learning can be used to analyze medical images, physiological signals, computer genomics, and more (Dash, Acharya, Mittal, Abraham, & Keleman, 2020; Rim, Sung, Min, & Hong, 2020). In clinical research, deep learning tools have been used to predict the risk of postoperative bleeding after colorectal surgery, diagnose childhood asthma, predict the time of treatment in patients with chronic lymphocytic anemia, predict the risk of mortality using public datasets, and predict the use and misuse of opioids (Chen et al., 2018).

Multiple layers of the network are explored in deep learning. Deep neural network (DNN), graph neural network (GNN), deep belief network (DBN), concurrent neural network (CNN), are some examples of network architecture (Alzubaidi et al., 2021). These types of deep learning models are critical for generating output results from raw dataset inputs. As previously discusses, these models have multiple layers and can be trained to perform difficult tasks. Furthermore, these models can be supervised or unsupervised. The significance of deep learning algorithms in biomedicine cannot be overemphasized. Studies have shown that they can detect lesions and enhance organ imaging to enhance the diagnosis and prognosis of disease conditions (Farhat, Sakr, & Kilany, 2020). They can be used to classify different stages of cancer (AlDubayan et al., 2020). With this wide range of applications, (Dash, Pani, Rodrigues, & Majhi, 2022), their multilayered capacity and complexity may continue to increase. It is likely that more deep architectures may emerge in the future to perform more rigorous, complex, and high-cognition tasks.

Data preprocessing techniques

Preprocessing involves changing raw data to understandable forms for statistical analysis. Real-world data are often inconsistent and may contain many errors (Tobore et al., 2019). Therefore, it is necessary to eliminate these errors before processing the data. Data cleaning, integration, transformation, and reduction are the basic techniques in data preprocessing (Malley, Ramazzotti, & Wu, 2016). These data processing steps help to ensure that quality datasets are processed. Data are collected with errors in most clinical research despite careful study design, conduct, and error-prevention strategies. Data cleaning can correct or minimize these errors to reduce their influence on the results obtained. Data cleaning methods include error deletion, missing data, and removing outliers, among others (Pani, Dash, dos Santos, Bukhari, & Flammini, 2022; Welch, von Recklinghausen, Taenzer, Savitz, & Weiss, 2017).

Furthermore, data may be collected from several sources with the aim of merging or integrating the data as one and making it accessible to researchers. These data could be used for diagnosis and therapeutic interventions. In the healthcare system, data are often collected, used, and reused, and often provide information on the epidemiology of disease (Carter, Shaw, & Macfarlane, 2002). These data can be shared with other health institutions. Several challenges may occur in the process of integrating data. For example, in genetic research, the sequencing of an individual's human genome can take up to 25 gigabytes of storage capacity. Hence the capacity to store this genetic information may be limited and therefore undermine the integration process (Dash, Pani, Balamurugan, & Abraham, 2021; López de Maturana et al., 2019). In addition, medical images are large files with several megabytes, hence storing and integrating these images can be difficult as well. Raw data can be transformed before subjecting it to statistical analysis. This is necessary for better analysis and interpretation of results. Examples of data transformation include the logarithmic transformation of H^+ concentration (pH) and the logarithmic transformation of the dose–response curve (Manikandan, 2010). However, when raw data are transformed, the summary statistics and the transformation must be clearly stated. To improve the storage efficiency of data, it is important to perform data reduction. The size and capacity of raw data may be reduced by summarizing and expunging invalid data. Data reduction may also include editing, scaling, encoding, sorting, collating, and producing a tabular data summary (Lee, 2020).

Early brain disease diagnosis using deep learning techniques

Several methods and techniques, including machine learning, have been used to support the diagnosis of brain diseases such as Alzheimer's disease and Parkinson's disease. Machine learning methods have shown appreciable success in the classification of this disease. Using MRI, the application of machine learning in Alzheimer's disease diagnosis can be organized into three primary stages: predetermination of regions-of-interest (ROIs), extracting features from ROIs, and developing a classification algorithm for the features. However, these features are extracted manually from the MRI data (Liu, Wee, Chen, & Shen, 2014; Suk, Lee, & Shen, 2016). Several techniques to solve this problem have been suggested due to the problem of extracting features and limited samples of radiological images. Among the suggested techniques is a sparse multitask learning framework that helps in feature extraction and learning vector quantization that models tissue distribution with a support vector machine (SVM) adopted as the classifier (Dash, Pani, Abraham, & Liang, 2021; Dash, Pani, Balamurugan, & Abraham, 2021; Liu et al., 2014; Ortiz, Górriz, Ramírez, & Martínez-Murcia, 2013), while other techniques have been used to train a classifier with low performance resulting from the complex relationship between feature extraction and classifier. Due to this problem, deep learning has shown better outcomes in the diagnosis of brain disease. Deep learning techniques exert better diagnostic value in that, as an alternative to extracting features from a classifier, they may learn the indices of neural networks from the images without the influence of human expertise. CNNs have shown good accuracy in feature classification from digital images and have been effectively employed in computer-aided diagnosis (Suk & Shen, 2015).

Deep learning uses a range of imaging protocols, including structural MRI, resting-state functional MRI (fMRI), PET, and multimodal procedures to classify brain disease, and this technique has shown promising results (Shen et al., 2017). Using the multimodal procedure, deep learning techniques have been applied to select features from neuroimaging data used to identify structural and functional biomarkers of brain disease. These markers include volume or thickness of the cortex in the ROI (e.g., hippocampus and entorhinal cortex), which is recognized as a feature to improve diagnostic accuracy. The neuroimages undergo various preprocessing steps, including registration and feature extraction, to select the appropriate features for diagnosis. Nevertheless, deep learning techniques can improve diagnosis using the initial neuroimaging data without feature selection (Jo, Nho, & Saykin, 2019).

Artificial intelligence and cognitive therapies and immunotherapies

Although, the conventional in-person therapeutic strategies remain the standard protocol for managing clinical symptoms of cognitive disorders such as depression, anxiety, and Alzheimer's disease, studies have proposed that computer-assisted cognitive therapies, including artificial intelligence, produce comparable results and effectively manage subminimal cognitive disease. The artificial intelligent approach to mental therapy is a noninvasive and less intensive technique to administer empirically validated therapies (Newman, Szkodny, Llera, & Przeworski, 2011; Warmerdam, Smit, Van Straten, Riper, & Cuijpers, 2010). Pieces of literature have shown that approximately 70% of patients showed interest in the use of mobile health (mHealth) apps to screen and manage their mental health, and these patients were shown to express themselves more freely when using the mHealth app compared to when they interact with their clinician (Rost et al., 2017; Torous et al., 2014). In addition, using artificial intelligence to administer cognitive therapies has been reported to be viable and effective in managing the symptoms of depression (Fitzpatrick, Darcy, & Vierhile, 2017; Frazier et al., 2016).

Nevertheless, the therapeutic response of using artificial intelligence to treat cognitive disorders is still in its infancy and needs further exploration despite the efficacy of computer-based therapy (CBT) in the treatment of anxiety and depression (Tolin, 2010). Artificial intelligence, via machine learning, plays a vital role in improving the administration of CBT. In this regard, the machine makes a decision based on the developed algorithm from the previous data training model, such that the machine can make a reasonable decision when a user provides a model with an input. Although this is a complex technique, it can undoubtedly improve accessibility to cognitive therapy, especially in vulnerable patients. By continuous practice, artificial intelligence will increasingly improve CBT administration while helping to reduce the cost, duration, and labor required for conventional therapeutic strategies (De Mello & De Souza, 2019). Also, the experience gathered may be used to modify and strategize treatment for several people on a large scale and reduce cost. In the field of immunotherapy, artificial intelligence via machine learning has been used to predict genotypes associated with poor prognosis (Yu et al., 2019). A machine learning algorithm has been used to predict immunopeptidome, which is a prerequisite in vaccine development. This algorithm may assist in developing an effective accessory for vaccines by evaluating epitopes in antigens (Moghram, Nabil, & Badr, 2018; Nagpal et al., 2015).

Summary

Several cognitive functions can be analyzed using artificial intelligence algorithms and data platforms. Neuronal degeneration is a major factor in neurological diseases causing reduced secretion of neurotransmitters, changes in biochemical activities such as increased inflammation, activation of reactive oxygen species, and abnormal firing level of neuronal activity. The human brain is made of billions of neurons connected with one another. The meeting point of neurons is called a synapse, which plays a vital role in cognition, memory, learning, language, and more. The complete mapping of neural connections in the brain is known as the connectome. The human connectome project represents the commitment of scientists toward understanding structural and functional connectivity network within healthy individuals that will create substantive data that will stimulate ground-breaking research in brain diseases. Clinical application transcriptome study typically generates a large number of *signature genes* that are associated with a disease. Gene signature may be useful in selecting treatments for patients through the application of artificial intelligence by selecting and analyzing appropriate medical images, physiological signals, and genomic data. The artificial intelligent approach to mental therapy is a noninvasive and less intensive technique to administer empirically validated therapies..

Conclusion and future perspectives

The significant interest and progress recorded by scientists on adopting and utilizing computer-aided techniques to diagnose different brain diseases in cognitive therapy has facilitated advancement and development in brain–computer interface in treating several neurological diseases. Recently, deep learning techniques have been projected to be very useful for the classification of different neuroimaging data associated with brain diseases. Thus, this chapter addressed demographic, environmental, imaging, clinical, and genetic factors linked to the understanding of complex algorithms in cognitive therapy for brain diseases utilizing deep learning models for precision medicine. It discussed the use of deep learning algorithms identifying disease biomarkers, connectome mapping, vasculature, electroencephalogram signals, postoperative seizure, and genetic signatures. Because using artificial intelligence to treat cognitive disorders is still in its infancy, and despite the fact that computer-based therapy has shown efficacy in the treatment of anxiety and depression, there is a need to further explore various techniques, algorithms, and platforms.

References

Adetunji, J. B., Adetunji, C. O., & Olaniyan, O. T. (2021). African walnuts: A natural depository of nutritional and bioactive compounds essential for food and nutritional security in Africa. In O. O. Babalola (Ed.), *Food security and safety*. Cham: Springer.

Adetunji, C. O., Ajayi, O. O., Akram, M., Olaniyan, O. T., Chishti, M. A., Inobeme, A., Olaniyan, S., Adetunji, J. B., Olaniyan, M., & Awotunde, S. O. (2021). Medicinal plants used in the treatment of influenza a virus infections. In K. Dua, S. Nammi, D. Chang, D. K. Chellappan, G. Gupta, & T. Collet (Eds.), *Medicinal plants for lung diseases*. Singapore: Springer.

Adetunji, C. O., Akram, M., Olaniyan, O. T., Ajayi, O. O., Inobeme, A., Olaniyan, S., Hameed, L., & Adetunji, J. B. (2021b). Medicinal plants used in the treatment of pulmonary hypertension. In K. Dua, S. Nammi, D. Chang, D. K. Chellappan, G. Gupta, & T. Collet (Eds.), *Medicinal plants for lung diseases*. Singapore: Springer. https://doi.org/10.1007/978-981-33-6850-7_14.

Adetunji, C. O., Akram, M., Olaniyan, O. T., Ajayi, O. O., Inobeme, A., Olaniyan, S., Hameed, L., & Adetunji, J. B. (2021c). Targeting SARS-CoV-2 novel Corona (COVID-19) virus infection using medicinal plants. In K. Dua, S. Nammi, D. Chang, D. K. Chellappan, G. Gupta, & T. Collet (Eds.), *Medicinal plants for lung diseases*. Singapore: Springer.

Adetunji, C. O., Inobeme, A., Olaniyan, O. T., Ajayi, O. O., Olaniyan, S., & Adetunji, J. B. (2021). Application of nanodrugs derived from active metabolites of medicinal plants for the treatment of inflammatory and lung diseases: Recent advances. In K. Dua, S. Nammi, D. Chang, D. K. Chellappan, G. Gupta, & T. Collet (Eds.), *Medicinal plants for lung diseases*. Singapore: Springer.

Adetunji, C. O., & Varma, A. (2020). Biotechnological application of trichoderma: A powerful fungal isolate with diverse potentials for the attainment of food safety, management of pest and diseases, healthy planet, and sustainable agriculture. In C. Manoharachary, H. B. Singh, & A. Varma (Eds.), *Vol. 61. Trichoderma: Agricultural applications and beyond. Soil biology*. Cham: Springer.

Akkus, Z., Galimzianova, A., Hoogi, A., Rubin, D. L., & Erickson, B. J. (2017). Deep learning for brain MRI segmentation: State of the art and future directions. *Journal of Digital Imaging*, *30*, 449–459. https://doi.org/10.1007/s10278-017-9983-4.

Akram, M., Jabeen, F., Daniyal, M., Zainab, R., Haq, U. U., Adetunji, C. O., Egbuna, C., Ephraim-Emmanuel, B. C., Patrick-Iwuanyanwu, K. C., & Ogbo, A. B. (2020). Genetic engineering of novel products of health significance: Recombinant DNA technology. In C. Egbuna, & T. G. Dable (Eds.), *Functional foods and nutraceuticals*. Cham: Springer. https://doi.org/10.1007/978-3-030-42319-3_26.

Alagiakrishnan, K., Sankaralingam, S., Ghosh, M., Mereu, L., & Senior, P. (2013). Antidiabetic drugs and their potential role in treating mild cognitive impairment and Alzheimer's disease. *Discovery Medicine, 16*(90), 277–286. PMID: 24333407.

AlDubayan, S. H., Conway, J. R., Camp, S. Y., Witkowski, L., Kofman, E., Reardon, B., Han, S., Moore, N., Elmarakeby, H., Salari, K., Choudhry, H., Al-Rubaish, A. M., Al-Sulaiman, A. A., Al-Ali, A. K., Taylor-Weiner, A., & Van Allen, E. M. (2020). Detection of pathogenic variants with germline genetic testing using deep learning vs standard methods in patients with prostate cancer and melanoma. *JAMA, 324*(19), 1957–1969.

Alzubaidi, L., Zhang, J., Humaidi, A. J., Al-Dujaili, A., Duan, Y., Al-Shamma, O., Santamaría, J., Fadhel, M. A., Al-Amidie, M., & Farhan, L. (2021). Review of deep learning: Concepts, CNN architectures, challenges, applications, future directions. *Journal of Big Data, 8*(1), 53.

Balafar, M. A., Ramli, A. R., Saripan, M. I., & Mashohor, S. (2010). Review of brain MRI image segmentation methods. *Artificial Intelligence Review, 33*, 261–274. https://doi.org/10.1007/s10462-010-9155-0.

Barnat, M., Capizzi, M., Aparicio, E., Boluda, S., Wennagel, D., Kacher, R., Kassem, R., Lenoir, S., Agasse, F., Braz, B. Y., & Liu, J.-P. (2020). Huntington's disease alters human neurodevelopment. *Science, 369*(6505), 787–793.

Cao, C., Liu, F., Tan, H., Song, D., Shu, W., Li, W., Zhou, Y., Bo, X., & Xie, Z. (2018). Deep learning and its applications in biomedicine. *Genomics, Proteomics & Bioinformatics, 16*(1), 17–32.

Carter, Y. H., Shaw, S., & Macfarlane, F. (2002). Primary care research team assessment (PCRTA): Development and evaluation. *Occasional Paper/Royal College of General Practitioners, 81*, iii–72.

Chen, D., Afzal, N., Sohn, S., Habermann, E. B., Naessens, J. M., Larson, D. W., & Liu, H. (2018). Postoperative bleeding risk prediction for patients undergoing colorectal surgery. *Surgery, 164*(6), 1209–1216.

Ching, T., Himmelstein, D. S., Beaulieu-Jones, B. K., Kalinin, A. A., Do, B. T., Way, G. P., Ferrero, E., Agapow, P. M., Zietz, M., Hoffman, M. M., Xie, W., Rosen, G. L., Lengerich, B. J., Israeli, J., Lanchantin, J., Woloszynek, S., Carpenter, A. E., Shrikumar, A., Xu, J., … Greene, C.S. (2018). Opportunities and obstacles for deep learning in biology and medicine. *Journal of the Royal Society, Interface, 15*(141), 20170387.

Crossman, A. R. (2000). Functional anatomy of movement disorders. *Journal of Anatomy, 196*(Pt 4), 519–525.

Dash, S., Acharya, B. R., Mittal, M., Abraham, A., & Keleman, A. (2020). Deep learning techniques for biomedical and health informatics. In *Studies in big data series* Springer. ISSN:2197-6511, ISBN:978-3-030-33966-1.

Dash, S., Pani, S. K., Abraham, A., & Liang, Y. (2021). *Advanced soft computing techniques in data science, Iot and cloud computing*. Springer International Publishing AG.

Dash, S., Pani, S. K., Balamurugan, K., & Abraham, A. (2021). *Information retrieval models for biomedical and health informatics, series: Artificial intelligence and soft computing for industrial transformation*. 436 pages, ISBN: 9781119711247 Wiley: Scrivener Publishing.

Dash, S., Pani, S. K., Rodrigues, J., & Majhi, B. (2022). *Deep learning, machine learning and IoT in biomedical and health informatics: Techniques and applications* (p. 382). Boca Raton: CRC Press. ISBN 9780367544256.

De Mello, F. L., & De Souza, S. A. (2019). Psychotherapy and artificial intelligence: A proposal for alignment. *Frontiers in Psychology*, 10.

Egbuna, C., Gupta, E., Ezzat, S. M., Jeevanandam, J., Mishra, N., Muhammad Akram, N., Sudharani, C. O., Adetunji, P. S., Ifemeje, J. C., Deepak, M., Bhavana, A., Angelo Mark, P., Walag, R. A., Adetunji, J. B., Laila, U., Olisah, M. C., & Onyekere, P. F. (2020). Aloe species as valuable sources of functional bioactives. In C. Egbuna, & G. Dable Tupas (Eds.), *Functional foods and nutraceuticals*. Cham: Springer.

Farhat, H., Sakr, G. E., & Kilany, R. (2020). Deep learning applications in pulmonary medical imaging: Recent updates and insights on COVID-19. *Machine Vision and Applications*, *31*(6), 53.

Fitzpatrick, K. K., Darcy, A., & Vierhile, M. (2017). Delivering cognitive behavior therapy to young adults with symptoms of depression and anxiety using a fully automated conversational agent (Woebot): A randomized controlled trial. *JMIR Mental Health*, *4*, e19.

Frazier, P., Richards, D., Mooney, J., Hofmann, S. G., Beidel, D., & Palmieri, P. A. (2016). Internet-delivered treatment for depression, anxiety, and stress in university students: A patient preference trail. *iproc*, *2*, e5.

Fu, Y., Patel, B. K., Wu, T., Li, J., & Gao, F. (2020). Advanced medical imaging analytics in breast cancer diagnosis. In *Women in industrial and systems engineering* (pp. 301–319). Springer.

Gudigar, A., Raghavendra, U., Hegde, A., Kalyani, M., Ciaccio, E. J., & Acharya, U. R. (2020). Brain pathology identification using computer aided diagnostic tool: A systematic review. *Computer Methods and Programs in Biomedicine*, *187*. https://doi.org/10.1016/j.cmpb.2019.105205, 105205.

Hassane, D. C., Guzman, M. L., Corbett, C., Li, X., Abboud, R., Young, F., Liesveld, J. L., Carroll, M., & Jordan, C. T. (2008). Discovery of agents that eradicate leukemia stem cells using an in silico screen of public gene expression data. *Blood*, *111*(12), 5654–5662.

Herculano-Houzel, S. (2009). The human brain in numbers: A linearly scaled-up primate brain. *Frontiers in Human Neuroscience*, *3*, 31. https://doi.org/10.3389/neuro.09.031.2009.

Hooper, N. M. (2005). Roles of proteolysis and lipid rafts in the processing of the amyloid precursor protein and prion protein. *Biochemical Society Transactions*, *33*(Pt 2), 335–338.

Jo, T., Nho, K., & Saykin, A. J. (2019). Deep learning in Alzheimer's disease: Diagnostic classification and prognostic prediction using neuroimaging data. *Frontiers in Aging Neuroscience*, *11*, 220.

Kaufman, D. M., & Milstein, M. J. (2013). Chapter 15 - Multiple sclerosis. In D. M. Kaufman, & M. J. Milstein (Eds.), *Kaufman's clinical neurology for psychiatrists* (7th ed., pp. 329–349). Philadelphia: W.B. Saunders. ISBN 978–0–7234-3748-2.

Kumar, A., & Sharma, S. (2021). Complex partial seizure [updated 2021 Jun 29]. In *StatPearls [Internet]*. Treasure Island (FL): StatPearls Publishing. Available from: https://www.ncbi.nlm.nih.gov/books/NBK519030/. (Accessed January 2022).

Lee, D. K. (2020). Data transformation: A focus on the interpretation. *Korean Journal of Anesthesiology*, *73*(6), 503–508.

Liu, Z., Wang, S., Di Dong, J. W., Fang, C., Zhou, X., Sun, K., Li, L., Li, B., Wang, M., & Tian, J. (2019). The applications of radiomics in precision diagnosis and treatment of oncology: Opportunities and challenges. *Theranostics*, *9*, 1303. https://doi.org/10.7150/thno.30309.

Liu, F., Wee, C. Y., Chen, H., & Shen, D. (2014). Inter-modality relationship constrained multi-modality multi-task feature selection for Alzheimer's disease and mild cognitive impairment identification. *NeuroImage*, *84*, 466–475.

López de Maturana, E., Alonso, L., Alarcón, P., Martín-Antoniano, I. A., Pineda, S., Piorno, L., & Malats, N. (2019). Challenges in the integration of omics and non-omics data. *Genes*, *10*(3), 238.

Malley, B., Ramazzotti, D., & Wu, J. T. (2016). Data pre-processing. In *Secondary analysis of electronic health records [internet]*. Cham: Springer. Chapter 12. Available from: https://www.ncbi.nlm.nih.gov/books/NBK543629/ https://doi.org/10.1007/978-3-319-43742-2_12.

Manaka, S., Ishijima, B., & Mayanagi, Y. (2003). Postoperative seizures: Epidemiology, pathology, and prophylaxis. *Neurologia Medico-Chirurgica (Tokyo)*, *43*(12), 589–600. https://doi.org/10.2176/nmc.43.589 (discussion 600; PMID: 14723265).

Manikandan, S. (2010). Data transformation. *The Journal of Pharmacy and Pharmacology*, *1*(2), 126–127.

McBee, M. P., Awan, O. A., Colucci, A. T., Ghobadi, C. W., Kadom, N., Kansagra, A. P., Tridandapani, S., & Auffermann, W. F. (2018). Deep learning in radiology. *Academic Radiology*, *25*, 1472–1480. https://doi.org/10.1016/j.acra.2018.02.018.

Moghram, B. A., Nabil, E., & Badr, A. (2018). Ab-initio conformational epitope structure prediction using genetic algorithm and SVM for vaccine design. *Computer Methods and Programs in Biomedicine*, *153*, 161–170.

Nagpal, G., Gupta, S., Chaudhary, K., Kumar Dhanda, S., Prakash, S., & Raghava, G. P. S. (2015). VaccineDA: Prediction, design and genome-wide screening of oligodeoxynucleotide-based vaccine adjuvants. *Scientific Reports*, *5*, 12478.

Newman, M. G., Szkodny, L. E., Llera, S. J., & Przeworski, A. (2011). A review of technology-assisted self-help and minimal contact therapies for anxiety and depression: Is human contact necessary for therapeutic efficacy? *Clinical Psychology Review*, *31*, 89–103.

Nguyen, H. G., Welty, C. J., & Cooperberg, M. R. (2015). "Diagnostic associations of gene expression signatures in prostate cancer tissue" *(PDF)*. *Current Opinion in Urology*, *25*(1), 65–70.

Ohnishi, S., & Takano, K. (2004). Amyloid fibrils from the viewpoint of protein folding. *Cellular and Molecular Life Sciences*, *61*(5), 511–524.

Ortiz, A., Górriz, J. M., Ramírez, J., & Martínez-Murcia, F. J. (2013). LVQ-SVM based CAD tool applied to structural MRI for the diagnosis of the Alzheimer's disease. *Pattern Recognition Letters*, *34*, 1725–1733.

Pani, S. K., Dash, S., dos Santos, W. P., Bukhari, S. A. C., & Flammini, F. (2022). *Assessing covid-19 and other pandemics and epidemics using computational modelling and data analysis*. Springer International Publishing. 431 pages, ISBN-13: 978-3030797522, Springer.

Priller, C., Bauer, T., Mitteregger, G., Krebs, B., Kretzschmar, H. A., & Herms, J. (2006). Synapse formation and function is modulated by the amyloid precursor protein. *The Journal of Neuroscience*, *26*(27), 7212–7221.

Purves, D., Augustine, G. A., Fitzpatrick, D., Hall, W., LaMantia, A. S., McNamara, J. O., & Williams, S. M. (2001). Modulation of movement by the basal ganglia—Box A. Huntington's disease. In D. Purves (Ed.), *Neuroscience* (2nd ed.). Sunderland, MA: Sinauer Associates. ISBN 978-0-87893-742-4.

Rim, B., Sung, N. J., Min, S., & Hong, M. (2020). Deep learning in physiological signal data: A survey. *Sensors (Basel, Switzerland)*, *20*(4), 969.

Roli, O. I., Adetunji, C. O., Mishra, P., Adetunji, J. B., & Mishra, R. R. (2020). Caffeine: Nutraceutical and health benefit of caffeine-containing commodities and products. In P. Mishra, R. R. Mishra, & C. O. Adetunji (Eds.), *Innovations in food technology*. Singapore: Springer.

Rost, T., Stein, J., Löbner, M., Kersting, A., Luck-Sikorski, C., & Riedel-Heller, S. G. (2017). User acceptance of computerized cognitive behavioral therapy for depression: Systematic review. *Journal of Medical Internet Research*, *19*, e309.

Sailer, A. M., van Zwam, W. H., Wildberger, J. E., & Grutters, J. P. (2015). Cost-effectiveness modelling in diagnostic imaging: A stepwise approach. *European Radiology*, *25*, 3629–3637. https://doi.org/10.1007/s00330-015-3770-8.

Sakai, K., & Yamada, K. (2019). Machine learning studies on major brain diseases: 5-year trends of 2014–2018. *Japanese Journal of Radiology*, *37*, 34–72. https://doi.org/10.1007/s11604-018-0794-4.

Senders, J. T., Arnaout, O., Karhade, A. V., Dasenbrock, H. H., Gormley, W. B., Broekman, M. L., & Smith, T. R. (2018). Natural and artificial intelligence in neurosurgery: A systematic review. *Neurosurgery*, *83*, 181–192. https://doi.org/10.1093/neuros/nyx384.

Shen, D., Wu, G., & Suk, H.-I. (2017). Deep learning in medical image analysis. *Annual Review of Biomedical Engineering, 19*, 221–248.

Stys, P. K., & Tsutsui, S. (2019). Recent advances in understanding multiple sclerosis. *F1000Research, 8*, 2100.

Suddhasuchi, D., Sharangi, A. B., Egbuna, C., Jeevanandam, J., Ezzat, S. M., Adetunji, C. O., Tijjani, H., Olisah, M. C., Patrick-Iwuanyanwu, K. C., Adetunji, J. B., Ifemeje, J. C., Akram, M., Moboladji, B. M., & Onyeike, P. C. (2020). Health benefits of isoflavones found exclusively of plants of the Fabaceae Family. In C. Egbuna, & T. G. Dable (Eds.), *Functional foods and nutraceuticals*. Cham: Springer. https://doi.org/10.1007/978-3-030-42319-3_22.

Suk, H. I., Lee, S. W., & Shen, D. (2016). Deep sparse multi-task learning for feature selection in Alzheimer's disease diagnosis. *Brain Structure & Function, 221*, 2569–2587.

Suk, H., & Shen, D. (2015). Alzheimer's Disease Neuroimaging Initiative (2015) deep learning in diagnosis of brain disorders. In S. Lee, H. Bülthoff, & K. Müller (Eds.), *Recent progress in brain and cognitive engineering. Trends in augmentation of human performance*. Dordrecht: Springer.

Tiraboschi, P., Hansen, L. A., Thal, L. J., & Corey-Bloom, J. (2004). The importance of neuritic plaques and tangles to the development and evolution of AD. *Neurology, 62*(11), 1984–1989.

Tobore, I., Li, J., Yuhang, L., Al-Handarish, Y., Kandwal, A., Nie, Z., & Wang, L. (2019). Deep learning intervention for health care challenges: Some biomedical domain considerations. *JMIR mHealth and uHealth*, 7(8), e11966.

Tolin, D. F. (2010). Is cognitive-behavioral therapy more effective than other therapies? A meta-analytic review. *Clinical Psychology Review, 30*, 710–720.

Torous, J., Chan, S. R., Yee-Marie Tan, S., Behrens, J., Mathew, I., Conrad, E. J., Hinton, L., Yellowlees, P., & Keshavan, M. (2014). Patient smartphone ownership and interest in mobile apps to monitor symptoms of mental health conditions: A survey in four geographically distinct psychiatric clinics. *JMIR Mental Health, 1*, e5.

Valliani, A. A.-A., Ranti, D., & Oermann, E. K. (2019). Deep learning and neurology: A systematic review. *Neurology and Therapy, 8*, 351. https://doi.org/10.1007/s40120-019-00153-8.

Varkey, J., Isas, J. M., Mizuno, N., Jensen, M. B., Bhatia, V. K., Jao, C. C., et al. (2010). Membrane curvature induction and tubulation are common features of synucleins and apolipoproteins. *The Journal of Biological Chemistry, 285*(42), 32486–32493.

Vieira, S., Pinaya, W. H., & Mechelli, A. (2017). Using deep learning to investigate the neuroimaging correlates of psychiatric and neurological disorders: Methods and applications. *Neuroscience and Biobehavioral Reviews, 74*, 58–75. https://doi.org/10.1016/j.neubiorev.2017.01.002.

Warmerdam, L., Smit, F., Van Straten, A., Riper, H., & Cuijpers, P. (2010). Cost-utility and cost-effectiveness of internet-based treatment for adults with depressive symptoms: Randomized trial. *Journal of Medical Internet Research, 12*, e53.

Warner, D. F. (2016). Defining a diagnostic gene signature for tuberculosis. *The Lancet. Respiratory Medicine, 4*(3), 170–171. https://doi.org/10.1016/s2213-2600(16)00063-1.

Welch, G., von Recklinghausen, F., Taenzer, A., Savitz, L., & Weiss, L. (2017). Data cleaning in the evaluation of a multi-site intervention project. *EGEMS (Washington, DC), 5*(3), 4.

Yasaka, K., & Abe, O. (2018). Deep learning and artificial intelligence in radiology: Current applications and future directions. *PLoS Medicine, 15*. https://doi.org/10.1371/journal.pmed.1002707, e1002707.

Yu, J., Hu, Y., Xu, Y., Wang, J., Kuang, J., Zhang, W., Shao, J., Guo, D., & Wang, Y. (2019). LUADpp: An effective prediction model on prognosis of lung adenocarcinomas based on somatic mutational features. *BMC Cancer, 19*, 263.

CHAPTER 12

Cognitive therapy for brain diseases using artificial intelligence models

Olugbemi T. Olaniyan[a], Charles O. Adetunji[b], Ayobami Dare[c], Olorunsola Adeyomoye[d], Mayowa J. Adeniyi[e], and Alex Enoch[f]

[a]Department of Physiology, Laboratory for Reproductive Biology and Developmental Programming, Rhema University, Aba, Nigeria
[b]Department of Microbiology, Applied Microbiology, Biotechnology and Nanotechnology Laboratory, Edo State University Uzairue, Iyamho, Nigeria
[c]Department of Physiology, School of Laboratory Medicine and Medical Sciences, College of Health Sciences, Westville Campus, University of KwaZulu-Natal, Durban, South Africa
[d]Department of Physiology, University of Medical Sciences, Ondo, Nigeria
[e]Department of Physiology, Edo State University Uzairue, Iyamho, Nigeria
[f]Department of Human Physiology, Ahmadu Bello University Zaria, Zaria, Nigeria

Introduction

In recent decades, the healthcare system has generated a large amount of data produced during practices like medical imaging, surgery, prognosis, laboratory testing, and physical examination (Senders et al., 2018). The analysis of this data has revolutionized the techniques used by healthcare providers over the past three decades to diagnose, understand, and manage brain diseases, detect risks, and respond to treatment modalities (Yasaka & Abe, 2018). More importantly, analysis of medical images has brought significant advancement to medical practice and facilitated robust, rapid, and noninvasive approaches to study, diagnose, and treat neurological disease. This noninvasive technique led to the development of more sophisticated and low-risk analysis of brain disease, including the application of magnetic resonance imaging (MRI), computed tomography (CT), and positron emission tomography (PET), all of which have improved the understanding of brain function, thereby assisting physicians in performing non-invasive examination of the brain anatomy and understanding the pathogenesis of different brain disorders (Cenek, Hu, York, & Dahl, 2018).

Artificial Intelligence for Neurological Disorders
https://doi.org/10.1016/B978-0-323-90277-9.00013-4

Although the study of brain disorders is an active area of research in neuroscience, most clinicians and scientists still diagnose and classify different neurological diseases subjectively rather than via a computer-assisted diagnosis (CAD) approach (Insel, 2014; Kalmady et al., 2019), thus increasing the incidence of misdiagnosis and inadequate management strategies. In addition, most medical data and brain scans are manually analyzed, which may be labor-intensive, time-consuming, and result in errors that may compromise interpretation. It has been reported that daily error rates and inconsistent radiological reports exceed 3%–5% (Brady, 2017). Thus, there is an urgent need to develop innovative approaches that efficiently analyze data and understand pathogenesis of brain diseases to assist clinicians in the adequate diagnosis and proper management of these disorders.

Recently, many traditional computer-based techniques and algorithms have been employed to improve clinical practice due to increased computational power and increased quality and quantity of neurological data (Dash, Shakyawar, Sharma, & Kaushik, 2019). However, the clinical efficiencies of most of these techniques are limited. For example, features like lesions and tumors may be too difficult to code with simple mathematical equations properly. Experts may often find it difficult to generate the exact protocol to apply for accurate analysis and management of disease conditions. Therefore, these innovative algorithms are generated from techniques such as machine learning (ML), deep learning (DL), and artificial neural networks (ANNs), which are subsets of artificial intelligence (AI) (Genç, 2019). ML is an AI technique that allows the system to "learn" and automatically extract features from qualitative and quantitative data from normal and abnormal brain structures necessary for accurate diagnosis. Using this technique, ML has been extensively applied to detect segmentation and precise localization of abnormal tissues needed for the diagnosis, analysis, and clinical management of brain diseases such as Alzheimer's disease (AD), schizophrenia, dementia, and other neurodegenerative diseases (Balafar, Ramli, Saripan, & Mashohor, 2010; Sakai & Yamada, 2019). The DL algorithm is currently transforming numerous processes in neuroscience, exerting greater benefits than other techniques in image analysis. In DL, important characteristics of images are extracted automatically from the raw data, thus reducing errors during manual computation and selection of potentially relevant features involved in ML (Gautam & Sharma, 2020; Goodfellow, Bengio, & Courville, 2016). Therefore, AI is a rapidly advancing technology that is presently transforming neuroscience to improve the management of brain diseases.

Brain diseases

AI-based applications have been designed to diagnose and manage brain diseases (Dwyer, Falkai, & Koutsouleris, 2018; Stead, 2018). These techniques are continuously used to monitor brain anatomy and physiology, which are the most essential biological phenotypes and major diagnostic biomarkers for brain disorders (Luxton, 2015). Therefore, different AI-related approaches that can generate comprehensive data to characterize several brain disorders may detect these diseases. Proper analysis of these data has been proposed to be the future direction in managing these diseases because important algorithms can be developed to assist in prompt, adequate, and early detection of brain disorders. We briefly discuss some of the major brain diseases to which AI has been extensively in the sections that follow.

Alzheimer's disease (AD) or mild cognitive impairment (MCI): MCI is regarded as an intermediate stage of dementia, while AD is the most common neurodegenerative disorder, with a high incidence among older persons. Due to the increased incidence of this brain disease, many data generated from brain images have assisted in the automatic classification of features from AD patients and normal controls using ML techniques and AI (Ni, Zhou, Ning, & Wang, 2016).

Schizophrenia: There is no specific confirmatory test for this brain disease. Thus, there is a paradigm shift toward the automatic diagnosis of this disorder using neuroimaging techniques (Arbabshirani, Plis, Sui, & Calhoun, 2017). Therefore, most studies using AI are designed to identify and classify specific features of schizophrenia from healthy controls and other disorders like bipolar disorder or obsessive–compulsive disorder (OCD) using data obtained from functional MRI (fMRI) (Anticevic et al., 2014; Schnack et al., 2014).

Brain tumor: The role of AI in brain tumors is to identify quantitative features on brain images, which are prerequisites for predicting diagnosis, prognosis, and treatment of specific brain cancers (Zhou et al., 2018). The AI-based approach used for brain tumors focuses mainly on tumor segmentation, tumor classification, and prediction of survival or genotype using images obtained from various MRI scans (Emblem et al., 2014; Li et al., 2018).

Depressive disorders: The major application of AI in major depressive disorder (MDD) is to classify and predict early-onset treatment response and score the clinical symptoms of this brain disease (Patel et al., 2015; Yoshida et al., 2017). The AI-based approach has helped in proper diagnosis, with adequate pharmacological intervention, which has shown improved efficacy in almost half of patients treated via this technique (Papakostas, 2009). This approach is dependent on the guidelines from the *Diagnosis and Statistical Manual of Mental Disorders* (DSM) and clinician judgment (Maj, 2013).

Parkinson's disease (PD): PD is a chronic neurodegenerative disease usually diagnosed by the presence of predominantly motor symptomatology, but it can also present with nonmotor hyposmia, paresthesia, depression, and pain (Chaudhuri, Yates, & Martinez-Martin, 2005; Nutt & Wooten, 2005). It is a general brain disease occurring in males and females of all ages with an incidence of 4.5–19/100,000 of the population (Marras & Tanner, 2004). The major aim of AI application in PD is to classify and differentiate PD from healthy controls by extracting features like progressive supranuclear palsy (PSP), as well as from subjects with scans without evidence of dopaminergic deficit (SWEDD) on fMRI images (Du et al., 2017).

Attention deficit hyperactivity disorder (ADHD): This brain disease frequently occurs in approximately 2.2%–17.8% of school-aged children but may extend to adulthood. It is characterized by a progressive inattention or hyperactivity-impulsivity that interferes with functioning or development (Ayano, Yohannes, & Abraha, 2020). In 2011, a global competition (called ADHD-200) was organized to use neuroimaging and phenotypic analysis to diagnose ADHD automatically. The major application of AI in ADHD is to identify imbalance in the dopaminergic and noradrenergic neurotransmission in the brain using data obtained from electroencephalography (EEG), MRI, and PET techniques (Kautzky et al., 2020; Pani, Dash, dos Santos, Bukhari, & Flammini, 2021).

Autism spectrum disorder (ASD): This is a neurodevelopmental disease associated with difficulties in social communication and interaction with restricted or repetitive patterns of behavior. It frequently affects youths and shares similar pathogenesis with other brain disorders. This disease has no distinct, identifiable biomarkers; present diagnostic criteria depend on behavioral observations by healthcare professionals (Falkmer, Anderson, Falkmer, & Horlin, 2013). However, the identification of unbiased and reliable biomarkers is highly important using AI-based neuroimaging techniques. AI can be used to classify this brain disease, especially by applying fMRI-based features and an estimation of ASD severity based on structure MRI (Song, Kim, Bong, Kim, & Yoo, 2019; Uddin, Dajani, Voorhies, Bednarz, & Kana, 2017).

Epilepsy: This is a chronic neurological disorder characterized by a continuous predisposition to epileptic seizures. At least one epileptic seizure is needed to confirm the disease (Fisher et al., 2005). Epilepsy has no age restriction, but risk increases with age. It occurs in both sexes with an incidence rate of approximately 23–190/100,000 of the population (Kotsopoulos, Van Merode, Kessels, De Krom, & Knottnerus, 2002). The main purposes of AI studies in epilepsy is to identify features that classify and differentiate epilepsy from other types of brain diseases, as well as to detect and predict treatment outcome using MRI techniques like T1W1 and DTI depending on the objectives (EL Azami et al., 2016; Munsell et al., 2015).

Multiple sclerosis (MS): This brain disorder is common among all neurological disorders and results from inflammatory demyelination of the nervous system. The pathogenesis of this brain disease is multifaceted, with initial characteristics of reversible neurological dysfunctions that become worse as the disease progresses (Goldenberg, 2012). There are no treatments available for this disease. Therefore, an appropriate diagnosis is crucial to reduce its occurrence. MS diagnosis depends on a comprehensive analysis of structural and functional characteristics of the brain structure using MRI to classify, detect, and segment MS lesions (Saccà et al., 2019; Zhao et al., 2017).

Stroke: Stroke is a clinical syndrome of cerebral deficit that lasts for more than 24 h without distinct cause. In industrialized countries, 75%–80% of strokes result from brain ischemia, while 10%–15% are correlated with intracerebral hemorrhage. Diagnosis fully depends on specialist examination of clinical symptoms (Hatano, 1976). AI techniques have been applied for stroke imaging to improve automatic or precise diagnosis of the disease and predict prognosis during therapy (Lee, Kim, Kim, & Kang, 2017) using information extracted from CT and MRI images (Bentley et al., 2014). The AI-based technique can help identify infarct or hemorrhage and stroke segmentation, classify stroke, detect blood vessel occlusion, score stroke severity, and assist with patient response to treatment. AI techniques such as convolutional neural networks (CNNs) can perform these image-based analyses efficiently and accurately (Chen, Bentley, & Rueckert, 2017).

Brain diseases and physiological signals

Evidence has shown the interplay between autonomic state and brain functions. For example, body feelings, especially those associated with emotions, strongly influence decision-making. Therefore, the body's internal condition regulates the mode of response to the external environment

(Critchley, 2009). It has been hypothesized that brain function might be closely related to neuro-modulatory brainstem activities. Thus, assessing the activity of the autonomic nervous system is indicative of the functional state (Samuels & Szabadi, 2008) and is important in neurological function. It has been reported that a close interaction exists between the pupil of the eye and brain function, reflecting the status of autonomic function and cognitive function. Thus, pupillometry has been proposed to assess the physiological function of the brain (Bradley, Miccoli, Escrig, & Lang, 2008). Pupils have been demonstrated to respond to different types of cognitive demand, like memory load, arithmetic tests, or speech processing. An increased level of brain activities causes a greater magnitude of pupillary response (Querino et al., 2015).

Another physiological signal that has been suggested as a marker for mental effort is the level of skin conductance (Mehler, Reimer, Coughlin, & Dusek, 2009; Reimer & Mehler, 2011). In competitive analyses, skin conductance and pupil size have been similarly suitable for indicating brain function, which is often altered in several brain diseases (Wickramasuriya, Amin, & Faghih, 2019).

Another physiological signal of brain disorder is immune dysregulation and inflammation. Emerging evidence have shown that peripheral growth factors, endocrine factors, pro-inflammatory cytokines, and metabolic markers play a significant role in the pathogenesis of major depressive disorders and antidepressant response (Schmidt, Shelton, & Duman, 2011). Alteration in the hypothalamic–pituitary–adrenal (HPA) axis is another signal associated with brain disease because the increased amount of glucocorticoids in the circulation varies directly with an increased incidence of depression in those with the metabolic disorder (Vogelzangs et al., 2009). These signals indicate that the body and the mind are inextricably connected with the evidence that these physiological signals originate from alteration in the autonomic nervous system and dysregulation of the HPA axis (HPA), both of which are affected by brain diseases.

Artificial intelligence

Artificial intelligence (AI) encompasses the science and engineering of creating intelligent machines that can think like humans and even surpass the level of human intelligence. AI is also a branch of computer science that performs tasks that would normally require human intelligence (Kayid, 2020; Sollini, Bartoli, Marciano, et al., 2020). AI encompasses a diverse set of methods, algorithms, and tools, including genetic algorithms, DL, and neural networks. These major areas are expanding rapidly and significantly impacting fields as diverse as robotics, health care, military, and space.

Currently, AI is also being applied in areas such as aviation, driving, medicine, image recognition, personal assistance, and online advertising. Furthermore, with the availability of a large amount of data, global connectivity, computing, and various algorithms at our disposal, AI will add a new level of effectiveness and complexity to technologies in the future.

Assistive AI may accomplish precision that may exceed that of human professionals. However, its adoption is frequently hampered by humans' lack of faith in assistive AI predictions. This is the major reason why researchers in the AI community have concentrated on making decisions regarding AI more clear by creating explainable artificial intelligence (XAI), in which the result of AI decisions can be explained and understood by humans (XAI Schmidt, Biessmann, & Teubner, 2020). One of the major goals of AI is to create a fully self-sufficient and intelligent agent that, like humans, interacts with their environment, learns optimal behaviors, and improves over time through trial and error.

AI is one of today's most contentious issues, and there appears to be little agreement on the similarities and differences between it and human intelligence. The pursuit of human-like intelligence characterizes many relevant discussions on the explainability and reliability of AI as its benchmark (Kayid, 2020).

Makridakis (2017) claimed that the AI revolution is on track and will bring extensive changes that will impact all aspects of our world by examining parallel discoveries of the industrial, digital, and AI revolutions. More so, it will have a significant effect on the rate of employment and how firms operate. Therefore, it will connecting great organizations with making decisions based on exploitation and analysis of their data and increased global competition among them (Makridakis, 2017).

People will be able to use the Internet to buy goods and obtain services from anywhere in the world and take advantage of the numerous additional benefits that will be available as a result of the extensive use of AI discoveries. The most difficult challenge for societies and businesses is to capitalize on the benefits of AI technologies, which provide massive opportunities for new products and services and enhanced productivity, therefore evading the perils of unemployment and greater inequalities in wealth (Makridakis, 2017).

Elegant AI technologies, providing increased human jurisdiction and computer automation, can improve the performance of humans, resulting in wider acceptance. Circumstances where full human control or full computer control is required, avoidance of the dangers of excessive human interference or increased computer autonomy, and finally, improvement of human performance by designing increased levels of human interference and high levels of computer automation were all outlined in the Human-Centered Artificial Intelligence (HCAI) framework. HCAI methodology is prone to yield harmless, reliable, and dependable schemes. Therefore, reaching these fits will significantly boost the performance of humans while also promoting sufficiency, efficiency, and innovation (Shneiderman, 2020).

Many lessons can be drawn from AI's previous successes and failures. To keep AI progressing, cogent and congruent interactions between projects specific to the application and visionary research ideas are required. To consider the associated legal and ethical issues, a clear approach is crucial to ensure that the evolution of AI benefits society as a whole and that its latent negative effects are abated from the start. Along with AI's unprecedented enthusiasm, there are also concerns about the sway of technology in our society. Such concerns should not stymie AI progress but rather motivate it to create a systematic framework in which future AI can thrive. Most importantly, it is critical to distinguish between practical reality science fiction, as AI is about to transform the future of our society, economy, and lives with sustained funding and investment (Kayid, 2020).

The most fruitful AI applications will be those that complement and compensate for the intrinsic cognitive and biological limitations of humans both in the workplace and in making policy. As a result, one of the most pressing issues is how to use it wisely. When is it safe to leave decision-making completely to AI, for what tasks and under what conditions, and when is human input required? How can we depend on the strengths of human intelligence and how can we effectively deploy AI systems to balance and compensate for the inborn limitations of human cognition? (Korteling, van de Boer-Visschedijk, Blankendaal, Boonekamp, & Eikelboom, 2021).

Artificial intelligence, neuroscience, and clinical practice

Artificial intelligence involves using computer technology to simulate and perform tasks. It contrasts with the natural intelligence displayed by humans, which depends on emotions and consciousness (Amisha, Malik, Pathania, & Rathaur, 2019). AI helps in planning, reasoning, learning, processing, representing, and manipulating objects (Amisha et al., 2019). It encompasses different fields of knowledge, such as mathematics, philosophy, computer science, linguistics, information engineering, and others. The awareness and use of AI to solve problems in neuroscience has continued to increase with current trends in digital globalization (Pedersen et al., 2020). AI is crucial in studying the etiology and pathogenesis of many neurodegenerative diseases. It is also useful for diagnosis and to administer therapeutic interventions. As a result of advances in technology, neuroscientists can collect large volumes of data with high precision using artificial intelligence models (Dash, Pani, Rodrigues, & Majhi, 2022; Niel & Bastard, 2019). However, the use of these data to understand neurological functions remains a big challenge. fMRI, CT, and single-photon emission computed tomography (SPECT) are some of the imaging techniques used to generate large data images to evaluate brain functions (Hosny, Parmar, Quackenbush, Schwartz, & Aerts, 2018). These are some of the techniques that utilize artificial intelligence to capture large volumes of scientific data in neuroscience. In the past, physicians were trained to understand, interpret, and give qualitative analysis of medical images, which is part of the basic requirements for becoming a neuro-specialist. At present, ML algorithms are being trained to understand the complex nature of imaging data and to quantitatively assess their characteristics (Currie, Hawk, Rohren, Vial, & Klein, 2019).

Furthermore, AI has brought a paradigm shift in clinical practice with increased capacity to generate and process large volumes of data. AI is all-encompassing, it has ability to accommodate both structured and unstructured data (Briganti & Le Moine, 2020). Many researchers have debated the possibility of replacing human physicians with artificial intelligence in the future. However, most believe that artificial intelligence may not completely replace human, but that it could help physicians to make better clinical decisions (Tai, 2020). Artificial intelligence has changed the narrative in the healthcare system, which now focuses on healthcare management rather than the treatment of disease (Adetunji, Adetunji, & Olaniyan, 2021; Adetunji, Ajay, et al., 2021; Adetunji, Akram, et al., 2021b; Adetunji, Akram, et al., 2021c; Adetunji, Inobeme, et al., 2021; Adetunji & Varma, 2020; Akram et al., 2020; Das et al., 2020; Egbuna et al., 2020; Roli, Adetunji, Mishra, Adetunji, & Mishra, 2020). This ensures early diagnosis of disease conditions, allowing people to remain healthy and strong (Davenport & Kalakota, 2019, Dash, Pani, Abraham, & Liang, 2021). AI also has application in healthcare administration and ensuring that there is proper clinical documentation. Information such as age, gender, geography, family history, metabolic and immune functions, and environment are easily documented and used in conjunction with other data sciences such as genetics to provide interventions and predict the prognosis of disease conditions (Dias &

Torkamani, 2019, Dash, Acharya, Mittal, Abraham, & Keleman, 2020). In addition, the process of drug discovery can be enhanced with the use of artificial intelligence in mining large chemistry data that could be useful in classifying potential drug compounds (Paul et al., 2021).

Data acquisition and image interpretation

Data acquisition involves using signals to convert real-world information to numerical values. In a data acquisition system, physical signals are emitted and detected by transducers and they undergo modification and are converted to values that are further processed by a computing system (Yang, Huang, Liu, Rong, & Lee, 2015). The trend in scientific research involves adopting a multidisciplinary approach where collaboration exists between a principal investigator, coordinator, data collector, database manager, and statisticians. This approach to scientific research is necessary to reduce data errors or problems that may occur in data acquisition, some of which include entry error, outliers, discrepancies, missing values, duplicates, and so on (Dash & Subudhi, 2016; Dash, Tripathy, & Rahman, 2017; Goldberg, Niemierko, & Turchin, 2008). These data errors may interfere with reproducibility, quality, and validity of research. A huge gap exists between data acquisition and analysis. However, with data preprocessing, most of these errors are eliminated. Data reprocessing techniques include cleaning, transformation, and reduction and they ensure the storage and management of quality data (Malley, Ramazzotti, & Wu, 2016). In the study of neurological disease, state-of-the-art equipment is used to extract large data. For instance, CT uses X-rays to illuminate parts of the body. These X-ray signals are detected by a photoelectric sensor that produces axial images that provide information about the part of the body under investigation (Taubmann, Berger, Bögel, et al., 2018). Physicians take advantage of this modern equipment to acquire big data. CT, SPECT, and X-rays are some of the equipment that provide image data in neurology, even though they are also used to study alteration in other body parts. The representation of these images can occur in various forms. Most times, representation refers to how these images are coded or stored. It may also include the format of the image file, capacity for compression, and the format of the meta-data (Rahman, Antani, Demner-Fushman, & Thoma, 2015). Visual content of images may also form part of how an image is being represented. In fact, studies have shown that stored image patches (small and specific image regions of interest) can be stored in a glossary and used for future retrieval of large images (Tagare,

Jaffe, & Duncan, 1997). At present, most medical images are interpreted by physicians, however, attention is shifting to the use artificial intelligence and ML approaches to provide quantitative and more reliable interpretation of medical images. For instance, in a study to evaluate acute ischemic stroke using medical images, ischemic scores, penumbra, and site of occlusion were quantified using ML (Kamal, Lopez, & Sheth, 2018). However, these techniques are used mainly to complement or support the physician's interpretation.

Artificial intelligence and cognitive behavioral therapy

Cognitive-behavioral therapy (CBT) is an intervention that helps in altering cognitive distortions and enhancing mental health. It also plays a vital role in developing self-dependent adaptive strategies and skills to overcome the prevailing problem. Studies have shown that it is comparably effective as the use of drugs (Gartlehner et al., 2017). CBT has been widely used to treat several psychological problems such as major depressive disorder, eating disorders, substance abuse, speech anxiety, social anxiety disorder, and more (Fitzpatrick, Darcy, & Vierhile, 2017). It is based on the work of Ivan Pavlov on classical conditioning in which exposure to extinct learned fear cues results in alleviation of stress, anxiety, and fear (Olatunji, Cisler, & Deacon, 2010). There is growing evidence that the amygdala discharges less during extinct fear learning (Ahs, Kragel, Zielinski, Brady, & LaBar, 2015).

One of the ways through which CBT is administered is through artificial intelligence. Established as an academic discipline in the 1950s, artificial intelligence is a technique that imitates human and animal functions. With artificial intelligence, machines can perform tasks that normally and customarily require human intelligence. Nowadays, artificial intelligence machines are designed to exhibit superb prowess to discern human speech and execute other executive functions like driving, gaming, and content delivery; artificial intelligence predicates the use of a set of specific instructions known as algorithms. Three types of artificial intelligence have been identified: vital or general artificial intelligence, weak or narrow artificial intelligence, and artificial superintelligence. Beyond this, artificial intelligence-enhanced CBT has been used to treat social anxiety disorder (Ngai, Tully, & Anderson, 2015) and generalized anxiety disorder (Aymerich-Franch & Bailenson, 2014; Poeschl, 2017).

Fulmer, Joerin, Gentile, Lakerink, and Rauws (2018) examined the possibility and feasibility of using an integrative psychological AI named Tess to reduce self-recognized symptoms of depression and anxiety in college students, a statistical reduction in depression symptoms indicating that AI is an accessible and cost-effective therapeutic. The researchers developed a CBT chatbot, an artificial intelligence technology that imitates normal conversational patterns. The chatbot was called "Woebot" and was used to deliver CBT. They discovered that the AI reduced anxiety and depression in students over 14 days. Virtual reality (VR) is a technology involving artificial intelligence. However, VR-coupled treatment could induce physiological reactions such as increased heart rate and skin conductance (Côté & Bouchard, 2005; Diemer, Alpers, Peperkorn, Shiban, & Mühlberger, 2015).

Challenges and pitfalls

One of the major challenges of artificial intelligence is the deficiency in trust, interpretability, and transparency. However, this can be overcome by using XAI (Adadi & Berrada, 2018; Došilović, Brčić, & Hlupić, 2018). Other challenges associated with artificial intelligence include data scarcity, knowledge limitation, and data security and privacy.

Another challenge is termed algorithm bias. Larson and Angwin reported an instance of such (2016) in a program called Correctional Offender Management Profiling for Alternative Sanctions (COMPAS), which is commonly used in US courts to predict recidivism risk Larson & Angwin, 2016. The researchers found that COMPAS claimed greater recidivism risk for Black defendants than it did for White defendants. With artificial intelligence, the need for human resources is reduced in the health sector, leading to an abysmal increase in unemployment and underemployment. The development of full artificial intelligence could put an end to humanity and then succeed it (Cellan-Jones, 2014).

Summary

Cognitive diseases are incurable disorders that hamper cognitive functions and abilities (Simpson, 2014). In addition to aging, several predisposing environmental factors can precipitate disease onset. Therefore, there are incontestable needs to develop technology-driven techniques for early diagnosis and prognosis, disease forecast, and misdiagnosis prevention. All studies presented in this chapter focused on advances in ML research algorithms and their applications for cognitive disease diagnosis, prognosis, and forecasting.

ML algorithms are useful in detecting, classifying, and predicting early pathological cognitive decline in older adults. Algorithms such as logistic regression designed by Ewers et al. (2012) claim to be useful in discriminating Alzheimer's disease and neurocognitive disease with Cox regression modeling the conversion time from mild cognitive disease to Alzheimer's. Sparse autoencoders and 3D CNN developed by Payan and Montana (2015) can predict Alzheimer's disease vs mild cognitive illness with accuracy of 95.39%. An ensemble-based method using a sparse representation-based classifier for predicting neurocognitive disease was developed by Liu, Zhang, Shen, Initiative, et al. (2012). Gomar et al. examined the tendency of cognitive markers and biomarkers to predict change from mild cognitive illness to Alzheimer's disease (Gomar et al., 2011). Zhang, Shen, Initiative, et al. (2012) proposed a multimodal, multitask learning system that can perform regression and classification simultaneously. In a bid to classify patients into mild cognitive impairment, Alzheimer's disease, and healthy categories, Moradi et al. were able to use data from the Alzheimer's Disease Neuroimaging Initiative (ADNI) to train the classifier with an accuracy of 77% (Moradi, Gaser, Huttunen, & Tohka, 2014).

Tierney et al. (1996) investigated the likelihood of developing a neuropsychological-based test to predict Alzheimer's disease onset. Using analysis of covariance, they analyzed group differences of neurocognitive disease and Alzheimer's disease. Using a dynamic Bayesian network, a causal relationship from fMRI was extracted and utilized to develop a neural anatomical network. With this, Burge, Clark, Lane, Link, and Qiu (2004) were able to classify patients into healthy and dementia categories with an accuracy of 73%. With the help of the support vector machine model, Challis et al. (2015) compared the Bayesian Gaussian process logistic model based on performance. They observed that by using fMRI as input data, models could comparably predict mild cognitive illness and neurocognitive illness with an accuracy of 80%.

Cognitive decline such as loneliness, agitation (Cheng, 2017), and social isolation (Meeks & Jeste, 2009), which are difficult to understand (Biddle et al., 2019), have been elucidated more clearly by artificial intelligence. In addition, with novel attributes such as handwriting (text), sensor (digital), speech (audio), and wearable device monitors, it is possible to identify new indicators of cognitive decline (Kourtis, Regele, Wright, & Jones, 2019). Wearable cognitive devices exhibit potential for continuous monitoring and tracking of changes in cognitive function.

Suk, Lee, and Shen (2014) developed a method of fusing data from MRI and PET via DL that involved integrating selected image-derived voxel patches into a deep Boltzmann machine. The fusion resulted in an improvement in classification, allowing brain regions associated with Alzheimer's disease to be confirmed. Fusion of biomarkers from cerebrospinal fluid, PET, and MRI using intermediate data integration (early and intermediate) and novel multimodal and multitask learning resulted in better classification of Alzheimer's disease compared to neurocognitive disease.

Brain disease is a global public health crisis (Gaugler, James, Johnson, et al., 2016) that has led to greater levels of mortality and decreased quality of life. Unfortunately, there are no adequate techniques for clinical management of brain diseases (Appleby, Nacopoulos, Milano, et al., 2013). Most of these diseases, such as Alzheimer's disease, might be a result of multidimensional factors associated with psychology, sociology, and physiology (Bemelmans, Tromp, Bunnik, et al., 2016; Cerman, Eraslan, & Cekic, 2015; Haass & Selkoe, 2007; Herrmann, Harimoto, Balshaw, et al., 2015; Kang, Weuve, & Grodstein, 2004; Robertson, 2013; Wang, Roe, Snyder, et al., 2012; Weuve, Hebert, Scherr, et al., 2015). The most common way of detecting Alzheimer's disease includes neuropsychological scales and clinical manifestations such as the Global Decline Scale (GDS), Mini-Mental State Examination (MMSE), Activities of Daily Living (ADL), and Montreal Cognitive Assessment (MoCA). Moreover, most of these approaches are meant for patients who have already developed Alzheimer's disease; they are of little help in early diagnosis and prevention. Most scientists around the world have applied some predication techniques that could help in resolving Alzheimer's disease, concentrating on risk factors and biomarkers. Biomarkers understood to be beneficial for recognition of Alzheimer's disease include amyloid beta (Aβ), P-tau, ApoE-ε4, T-tau, and Alzheimer-associated neuronal thread protein (AD7c-NTP) (Aggarwal, Shah, & Bennett, 2015; Almeida, Schultz, Austin, et al., 2015a, 2015b; Lim, Villemagne, Pietrzak, et al., 2015).

Interestingly, one of the most current techniques for the identification of Alzheimer's disease includes pathological detections, MRI of the brain, and genetic detections of some proteins. However, these methodologies have several drawbacks, such as being high cost and time-intensive. Therefore, the application of mathematical models could help in early predication and diagnosis of several brain diseases (Wang, Du, Zhang, et al., 2017; Wang, Zhang, Liu, et al., 2015; Zhang, Dong, Phillips, et al., 2015). Furthermore, ANNs have been identified as nonlinear adaptive processing systems that contain numerous artificial neurons that are available in a way that joined to each other. This plays a greater role in resolving several medical challenges because they portends the capacity to approximate any role with arbitrary precision (Hwang, Lee, Kim, et al., 2015; Tang, Wu, Huang, et al., 2013).

Conclusion and future direction

Neurocognitive diseases are chronic conditions affecting several individuals. As such, there is a need for development and deployment of AI technology in the early diagnosis, prognosis, and disease prediction. This chapter outlined various advances in ML research algorithms, platforms and several applications for cognitive disease diagnosis, prediction, risk assessment, classification, diagnosis, segmentation tasks, prognosis, and forecasting. It also addressed challenges linked with the use of ML to classify and group patients with neurocognitive diseases. In addition, the chapter discussed several research works that investigated the roles of data fusion. Although there is immense data available on using ML algorithms for the classification of neurocognitive diseases, more research is needed to investigate their applicability to other neurological conditions like delirium and dementia.

References

Adadi, A., & Berrada, M. (2018). Peeking inside the black-box: A survey on Explainable Artificial Intelligence (XAI). *IEEE Access*, *6*, 52138–52160.

Adetunji, J. B., Adetunji, C. O., & Olaniyan, O. T. (2021). African walnuts: a natural depository of nutritional and bioactive compounds essential for food and nutritional security in Africa. In O. O. Babalola (Ed.), *Food security and safety*. Cham: Springer.

Adetunji, C. O., Ajayi, O. O., Akram, M., Olaniyan, O. T., Chishti, M. A., Inobeme, A., Olaniyan, S., Adetunji, J. B., Olaniyan, M., & Awotunde, S. O. (2021). Medicinal plants used in the treatment of influenza a virus infections. In K. Dua, S. Nammi, D. Chang, D. K. Chellappan, G. Gupta, & T. Collet (Eds.), *Medicinal plants for lung diseases*. Singapore: Springer.

Adetunji, C. O., Akram, M., Olaniyan, O. T., Ajayi, O. O., Inobeme, A., Olaniyan, S., Hameed, L., & Adetunji, J. B. (2021b). Medicinal plants used in the treatment of pulmonary hypertension. In K. Dua, S. Nammi, D. Chang, D. K. Chellappan, G. Gupta, & T. Collet (Eds.), *Medicinal plants for lung diseases*. Singapore: Springer. https://doi.org/10.1007/978-981-33-6850-7_14.

Adetunji, C. O., Akram, M., Olaniyan, O. T., Ajayi, O. O., Inobeme, A., Olaniyan, S., Hameed, L., & Adetunji, J. B. (2021c). Targeting SARS-CoV-2 novel corona (COVID-19) virus infection using medicinal plants. In K. Dua, S. Nammi, D. Chang, D. K. Chellappan, G. Gupta, & T. Collet (Eds.), *Medicinal plants for lung diseases*. Singapore: Springer.

Adetunji, C. O., Inobeme, A., Olaniyan, O. T., Ajayi, O. O., Olaniyan, S., & Adetunji, J. B. (2021). Application of nanodrugs derived from active metabolites of medicinal plants for the treatment of inflammatory and lung diseases: recent advances. In K. Dua, S. Nammi, D. Chang, D. K. Chellappan, G. Gupta, & T. Collet (Eds.), *Medicinal plants for lung diseases*. Singapore: Springer.

Adetunji, C. O., & Varma, A. (2020). Biotechnological application of trichoderma: a powerful fungal isolate with diverse potentials for the attainment of food safety, management of pest and diseases, healthy planet, and sustainable agriculture. In C. Manoharachary, H. B. Singh, & A. Varma (Eds.), *Vol. 61. Trichoderma: Agricultural applications and beyond. soil biology*. Cham: Springer.

Aggarwal, N. T., Shah, R. C., & Bennett, D. A. (2015). Alzheimer's disease: unique markers for diagnosis & new treatment modalities. *The Indian Journal of Medical Research, 142*(4), 369–382.

Ahs, F., Kragel, P. A., Zielinski, D. J., Brady, R., & LaBar, K. S. (2015). Medial prefrontal pathways for the contextual regulation of extinguished fear in humans. *NeuroImage, 122*, 262–271.

Akram, M., Jabeen, F., Daniyal, M., Zainab, R., Haq, U. U., Adetunji, C. O., Egbuna, C., Ephraim-Emmanuel, B. C., Patrick-Iwuanyanwu, K. C., & Ogbo, A. B. (2020). Genetic engineering of novel products of health significance: Recombinant DNA technology. In C. Egbuna, & T. G. Dable (Eds.), *Functional foods and nutraceuticals*. Cham: Springer. https://doi.org/10.1007/978-3-030-42319-3_26.

Almeida, R. P., Schultz, S. A., Austin, B. P., et al. (2015a). Effect of cognitive reserve on age-related changes in cerebrospinal fluid biomarkers of Alzheimer disease. *JAMA Neurology, 72*(6), 699–706.

Almeida, R. P., Schultz, S. A., Austin, B. P., et al. (2015b). Cognitive reserve and age-related changes in Alzheimer disease. *JAMA Neurology, 72*(6), 935–938.

Amisha, Malik, P., Pathania, M., & Rathaur, V. K. (2019). Overview of artificial intelligence in medicine. *Journal of Family Medicine and Primary Care, 8*(7), 2328–2331.

Anticevic, A., Cole, M. W., Repovs, G., Murray, J. D., Brumbaugh, M. S., Winkler, A. M., Savic, A., Krystal, J. H., Pearlson, G. D., & Glahn, D. C. (2014). Characterizing thalamo-cortical disturbances in schizophrenia and bipolar illness. *Cerebral Cortex, 24*, 3116–3130.

Appleby, B. S., Nacopoulos, D., Milano, N., et al. (2013). A review: treatment of Alzheimer's disease discovered in repurposed agents. *Dementia and Geriatric Cognitive Disorders, 35*(1–2), 1–22.

Arbabshirani, M. R., Plis, S., Sui, J., & Calhoun, V. D. (2017). Single subject prediction of brain disorders in neuroimaging: Promises and pitfalls. *NeuroImage, 145*, 137–165.

Ayano, G., Yohannes, K., & Abraha, M. (2020). Epidemiology of attention-deficit/hyperactivity disorder (ADHD) in children and adolescents in Africa: A systematic review and meta-analysis. *Annals of General Psychiatry, 19*, 21.

Aymerich-Franch, L., & Bailenson, J. (2014). The use of doppelgangers in virtual reality to treat public speaking anxiety: a gender comparison. In *Proceedings of the international society for presence research annual conference* (pp. 173–186).

Balafar, M. A., Ramli, A. R., Saripan, M. I., & Mashohor, S. (2010). Review of brain MRI image segmentation methods. *Artificial Intelligence Review, 33*, 261–274.

Bemelmans, S. A. S. A., Tromp, K., Bunnik, E. M., et al. (2016). Psychological, behavioral and social effects of disclosing Alzheimer's disease biomarkers to research participants: A systematic review. *Alzheimer's Research & Therapy, 8*, 46.

Bentley, P., Ganesalingam, J., Carlton Jones, A. L., Mahady, K., Epton, S., Rinne, P., Sharma, P., Halse, O., Mehta, A., & Rueckert, D. (2014). Prediction of stroke thrombolysis outcome using CT brain machine learning. *NeuroImage: Clinical, 4*, 635–640.

Biddle, K. D., Uquillas, O., Jacobs, H. I. L., Zide, B., Kirn, D. R., Rentz, D. M., Psy, D., Johnson, K. A., Sperling, R. A., & Donovan, N. J. (2019). Social engagement and Amyloid-B-related cognitive decline incognitively normal older adults. *The American Journal of Geriatric Psychiatry, 27*, 1247–1256.

Bradley, M. M., Miccoli, L., Escrig, M. A., & Lang, P. J. (2008). The pupil as a measure of emotional arousal and autonomic activation. *Psychophysiology, 45*, 602–607.

Brady, A. P. (2017). Error and discrepancy in radiology: inevitable or avoidable? *Insights Into Imaging, 8*, 171–182.

Briganti, G., & Le Moine, O. (2020). Artificial intelligence in medicine: Today and tomorrow. *Frontiers in Medicine, 7*, 27.

Burge, J., Clark, V. P., Lane, T., Link, H., & Qiu, S. (2004). Bayesian classification of FMRI data: Evidence for altered neural networks in dementia. In *University of New Mexico, Tech. Rep. TR-CS-2004-28*.

Cellan-Jones, R. (*2014*). *Stephen Hawking warns artificial intelligence could end mankind*. BBC News. Archived from the original on 30 October 2015. Retrieved 30 October 2015.

Cenek, M., Hu, M., York, G., & Dahl, S. (2018). Survey of image processing techniques for brain pathology diagnosis: Challenges and opportunities. *Frontiers in Robotics and AI*, 5.

Cerman, E., Eraslan, M., & Cekic, O. (2015). Age-related macular degeneration and Alzheimer disease. *Turkish Journal of Medical Sciences*, *45*(5), 1004–1009.

Challis, E., Hurley, P., Serra, L., Bozzali, M., Oliver, S., & Cercignani, M. (2015). Gaussian process classification of Alzheimer's disease and mild cognitive impairment from resting-state FMRI. *NeuroImage*, *112*, 232–243.

Chaudhuri, K. R., Yates, L., & Martinez-Martin, P. (2005). The non-motor symptom complex of Parkinson's disease: a comprehensive assessment is essential. *Current Neurology and Neuroscience Reports*, *5*, 275–283.

Chen, L., Bentley, P., & Rueckert, D. (2017). Fully automatic acute ischemic lesion segmentation in DWI using convolutional neural networks. *NeuroImage: Clinical*, *15*, 633–643.

Cheng, S. T. (2017). Dementia caregiver burden: A research update and critical analysis. *Current Psychiatry Reports*, *19*(9), 64. https://doi.org/10.1007/s11920-017-0818-2. PMID: 28795386. PMCID: PMC5550537.

Côté, S., & Bouchard, S. (2005). Documenting the efficacy of virtual reality exposure with psychophysiological and information processing measures. *Applied Psychophysiology and Biofeedback*, *30*(3), 217–232.

Critchley, H. D. (2009). Psychophysiology of neural, cognitive and affective integration: fMRI and autonomic indicants. *International Journal of Psychophysiology*, *73*, 88–94.

Currie, G., Hawk, K. E., Rohren, E., Vial, A., & Klein, R. (2019). Machine learning and deep learning in medical imaging: Intelligent imaging. *Journal of Medical Imaging and Radiation Sciences*, *50*(4), 477–487.

Das, S., Sharangi, A. B., Egbuna, C., Jeevanandam, J., Ezzat, S. M., Adetunji, C. O., Tijjani, H., Olisah, M. C., Patrick-Iwuanyanwu, K. C., Adetunji, J. B., Ifemeje, J. C., Akram, M., Moboladji, B. M., & Onyeike, P. C. (2020). Health benefits of isoflavones found exclusively of plants of the Fabaceae Family. In C. Egbuna, & T. G. Dable (Eds.), *Functional foods and nutraceuticals*. Cham: Springer. https://doi.org/10.1007/978-3-030-42319-3_22.

Dash, S., Acharya, B. R., Mittal, M., Abraham, A., & Keleman, A. (2020). Deep learning techniques for biomedical and health informatics. In *Studies in big data series* Springer. ISSN:2197-6511, ISBN:978-3-030-33966-1.

Dash, S., Pani, S., Abraham, A., & Liang, Y. (2021). *Advanced soft computing techniques in data science, IoT and cloud computing, series: Studies in big data. 89* (p. 443). Springer. ISSN: 2197-6503, ISBN 978-3-030-75656-7, ISBN 978-3-030-75657-4 (eBook) https://doi.org/10.1007/978-3-030-75657-4.

Dash, S., Pani, S. K., Rodrigues, J. P. C., & Majhi, B. (2022). *Deep learning, machine learning and IoT in biomedical and health informatics: techniques and applications* (p. 382). CRC Press, Taylor & Francis group. ISBN 9780367544256.

Dash, S., Shakyawar, S. K., Sharma, M., & Kaushik, S. (2019). Big data in healthcare: management, analysis and future prospects. *Journal of Big Data*, *6*, 54.

Dash, S., & Subudhi, B. (2016). *Computational intelligence applications in bioinformatics* (pp. 1–514). IGI Global. ISBN13: 9781522504276, ISBN10: 1522504273, EISBN13: 9781522504283 https://doi.org/10.4018/978-1-5225-0427-6. USA.

Dash, S., Tripathy, B. K., & Rahman, A. (2017). *Modeling, analysis and application on nature-inspired metaheuristic algorithms* (pp. 1–538). IGI Global USA. https://doi.org/10.4018/978-1-5225-2857-9. ISBN13: 9781522528579.

Davenport, T., & Kalakota, R. (2019). The potential for artificial intelligence in healthcare. *Future Healthcare Journal*, *6*(2), 94–98.

Dias, R., & Torkamani, A. (2019). Artificial intelligence in clinical and genomic diagnostics. *Genome Medicine*, *11*(1), 70.

Diemer, J., Alpers, G. W., Peperkorn, H. M., Shiban, Y., & Mühlberger, A. (2015). The impact of perception and presence on emotional reactions: A review of research in virtual reality. *Frontiers in Psychology, 6*, 26.

Došilović, F. K., Brčić, M., & Hlupić, N. (2018). Explainable artificial intelligence: A survey. In *2018 41st International convention on information and communication technology, electronics and microelectronics (MIPRO)* (pp. 0210–0215). IEEE.

Du, G., Lewis, M. M., Kanekar, S., Sterling, N. W., He, L., Kong, L., Li, R., & Huang, X. (2017). Combined diffusion tensor imaging and apparent transverse relaxation rate differentiate Parkinson disease and atypical parkinsonism. *AJNR. American Journal of Neuroradiology, 38*, 966–972.

Dwyer, D. B., Falkai, P., & Koutsouleris, N. (2018). Machine learning approaches for clinical psychology and psychiatry. *Annual Review of Clinical Psychology, 14*, 91–118.

Egbuna, C., Gupta, E., Ezzat, S. M., Jeevanandam, J., Mishra, N., Muhammad Akram, N., Sudharani, C. O., Adetunji, P. S., Ifemeje, J. C., Deepak, M., Bhavana, A., Angelo Mark, P., Walag, R. A., Adetunji, J. B., Laila, U., Olisah, M. C., & Onyekere, P. F. (2020). Aloe species as valuable sources of functional bioactives. In C. Egbuna, & G. Dable Tupas (Eds.), *Functional foods and nutraceuticals*. Cham: Springer.

EL Azami, M., Hammers, A., Jung, J., Costes, N., Bouet, R., & Lartizien, C. (2016). Detection of lesions underlying intractable epilepsy on T1-weighted MRI as an outlier detection problem. *PLoS One, 11*, e0161498.

Emblem, K. E., Due-Tonnessen, P., Hald, J. K., Bjornerud, A., Pinho, M. C., Scheie, D., Schad, L. R., Meling, T. R., & Zoellner, F. G. (2014). Machine learning in preoperative glioma MRI: survival associations by perfusion-based support vector machine outperforms traditional MRI. *Journal of Magnetic Resonance Imaging, 40*, 47–54.

Ewers, M., Walsh, C., Trojanowski, J. Q., Shaw, L. M., Petersen, R. C., Jack, C. R., Feldman, H. H., Bokde, A. L., Alexander, G. E., Scheltens, P., et al. (2012). Prediction of conversion from mild cognitive impairment to Alzheimer's disease dementia based upon biomarkers and neuropsychological test performance. *Neurobiology of Aging, 33*(7), 1203–1214.

Falkmer, T., Anderson, K., Falkmer, M., & Horlin, C. (2013). Diagnostic procedures in autism spectrum disorders: A systematic literature review. *European Child & Adolescent Psychiatry, 22*, 329–340.

Fisher, R. S., Van Emde Boas, W., Blume, W., Elger, C., Genton, P., Lee, P., & EngeL, J., JR. (2005). Epileptic seizures and epilepsy: Definitions proposed by the International League Against Epilepsy (ILAE) and the International Bureau for Epilepsy (IBE). *Epilepsia, 46*, 470–472.

Fitzpatrick, K. K., Darcy, A., & Vierhile, M. (2017). Delivering cognitive behavior therapy to young adults with symptoms of depression and anxiety using a fully automated conversational agent (Woebot): A randomized controlled trial. *JMIR Mental Health, 4*, e19.

Fulmer, R., Joerin, A., Gentile, B., Lakerink, L., & Rauws, M. (2018). Using psychological artificial intelligence (Tess) to relieve symptoms of depression and anxiety: Randomized controlled trial. *JMIR Mental Health, 5*(4). https://doi.org/10.2196/mental.9782, e64. 30545815. PMC6315222.

Gartlehner, G., Wagner, G., Matyas, N., Titscher, V., Greimel, J., Lux, L., Gaynes, B. N., Viswanathan, M., & Patel, S. (*2017*). Pharmacological and non-pharmacological treatments for major depressive disorder: Review of systematic reviews. *BMJ Open*, 7(*6*), e014912.

Gaugler, J., James, B., Johnson, T., et al. (2016). Alzheimer's disease facts and figures. *Alzheimers Dement, 12*(4), 459–509.

Gautam, R., & Sharma, M. (2020). Prevalence and diagnosis of neurological disorders using different deep learning techniques: A meta-analysis. *Journal of Medical Systems, 44*, 49.

Genç, Ö. (2019). *Notes on Artificial Intelligence, Machine Learning and Deep Learning for curious people. Available at* https://towardsdatascience.com/notes-on-artificial-intelligence-ai-machine-learning-ml-and-deep-learning-dl-for-56e51a2071c2.

Goldberg, S. I., Niemierko, A., & Turchin, A. (2008). Analysis of data errors in clinical research databases. In *2008. AMIA. Annual Symposium proceedings. AMIA Symposium* (pp. 242–246).

Goldenberg, M. M. (2012). Multiple sclerosis review. *P & T: A Peer-Reviewed Journal for Formulary Management, 37*, 175–184.

Gomar, J. J., Bobes-Bascaran, M. T., Conejero-Goldberg, C., Davies, P., Goldberg, T. E., Initiative, A. D. N., et al. (2011). Utility of combinations of biomarkers, cognitive markers, and risk factors to predict conversion from mild cognitive impairment to Alzheimer disease in patients in the Alzheimer's disease neuroimaging initiative. *Archives of General Psychiatry, 68*(9), 961–969.

Goodfellow, I., Bengio, Y., & Courville, A. (2016). Deep learning. In P. Kantor (Ed.), *Foundations of statistical natural language processing* (1st ed.). MIT Press.

Haass, C., & Selkoe, D. J. (2007). Soluble protein oligomers in neurodegeneration: lessons from the Alzheimer' s amyloid beta-peptide. *Nature Reviews. Molecular Cell Biology, 8*(2), 101–112.

Hatano, S. (1976). Experience from a multicentre stroke register: a preliminary report. *Bulletin of the World Health Organization, 54*, 541–553.

Herrmann, N., Harimoto, T., Balshaw, R., et al. (2015). Risk factors for progression of Alzheimer disease in a Canadian population: The Canadian outcomes study in dementia (COSID). *Canadian Journal of Psychiatry, 60*(4), 189–199.

Hosny, A., Parmar, C., Quackenbush, J., Schwartz, L. H., & Aerts, H. (2018). Artificial intelligence in radiology. *Nature reviews. Cancer, 18*(8), 500–510.

Hwang, Y. N., Lee, J. H., Kim, G. Y., et al. (2015). Classification of focal liver lesions on ultrasound images by extracting hybrid textural features and using an artificial neural network. *Bio-medical Materials and Engineering, 26*(s1), S1599–S1611.

Insel, T. R. (2014). The NIMH Research Domain Criteria (RDoC) Project: Precision medicine for psychiatry. *The American Journal of Psychiatry, 171*, 395–397.

Kalmady, S. V., Greiner, R., Agrawal, R., Shivakumar, V., Narayanaswamy, J. C., Brown, M. R. G., Greenshaw, A. J., Dursun, S. M., & Venkatasubramanian, G. (2019). Towards artificial intelligence in mental health by improving schizophrenia prediction with multiple brain parcellation ensemble-learning. *NPJ Schizophrenia, 5*, 2.

Kamal, H., Lopez, V., & Sheth, S. A. (2018). Machine learning in acute ischemic stroke neuroimaging. *Frontiers in Neurology, 9*, 945.

Kang, J. H., Weuve, J., & Grodstein, F. (2004). Postmenopausal hormone therapy and risk of cognitive decline in community-dwelling aging women. *Neurology, 63*(1), 101–107.

Kautzky, A., Vanicek, T., Philippe, C., Kranz, G. S., Wadsak, W., Mitterhauser, M., Hartmann, A., Hahn, A., Hacker, M., Rujescu, D., Kasper, S., & Lanzenberger, R. (2020). Machine learning classification of ADHD and HC by multimodal serotonergic data. *Translational Psychiatry, 10*, 104.

Kayid, A. (2020). *The role of Artificial Intelligence in future technology*. Research gate. https://doi.org/10.13140/RG.2.2.12799.23201.

Korteling, J. E., van de Boer-Visschedijk, G. C., Blankendaal, R. A. M., Boonekamp, C., & Eikelboom, A. R. (2021). Human- versus artificial intelligence. *Frontiers in Artificial Intelligence, 2021*. https://doi.org/10.3389/frai.2021.622364.

Kotsopoulos, I. A., Van Merode, T., Kessels, F. G., De Krom, M. C., & Knottnerus, J. A. (2002). Systematic review and meta-analysis of incidence studies of epilepsy and unprovoked seizures. *Epilepsia, 43*, 1402–1409.

Kourtis, L. C., Regele, O. B., Wright, J. M., & Jones, G. B. (2019). Digital biomarkers for Alzheimer's disease: The mobile/wearable devices opportunity. *NPJ Digital Medicine*, 1–9.

Larson, J., & Angwin, J. (*2016*). *How we analyzed the COMPAS recidivism algorithm. ProPublica.* Archived from the original on 29 April 2019. Retrieved 19 June 2020.
Lee, E.-J., Kim, Y.-H., Kim, N., & Kang, D.-W. (2017). Deep into the brain: Artificial intelligence in stroke imaging. *Journal of Stroke*, *19*, 277–285.
Li, Y., Qian, Z., Xu, K., Wang, K., Fan, X., Li, S., Jiang, T., Liu, X., & Wang, Y. (2018). MRI features predict p53 status in lower-grade gliomas via a machine-learning approach. *NeuroImage: Clinical*, *17*, 306–311.
Lim, Y. Y., Villemagne, V. L., Pietrzak, R. H., et al. (2015). APOE ε4 moderates amyloid-related memory decline in preclinical Alzheimer's disease. *Neurobiology of Aging*, *36*(3), 1239–1244.
Liu, M., Zhang, D., Shen, D., Initiative, A. D. N., et al. (2012). Ensemble sparse classification of Alzheimer's disease. *NeuroImage*, *60*(2), 1106–1116.
Luxton, D. D. (Ed.). (2015). *Artificial intelligence in behavioral and mental health care. London: Academic Press.*
Maj, M. (2013). "Clinical judgment" and the DSM-5 diagnosis of major depression. *World Psychiatry: Official Journal of the World Psychiatric Association (WPA)*, *12*, 89–91.
Makridakis, S. (2017). The forthcoming Artificial Intelligence (AI) revolution: Its impact on society and firms. *Futures*, *90*, 46–60. https://doi.org/10.1016/j.futures.2017.03.006.
Malley, B., Ramazzotti, D., & Wu, J. T. (2016). Data pre-processing. In *Secondary analysis of electronic health records [internet]*. Cham (CH): Springer. https://doi.org/10.1007/978-3-319-43742-2_12. Chapter 12. Available from: https://www.ncbi.nlm.nih.gov/books/NBK543629/.
Marras, C., & Tanner, C. (2004). Epidemiology of Parkinson's disease. In R. Watts, & W. E. Koller (Eds.), *Movement disorders, neurologic principles and practice* (*2nd ed.*). New York: McGraw Hill.
Meeks, T. W., & Jeste, D. V. (2009). Neurobiology of wisdom. *Archives of General Psychiatry*, *66*, 355–365.
Mehler, B., Reimer, B., Coughlin, J., & Dusek, J. (2009). Impact of incremental increases in cognitive workload on physiological arousal and performance in young adult drivers. *Transportation Research Record Journal of the Transportation Research Board*, *2138*, 6–12.
Moradi, E., Gaser, C., Huttunen, H., Tohka, J., & Alzheimerr's, D. N. I. (2014). MRI based dementia classification using semi-supervised learning and domain adaption. *MICCAI 2014 workshop proceedings, challange on computer-aided diagnosis of dementia, based on structural MRI data, September 14-18 2014, Boston, USA* (pp. 65–73). http://caddementia.grand-challenge.org/workshop/.
Munsell, B. C., Wee, C. Y., Keller, S. S., Weber, B., Elger, C., DA Silva, L. A., Nesland, T., Styner, M., Shen, D., & Bonilha, L. (2015). Evaluation of machine learning algorithms for treatment outcome prediction in patients with epilepsy based on structural connectome data. *NeuroImage*, *118*, 219–230.
Ngai, I., Tully, E. C., & Anderson, P. L. (2015). The course of the working alliance during virtual reality and exposure group therapy for social anxiety disorder. *Behavioural and Cognitive Psychotherapy*, *43*(2), 167–181.
Ni, H., Zhou, L., Ning, X., & Wang, L. (2016). Exploring multifractal-based features for mild Alzheimer's disease classification. *Magnetic Resonance in Medicine*, *76*, 259–269.
Niel, O., & Bastard, P. (2019). Artificial intelligence in nephrology: core concepts, clinical applications, and perspectives. *American Journal of Kidney Diseases: The Official Journal of the National Kidney Foundation*, *74*(6), 803–810.
Nutt, J. G., & Wooten, G. F. (2005). Clinical practice. Diagnosis and initial management of Parkinson's disease. *The New England Journal of Medicine*, *353*, 1021–1027.
Olatunji, B. O., Cisler, J. M., & Deacon, B. J. (2010). Efficacy of cognitive behavioral therapy for anxiety disorders: a review of meta-analytic findings. *Psychiatric Clinics*, *33*(3), 557–577.

Pani, S., Dash, S., dos Santos, W., Bukhari, S. A. C., & Flammini, F. (2021). *Assessing covid-19 and other pandemics and epidemics using computational modelling and data analysis* (p. 431). Springer. ISBN-13: 978-3030797522.

Papakostas, G. I. (2009). Managing partial response or nonresponse: switching, augmentation, and combination strategies for major depressive disorder. *The Journal of Clinical Psychiatry, 70*(Suppl 6), 16–25.

Patel, M. J., Andreescu, C., Price, J. C., Edelman, K. L., Reynolds, C. F., 3rd, & Aizenstein, H. J. (2015). Machine learning approaches for integrating clinical and imaging features in late-life depression classification and response prediction. *International Journal of Geriatric Psychiatry, 30*, 1056–1067.

Paul, D., Sanap, G., Shenoy, S., Kalyane, D., Kalia, K., & Tekade, R. K. (2021). Artificial intelligence in drug discovery and development. *Drug Discovery Today, 26*(1), 80–93.

Payan, A., & Montana, G. (2015). *Predicting Alzheimer's disease: a neuroimaging study with 3d convolutional neural networks*. arXiv preprint arXiv:1502.02506.

Pedersen, M., Verspoor, K., Jenkinson, M., Law, M., Abbott, D. F., & Jackson, G. D. (2020). Artificial intelligence for clinical decision support in neurology. *Brain Communications, 2*(2), fcaa096.

Poeschl, S. (2017). Virtual reality training for public speaking—A QUEST-VR framework validation. *Frontiers in ICT, 4*, 13.

Querino, E., Dos Santos, L., Ginani, G., Nicolau, E., Miranda, D., Romano-Silva, M., & Malloy-Diniz, L. (2015). Cognitive effort and pupil dilation in controlled and automatic processes. *Translational Neuroscience, 6*, 168–173.

Rahman, M. M., Antani, S. K., Demner-Fushman, D., & Thoma, G. R. (2015). Biomedical image representation approach using visualness and spatial information in a concept feature space for interactive region-of-interest-based retrieval. *Journal of Medical Imaging, 2*(4), 046502.

Reimer, B., & Mehler, B. (2011). The impact of cognitive workload on physiological arousal in young adult drivers: A field study and simulation validation. *Ergonomics, 54*, 932–942.

Robertson, I. H. (2013). A noradrenergic theory of cognitive reserve: implications for Alzheimer's disease. *Neurobiology of Aging, 34*(1), 298–308.

Roli, O. I., Adetunji, C. O., Mishra, P., Adetunji, J. B., & Mishra, R. R. (2020). Caffeine: Nutraceutical and health benefit of caffeine-containing commodities and products. In P. Mishra, R. R. Mishra, & C. O. Adetunji (Eds.), *Innovations in food technology*. Singapore: Springer.

Saccà, V., Sarica, A., Novellino, F., Barone, S., Tallarico, T., Filippelli, E., Granata, A., Chiriaco, C., Bruno Bossio, R., Valentino, P., & Quattrone, A. (2019). Evaluation of machine learning algorithms performance for the prediction of early multiple sclerosis from resting-state FMRI connectivity data. *Brain Imaging and Behavior, 13*, 1103–1114.

Sakai, K., & Yamada, K. (2019). Machine learning studies on major brain diseases: 5-year trends of 2014-2018. *Japanese Journal of Radiology, 37*, 34–72.

Samuels, E. R., & Szabadi, E. (2008). Functional neuroanatomy of the noradrenergic locus coeruleus: its roles in the regulation of arousal and autonomic function part II: Physiological and pharmacological manipulations and pathological alterations of locus coeruleus activity in humans. *Current Neuropharmacology, 6*, 254–285.

Schmidt, P., Biessmann, F., & Teubner, T. (2020). Transparency and trust in artificial intelligence systems. *Journal of Decision Systems, 29*(4), 260–278. https://doi.org/10.1080/12460125.2020.1819094.

Schmidt, H. D., Shelton, R. C., & Duman, R. S. (2011). Functional biomarkers of depression: diagnosis, treatment, and pathophysiology. *Neuropsychopharmacology, 36*, 2375–2394.

Schnack, H. G., Nieuwenhuis, M., Van Haren, N. E., Abramovic, L., Scheewe, T. W., Brouwer, R. M., Hulshoff Pol, H. E., & Kahn, R. S. (2014). Can structural MRI aid in clinical classification? A machine learning study in two independent samples of patients with schizophrenia, bipolar disorder and healthy subjects. *NeuroImage, 84*, 299–306.

Senders, J. T., Arnaout, O., Karhade, A. V., Dasenbrock, H. H., Gormley, W. B., Broekman, M. L., & Smith, T. R. (2018). Natural and artificial intelligence in neurosurgery: A systematic review. *Neurosurgery, 83*, 181–192.

Shneiderman, B. (2020). Human-centered artificial intelligence: reliable, safe & trustworthy. *International Journal of Human Computer Interaction, 36*(6), 495–504. https://doi.org/10.1080/10447318.2020.1741118.

Simpson, J. R. (*2014*). DSM-5 and neurocognitive disorders. *The Journal of the American Academy of Psychiatry and the Law, **42**(2), 159–164.*

Sollini, M., Bartoli, F., Marciano, A., et al. (2020). Artificial intelligence and hybrid imaging: The best match for personalized medicine in oncology. *European Journal of Hybrid Imaging, 4*(24). https://doi.org/10.1186/s41824-020-00094-8.

Song, D. Y., Kim, S. Y., Bong, G., Kim, J. M., & Yoo, H. J. (2019). The use of artificial intelligence in screening and diagnosis of autism spectrum disorder: A literature review. *Soa Chongsonyon Chongsin Uihak, 30*, 145–152.

Stead, W. W. (2018). Clinical implications and challenges of artificial intelligence and deep learning. *JAMA, 320*, 1107–1108.

Suk, H. I., Lee, S. W., & Shen, D. (2014). Hierarchical feature representation and multimodal fusion with deep learning for AD/MCI diagnosis. *Neuroimage, 101*, 569–582. https://doi.org/10.1016/j.neuroimage.2014.06.077.

Tagare, H. D., Jaffe, C. C., & Duncan, J. (1997). Medical image databases: a content-based retrieval approach. *Journal of the American Medical Informatics Association, 4*(3), 184–198.

Tai, M. C. (2020). The impact of artificial intelligence on human society and bioethics. *Tzu Chi Medical Journal, 32*(4), 339–343.

Tang, J., Wu, L., Huang, H., et al. (2013). Back propagation artificial neural network for community Alzheimer's disease screening in China. *Neural Regeneration Research, 8*(3), 270–276.

Taubmann, O., Berger, M., Bögel, M., et al. (2018). Computed tomography. 2018 Aug 3. In A. Maier, S. Steidl, V. Christlein, et al. (Eds.), *Medical imaging systems: An introductory guide [internet]*. Cham (CH): Springer. https://doi.org/10.1007/978-3-319-96520-8_8. Chapter 8. Available from: https://www.ncbi.nlm.nih.gov/books/NBK546157/.

Tierney, M., Szalai, J., Snow, W., Fisher, R., Nores, A., Nadon, G., Dunn, E., & George-Hyslop, P. S. (1996). Prediction of probable Alzheimer's disease in memory-impaired patients Aprospective longitudinal study. *Neurology, 46*(3), 661–665.

Uddin, L. Q., Dajani, D. R., Voorhies, W., Bednarz, H., & Kana, R. K. (2017). Progress and roadblocks in the search for brain-based biomarkers of autism and attention-deficit/hyperactivity disorder. *Translational Psychiatry*, 7, e1218.

Vogelzangs, N., Beekman, A. T., Dik, M. G., Bremmer, M. A., Comijs, H. C., Hoogendijk, W. J., Deeg, D. J., & Penninx, B. W. (2009). Late-life depression, cortisol, and the metabolic syndrome. *The American Journal of Geriatric Psychiatry, 17*, 716–721.

Wang, S. H., Du, S., Zhang, Y., et al. (2017). Alzheimer's disease detection by Pseudo Zernike moment and linear regression classification. *CNS & Neurological Disorders Drug Targets, 16*(1), 11–15.

Wang, L., Roe, C. M., Snyder, A. Z., et al. (2012). Alzheimer disease family history impacts resting state unctional connectivity. *Annals of Neurology, 72*(4), 571–577.

Wang, S., Zhang, Y., Liu, G., et al. (2015). Detection of Alzheimer's disease by three-dimensional displacement field estimation in structural magnetic resonance imaging. *Journal of Alzheimers Disease, 50*(1), 233–248.

Weuve, J., Hebert, L. E., Scherr, P. A., et al. (2015). Prevalence of Alzheimer disease in US states. *Epidemiology, 26*(1), e4–e6.

Wickramasuriya, D. S., Amin, M. R., & Faghih, R. T. (2019). Skin conductance as a viable alternative for closing the deep brain stimulation loop in neuropsychiatric disorders. *Frontiers in Neuroscience*, *13*.

Yang, C. H., Huang, Y. J., Liu, A., Rong, Y., & Lee, T. F. (2015). Data acquisition and processing in biology and medicine. *BioMed Research International*, *2015*, 538031.

Yasaka, K., & Abe, O. (2018). Deep learning and artificial intelligence in radiology: Current applications and future directions. *PLoS Medicine*, *15*, e1002707.

Yoshida, K., Shimizu, Y., Yoshimoto, J., Takamura, M., Okada, G., Okamoto, Y., Yamawaki, S., & Doya, K. (2017). Prediction of clinical depression scores and detection of changes in whole-brain using resting-state functional MRI data with partial least squares regression. *PLoS One*, *12*, e0179638.

Zhang, Y., Dong, Z., Phillips, P., et al. (2015). Detection of subjects and brain regions related to Alzheimer's disease using 3D MRI scans based on eigenbrain and machine learning. *Frontiers in Computational Neuroscience*, *9*(9), 66.

Zhang, D., Shen, D., Initiative, A. D. N., et al. (2012). Multi-modal multi-task learning for joint prediction of multiple regression and classification variables in Alzheimer's disease. *NeuroImage*, *59*(2), 895–907.

Zhao, Y., Healy, B. C., Rotstein, D., Guttmann, C. R., Bakshi, R., Weiner, H. L., Brodley, C. E., & Chitnis, T. (2017). Exploration of machine learning techniques in predicting multiple sclerosis disease course. *PLoS One*, *12*, e0174866.

Zhou, M., Scott, J., Chaudhury, B., Hall, L., Goldgof, D., Yeom, K. W., Iv, M., OU, Y., Kalpathy-Cramer, J., NapeL, S., Gillies, R., Gevaert, O., & Gatenby, R. (2018). Radiomics in brain tumor: image assessment, quantitative feature descriptors, and machine-learning approaches. *AJNR. American Journal of Neuroradiology*, *39*, 208–216.

CHAPTER 13

Clinical applications of deep learning in neurology and its enhancements with future predictions

Olugbemi T. Olaniyan[a], Charles O. Adetunji[b], Ayobami Dare[c], Olorunsola Adeyomoye[d], Mayowa J. Adeniyi[e], and Alex Enoch[f]
[a]Department of Physiology, Laboratory for Reproductive Biology and Developmental Programming, Rhema University, Aba, Nigeria
[b]Department of Microbiology, Applied Microbiology, Biotechnology and Nanotechnology Laboratory, Edo State University Uzairue, Iyamho, Nigeria
[c]Department of Physiology, School of Laboratory Medicine and Medical Sciences, College of Health Sciences, Westville Campus, University of KwaZulu-Natal, Durban, South Africa
[d]Department of Physiology, University of Medical Sciences, Ondo, Nigeria
[e]Department of Physiology, Edo State University Uzairue, Iyamho, Nigeria
[f]Department of Human Physiology, Ahmadu Bello University Zaria, Zaria, Nigeria

Introduction

In the last few decades, there have been several technological developments in health care, including growth of biomedical data and the introduction of novel, high-performance computing tools to assist in evaluating this large volume of data. This has led to revolution in the work performed by healthcare professionals. Computer systems with increased power and speed present both challenges and opportunities for healthcare professionals and researchers. Machine learning techniques play a vital role in disease identification, evaluation, and analysis and their eventual translation into novel data-driven medical tools can enhance speed and quality of care (Dash, Chakraborty, Giri, Pani, & Frnda, 2021; Jensen, Jensen, & Brunak, 2012; Kohli, Summers, & Geis, 2017; Luo, Wu, Gopukumar, & Zhao, 2016).

Moreover, recent representation learning techniques have allowed the automatic identification of representations required to derive new knowledge from raw data (Bengio, Courville, & Vincent, 2013). For example, deep learning algorithms entail representation learning that uses nonlinear functions to transform raw input data into more advanced structures that allow for the recognition of novel patterns (LeCun, Bengio, & Hinton, 2015). The great bedrock of a lot of success in modern engineering advances have

Artificial Intelligence for Neurological Disorders
https://doi.org/10.1016/B978-0-323-90277-9.00016-X

been witnessed from self-driving cars and face recognition to language translation and chat-bots toward driving scene understanding: a dataset for learning driver (Gilani & Mian, 2017; Li, Lin, Shen, Brandt, & Hua, 2015; Maqueda, Loquercio, Gallego, Garcia, & Scaramuzza, 2018; Ramanishka, Chen, Misu, & Saenko, 2018; Wu, Schuster, Chen, et al., 2016; Zhang et al., 2018).

Interestingly, the application of deep learning algorithms has played a crucial role in several developments in medicine and other commercial research (Adetunji & Varma, 2020; Adetunji, Adetunji, & Olaniyan, 2021; Adetunji, Ajayi, et al., 2021; Adetunji et al., 2021a, 2021b; Adetunji, Inobeme et al., 2021; Akram et al., 2020; Egbuna et al., 2020; Roli, Adetunji, Mishra, Adetunji, & Mishra, 2020). This chapter gives a detailed account of advances in the adoption of artificial intelligence (AI) applications to enhance neurological techniques in clinical settings, such as medical imaging, translational bioinformatics, cheminformatics, pervasive sensing, public health, and medical informatics. It also provides a critical analysis of potential pitfalls of these techniques.

Neural network systems, biomarkers, and physiological signals

Several machine learning techniques have been designed to analyze data in medicine. For example, high-throughput-based approaches have been designed for cancer research. Wong, Zhu, and Yip (2009) proposed a model based on Biomarker Association Networks to classify cancer. Their model was comparable to existing ones such as K-nearest neighborhood, Fisher discriminant analysis, radial basis function kernel, Bayesian network, and support vector machines (SVMs) with linear kernel.

Alzheimer's disease is the most prevalent neurodegenerative disease and has a high rate of misdiagnosis. This necessitates highly intensified research efforts in a bid to surmount diagnostic challenges. For example, to simplify the classification of Alzheimer's disease, Aljović, Badnjević, and Gurbeta (2016) designed and presented an artificial neural network (ANN) system. This system uses biomarkers from cerebrospinal fluid such as tau-phospho, Aβ40, albumin ratio, Aβ42, and tau-total.

In an attempt to classify contact dermatitis signature genes (Fortino et al., 2020), 89 patch-test positive reaction tissue samples to two irritants and four contact allergens were investigated and analyzed via random forest classification and co-expression network analysis with the aim of discovering potential biomarkers. The results showed that random forest classification was able to identify *PKMYT*, *CD47*, *HISTH1A*, *BATF*, *FASLG*,

RGS16, SYNPO, SELE, PTPN7, WARS, PRC1, EXO1, RRM2, PBK, RAD54L, KIFC1, SPC25, TPX2, TPX2, DLGAP5, IL37, and *CH25H* as potential biomarkers that can be used to distinguish allergic and irritant contact dermatitis in human skin.

Tibshirani, Hastie, Narasimhan, and Chu (2002) developed a nearest shrunken centroid model to identify subsets of genes that typify leukemias and small round blue cell tumors. Khan et al. (2001) designed an approach for classifying cancer to strata based on their gene expression. Training of ANNs with the aid of small round blue cell tumor indicates the promising role of these methods in diagnosing and treating tumors.

Neurological techniques, biomedical informatics, and computational neurophysiology

Neurological techniques

There are several neurological techniques available. Some of them are non-invasive, including electroencephalography, which is a non-invasive technique that measures brain electrical activity from the surface. Electrical phenomenon taking place in the brain is converted to discernible signals. Electroencephalography is used both physiologically and diagnostically. Physiologically, electroencephalography can be used to monitor changes in the brain during different life phases. For instance, electroencephalographic sleep analysis can deduce brain development timing (Campbell, Kraus, Burright, & Feinberg, 2016). With electroencephalography, a non-epileptic seizure can be differentiated from an epileptic seizure, and catatonia can be distinguished from delirium. Other conditions that the technique can diagnose include sleep disorder, epilepsy, brain dysfunction, brain damage resulting from a head injury, encephalitis, and more.

Forms of electroencephalography are quantitative electroencephalography and stereo electroencephalography. The former deals with the analysis of electroencephalography data numerically (Kececi & Degirmenci, 2008). Stereo electroencephalography is concerned with the use of depth electrodes to measure brain electrical activities. Insertion of electrodes directly into the brain through the sectioned skull may be required, especially in evaluating epilepsy surgery. This craniotomy-related technique is known as electrocorticography. Unlike electroencephalography, electrocorticography is an invasive electrophysiological technique that measures cortical electrical activity. The technique was first introduced by Surgeons Wilser Penfiels and Herbert Jasper in 1950 (Palmini, 2006).

Functional magnetic resonance imaging (fMRI) is another non-invasive method for discerning brain electrical activity. This technique uses a magnetic field and is based on the change in blood flow (Logothetis, Pauls, Auguth, Trinath, & Oeltermann, 2001). Positron emission tomography (PET) uses radioactivity referred to as radiotracers to record changes in metabolic activities (Bailey, Townsend, Valk, & Maisy, 2005). Magnetoencephalography can record brain electrical activity using magnetic fields produced by electric currents (Boto et al., 2018). Apart from transcranial direct current stimulation and transcranial magnetic stimulation, single-unit recording is another invasive method that uses microelectrodes inserted into the brain to record electrical activities.

Biomedical informatics

Telemedicine, the dissemination of health information through telecommunication systems, is increasingly being used. Patients in remote locations can now access quality health care and prescriptions without necessarily traveling to hospitals or clinics. Biomedical informatics is an integral part of telemedicine. It is an interdisciplinary field that deals with the improvement of health using computation models. It is a rallying point for disciplines such as medicine, computer science, engineering, and mathematics (Nadri, Rahimi, Timpka, & Sedghi, 2017). Biomedical informatics is now being used in neurophysiology, neurodiagnosis, brain mapping, and simulation. Medical informatics also includes modern applications of neuroinformatic and cognitive informatics in the fields of brain mapping and emulation, where the concept is used in the design and development of computational models of the nervous system and software for diagnosis of neurological disorders and formation of patient information databases.

Computational neurophysiology

Computational neurophysiology is a multidisciplinary field that studies the brain's functions using the information processing attributes of the structures that make up the nervous system. It is concerned with the analysis of brain data to produce practical applications. Types of brain models are realistic brain models and simplifying brain models. Realistic brain models represent every aspect of the brain up to the cellular level, and as such, they are costly to create (Sejnowski, Koch, & Churchland, 1988). Simplifying brain models regulate model scope to evaluate physical attributes of the nervous system (Sejnowski et al., 1988).

Data and image acquisition

Data acquisition for deep learning techniques encompasses collecting, exploring, and cleaning the data required to train the system and develop the algorithm necessary to execute a specific task. These processes are needed to normalize and increase data quality for neural network training. The data are usually qualitative (e.g., images) or quantitative (e.g., pressure) (Chennubhotla et al., 2017). Data acquisition is a time-intensive step in AI, but it is vital to any classical training. However, some strategies have been suggested to promote quality datasets (Lambin et al., 2012) while significant advances are being made to ensure that data curation is done automatically (Thirumuruganathan, Tang, & Ouzzani, 2018). Data acquisition involves the selection of relevant features (such as sequences, phases, planes) and removing artifacts from images from a picture and archiving communication system (PACS). Data exploration evaluates the qualitative and quantitative characteristics of raw data to ensure that they possess specific features needed to develop a relevant model (Montagnon et al., 2020).

AI systems have brought significant improvements to radiological imaging by producing high-quality images that provide data used to train the system. AI contributes by reducing ionizing radiation dose, which is required for good output, for imaging procedures such as X-ray, computerized tomography (CT) scan, and nuclear imaging. Deep learning helps to improve accuracy and reduce the time taken to interpret a scan by replicating and automating the visual tasks done by the radiologist, thereby improving diagnosis and clinical decisions. Deep learning is currently being used for image reconstruction from sparse and under-sampled data to reduce inherent noise (artifacts) that occur during physical acquisition of images and maintain a better representation of the anatomical structures and abnormalities. AI is important in image acquisition for distinguishing a higher-resolution image from a lower-resolution image. To achieve this, images obtained by low radiation or sparsely acquired scan can be improved by using neural networks to generate full radiation equivalent images or fully acquired scans. However, this procedure needs to be validated (Harvey & Topol, 2020).

Artificial intelligence and deep learning

The medical discipline is rapidly becoming data-intensive, resulting from the large amount of information derived from several physiology and pathological indices generated during diagnosis and treatment. The high volume and heterogeneous nature of these data pose significant benefits. They can be used to extract reasonable information and develop models that could change the mode of disease diagnosis and progression, characterize disease conditions, and even treat diseases, all within the privacy-critical environment. The effective use of data to perform a specific task without human interference brought about the concept of AI, which includes deep learning and machine learning.

AI is a branch of science that uses a computer to execute tasks regarded as uniquely human (i.e., using the computer to mimic human behavior). The concept of AI was first described in 1956 by John McCarthy as the science of developing intelligent machines, and this concept has been widely applied in medical practice (CB Insights Research, 2017). AI has led to a paradigm shift in medical practice due to the large volume of data and rapid advancement in analytical techniques. AI is not meant to replace physicians but to improve decision-making in clinical practice. Different AI techniques can reveal clinically relevant ideas embedded in huge amounts of data, which can assist the physician in making proper medical decisions (1–3). The common AI techniques used for medical application include machine learning techniques, which are used to analyze structured data such as images (e.g., classical SVM, neural network), deep learning which is a recent form of machine learning, and natural language processing, which is used for extracting information from unstructured data (e.g., medical notes) to improved structured data (Darcy, Louie, & Roberts, 2016; Goodfellow, Bengio, & Courville, 2016).

The use of AI in health care (e.g., neurology) has been associated with significant benefits because AI utilizes advanced algorithms to learn features from many neurological data and later uses the information obtained to improve clinical practice. These vast benefits result from the adequate training of AI systems with data (in the form of, but not limited to, images, physical evaluation, and laboratory tests) to identify similar groups of subjects, the relationship between subject features, and relevant results (Administration UFaD, 2013). An AI system can minimize diagnostic and treatment errors inevitable in human clinical practice (Jha & Topol, 2016; Winters et al., 2012). The system can collect helpful information from a large patient population to assist in making real-time inferences for health risk alerts and health outcome predictions (Neill, 2013).

Deep learning refers to new techniques (an extension of the classical neural network) that in recent years have exhibited appreciable gain compared to currently available machine learning algorithms. For instance, deep learning techniques have transformed image classification and speech recognition due to their high flexibility and accuracy (LeCun et al., 2015). Recently, the deep learning technique has been effectively applied in neurology, producing better results than other techniques and is currently being used on a large scale to improve imaging analysis. In deep learning, the computer automatically learns features from the images and other raw data via a convoluted neural network (CNN), whereas machine learning uses a manual method (Lecun, Bottou, Bengio, & Haffner, 1998). Deep learning can assess more intricate patterns in the data because it can accommodate the increasing volume and complexity of data (Ravi et al., 2017). The frequently used deep learning algorithms in medical applications include recurrent neural networks, deep belief networks, CNNs, and deep neural networks.

Artificial intelligence and neurological disease prediction

In the last decade, studies in the field of AI have been growing at a rapid pace. Researchers have argued that a computer-aided diagnosis (CAD) system trained using a large amount of patient data, physiological signals, and images, as well as integration of advanced signal processing and AI techniques, can assist neurologists, neurosurgeons, radiologists, and other health providers in making diagnoses (Acharya, Hagiwara, & Adeli, 2018; Bairy et al., 2017; Bhat, Acharya, Adeli, Bairy, & Adeli, 2014; Sridhar, Bhat, Acharya, Adeli, & Bairy, 2017).

Support tools focused on machine learning promise to improve the precision, reliability, and quality of clinical decision-making and evaluation. Since accurate diagnostics and early-stage diagnosis are top priorities in medical practice, a lot of work is being put into developing new diagnostic support methods that will improve accuracy while reducing time and resources (Berardelli et al., 2013).

The standard of living for patients and their relatives or caregivers can be improved by early diagnosis of neurological conditions. Furthermore, early diagnosis can aid scientists in the creation of appropriate treatments. However, since mental ailments are a major public health risk with an increased socioeconomic cost to society, advances for early diagnosis of these disorders is an economic necessity (Papakostas, Savio, Gran, & Kaburlasos, 2015).

AI, especially machine learning algorithms, can recognize important trends and learn the underlying features of massive, heterogenous, and multivariate data representations. This knowledge can later be utilized to develop models comprised of mathematical and statistical instructions that can perform recognition tasks. Recent significant improvements in computing capacity, combined with the increasing availability of vast datasets from which to learn, have resulted in AI-based technologies approaching, if not exceeding, human capabilities in various data analytic tasks (Cummins, Matcham, Klapper, & Schuller, 2020; Dash, 2014).

Clinical tests are also a complex inferential process by which a clinician must evaluate a detailed classification of symptoms and various circumstantial factors to better determine the specific condition that may be present in a particular patient. Evidence from the literature on machine learning and affective computing suggest that AI technologies can help with the diagnosis of several neurological conditions, including bipolar disorder (BD) and major depressive disorder (MDD) (Khorram, Gideon, McInnis, & Provost, 2016; Pampouchidou et al., 2017; Ringeval et al., 2017, 2018).

Electroencephalography readings, which include information about the brain's electrical activity, are a valuable diagnostic tool in neurology. Many machine learning methods have been used to interpret these signals and make predictions. Subasi, Kevric, and Abdullah and Canbaz, M. (2019) suggested an algorithm that used the machine learning techniques of SVM and genetic algorithms (GAs) to diagnose epileptic seizures in EEG documents. Results showed these methods obtained accuracy of 99.38% on the EEG dataset used. Avcu, Zhang, and Chan (2019) introduced CNNs that can detect seizures with 93.3% accuracy using only two channels.

Connectome mapping and machine learning are both well adapted to epilepsy diagnosis. Epilepsy is a condition characterized by abnormal alterations in brain organization that result in hyperactivation of specific brain regions and seizures. Using anatomical MRI images combined with diffusion tensor imaging (DTI) tractography, machine learning has demonstrated substantial improvements over previous methods in identifying epileptiform structures and movement as well as predicting surgical outcomes. Munsell et al. used SVMs in a 2015 report, consistent with previous research (Subasi et al., 2019). However, they did use a feature selection and regularization method to limit the clinical characteristics used in predictive tasks. By doing so, they overcame a key obstacle in machine learning investigation in neurology: a limited number of test subjects compared to the large number of features identified by connectome mapping. The SVM was able to distinguish epileptic patients from healthy controls with an accuracy of 80% and predict the outcome of surgery with an accuracy of 70% using these methods, findings that are comparable to professional clinical decision-making (Munsell, Wee, et al., 2015).

The study of connectome mapping has been a focus of machine learning science. Scientists and clinicians will see a wiring illustration of the medium and wide connections in the brain using fMRI methods combined with DTI and white matter fiber tractography. This image of brain connectivity can be used to quantify the relationships between functionally and physically linked objects as well as to predict changes in the brain's resting state. Although genetic profiles may reveal a predisposition to microscopic changes in brain structure, connectome mapping models the brain as a network of interconnected regions to display the macroscopic manifestations of these changes (Mohammad-Rezazadeh, Frohlich, Loo, & Jeste, 2016).

Current imaging technologies and throughput capabilities have restricted connectome mapping to a macroscopic level of brain wiring, with 1010 neurons interconnected by 1014 connections. When coupled with machine learning, these techniques display promise in discovering novel characteristics for detecting epilepsy, attention deficit hyperactivity disorder (ADHD), autism spectrum disorders (ASDs), and other neurological conditions.

To see if machine learning methods have helped neurologists in other ways, we will look at a case study of Parkinson's disease (PD) (Dash, Abraham, & Rehman, 2018). Starting with the initial stages of PD, Prashanth, Dutta-Roy, Mandal, and Ghosh (2016) presented a machine learning method that can reliably predict PD with a 96.40% accuracy rate. Nonmotor features and olfactory loss, cerebrospinal fluid (CSF) measurements, and data from dopaminergic imaging markers are used in the process. Another machine learning system can now be used to predict disease progression after the disease has been diagnosed. Rastegar, Ho, Halliday, and Dzamko (2019) introduced a method that used serum cytokines from a baseline point to predict the result for 2 years after 1 year.

Non-clinical health-related applications

Deep learning is a branch of machine learning that works based on ANNs. It can be supervised, semi-supervised, or unsupervised (Cao et al., 2018; Dash, Patra, & Tripathy, 2012). Advancements in medical technologies have generated large volumes of data used for interpreting medical imaging, electroencephalography, and more. Learning from these datasets has made it possible to understand the pathogenesis of many disease conditions. Furthermore, evidence has shown that deep learning has several other applications outside biomedicine (Chen et al., 2019). Deep learning can be applied for processing electronic health data involving diagnosis, medications, and laboratory investigations (Rajkomar et al., 2018). A healthcare center can process these data using a deep learning system to obtain results better than those obtained by conventional methods. Molecular genetics is another field currently exploring the use of deep learning for DNA and RNA sequencing (Koumakis, 2020). It helps improve performance over traditional models, enhances interpretation, and allows for better understanding of the structure of biological data. Deep learning is also relevant in analyzing smartphone data, which helps identify health risk factors, take preventive care, and manage illnesses. Millions of datasets are used to build a system model that trains machines to learn and test the results in a safe environment. In view of this, a large set of data from sensors, geomapping, and cameras are used to build automated cars that can self-navigate through traffic; however, the major limitation of using these automobiles is their capacity to handle unprecedented events (Zhang et al., 2019). Language learning is a cognitive skill that individuals build through constant or repeated exposure to a language and develop appropriate responses over the years. The use of deep learning in natural language processing is gaining momentum (Wu et al., 2020). Learning from a complex volume of language data can now be used to build responses that mimic natural language. A deep learning system can be used to interpret language, which is a complex human task. In addition, answering questions, classifying text, and sentiment analysis are also components of deep learning (Burns, Li, & Peng, 2019). Deep learning can be used as a visual assistant that provides access to online information and provides interaction with other people. Furthermore, with deep learning models, speech can directly be translated to text, used to make notes, and book appointments. Other deep learning applications include detecting fraud, playing games, picture description, and electronic demographic predictions.

Challenges and potential pitfalls of neurological techniques

Neurological techniques are methods used for conducting research, diagnosing disease conditions, and assessing neurological functions. CT, PET, fMRI, electroencephalography, and magnetoencephalography are some of the techniques used in the field of neuroscience. Over the years, technological advancement in neurological techniques has made it possible to assess cognitive and other brain functions. Nevertheless, these techniques pose several challenges, limitations, or pitfalls. PET has become relevant in managing neurological disorders (Hooker & Carson, 2019). During PET scanning, 18 F-fluorodeoxyglucose (FDG) is commonly used, and evidence has shown that FDG is often associated with an artifact that interferes with interpreting the results (Almuhaideb, Papathanasiou, & Bomanji, 2011).

Furthermore, the combined use of PET/CT scanners has been described with artifacts on many clinical images. The shortage of well-trained professionals to handle neurological equipment can pose challenges to the field of neuroscience. High-energy X-rays to assess neurological functions could result in loss of electronic equilibrium, thereby causing a high beam of photon energy, damaging cellular structure and functions (Sharma, 2008). Evidence has shown that many computer programs for radiation therapy do not consider the changes in beam profile (Weissman & Feinstein, 1971). The use of CT imaging has gained momentum and can replace the traditional X-ray during autopsies. However, many experts who are not familiar with these imaging techniques might misinterpret the results obtained. CT is the most widely used imaging technique to assess neurologic and other body functions. It is often used with tracers, which can emit high-energy beams capable of damaging cellular structures (Anand, Singh, & Dash, 2009). fMRI has also contributed significantly to advances in neuroscience. Studies have shown that taking medications could interfere with imaging results (Brammer, 2009). In addition, alcohol, caffeine, and several other substances can alter images, and therefore clinicians and other professionals should be aware of these challenges to ensure that imaging results are accurate and well interpreted.

Conclusion and future directions

This chapter provided an overview of the role of machine learning in medical sciences and the advancements made in adopting AI to enhance neurological techniques in clinical settings. Also, it highlighted some critical analysis of potential pitfalls of these techniques. Non-invasive machine learning techniques measure brain electrical activity from the surface by generating data and convert this electrical phenomenon to discernible signals that are further analyzed utilizing algorithms and AI techniques. Data acquisition for deep learning techniques encompasses collecting, exploring, and cleaning the data required to train the system and develop the algorithm necessary to execute a specific task. Data exploration evaluates the qualitative and quantitative characteristics of raw data, which are assessed to ensure that they possess specific features needed to develop a relevant model.

References

Acharya, U. R., Hagiwara, Y., & Adeli, H. (2018). Automated seizure prediction. *Epilepsy and Behaviour, 88*, 251–261.

Adetunji, J. B., Adetunji, C. O., & Olaniyan, O. T. (2021). African walnuts: A natural depository of nutritional and bioactive compounds essential for food and nutritional security in Africa. In O. O. Babalola (Ed.), *Food security and safety*. Cham: Springer.

Adetunji, C. O., Ajayi, O. O., Akram, M., Olaniyan, O. T., Chishti, M. A., Inobeme, A., et al. (2021). Medicinal plants used in the treatment of influenza A virus infections. In K. Dua, S. Nammi, D. Chang, D. K. Chellappan, G. Gupta, & T. Collet (Eds.), *Medicinal plants for lung diseases*. Singapore: Springer.

Adetunji, C. O., Akram, M., Olaniyan, O. T., Ajayi, O. O., Inobeme, A., Olaniyan, S., et al. (2021a). Medicinal plants used in the treatment of pulmonary hypertension. In K. Dua, S. Nammi, D. Chang, D. K. Chellappan, G. Gupta, & T. Collet (Eds.), *Medicinal plants for lung diseases*. Singapore: Springer. https://doi.org/10.1007/978-981-33-6850-7_14.

Adetunji, C. O., Akram, M., Olaniyan, O. T., Ajayi, O. O., Inobeme, A., Olaniyan, S., et al. (2021b). Targeting SARS-CoV-2 novel corona (COVID-19) virus infection using medicinal plants. In K. Dua, S. Nammi, D. Chang, D. K. Chellappan, G. Gupta, & T. Collet (Eds.), *Medicinal plants for lung diseases*. Singapore: Springer.

Adetunji, C. O., Inobeme, A., Olaniyan, O. T., Ajayi, O. O., Olaniyan, S., & Adetunji, J. B. (2021). Application of nanodrugs derived from active metabolites of medicinal plants for the treatment of inflammatory and lung diseases: Recent advances. In K. Dua, S. Nammi, D. Chang, D. K. Chellappan, G. Gupta, & T. Collet (Eds.), *Medicinal plants for lung diseases*. Singapore: Springer.

Adetunji, C. O., & Varma, A. (2020). Biotechnological application of trichoderma: a powerful fungal isolate with diverse potentials for the attainment of food safety, management of pest and diseases, healthy planet, and sustainable agriculture. In C. Manoharachary, H. B. Singh, & A. Varma (Eds.), *Vol. 61. Trichoderma: Agricultural applications and beyond. Soil biology*. Cham: Springer.

Administration Ufad. (2013). *Guidance for industry: electronic source data in clinical investigations*. https://www.fda.gov/downloads/drugs/guidances/ucm328691.pdf. (Accessed 1 June 2017).

Akram, M., Jabeen, F., Daniyal, M., Zainab, R., Haq, U. U., Adetunji, C. O., et al. (2020). Genetic engineering of novel products of health significance: Recombinant DNA technology. In C. Egbuna, & T. G. Dable (Eds.), *Functional foods and nutraceuticals*. Cham: Springer. https://doi.org/10.1007/978-3-030-42319-3_26.

Aljović, A., Badnjević, A., & Gurbeta, L. (2016). Artificial neural networks in the discrimination of Alzheimer's disease using biomarkers data. In *2016 5th mediterranean conference on embedded computing (MECO), Bar, Montenegro* (pp. 286–289). https://doi.org/10.1109/MECO.2016.7525762.

Almuhaideb, A., Papathanasiou, N., & Bomanji, J. (2011). 18F-FDG PET/CT imaging in oncology. *Annals of Saudi Medicine, 31*(1), 3–13.

Anand, S. S., Singh, H., & Dash, A. K. (2009). Clinical applications of PET and PET-CT. *Medical Journal, Armed Forces India, 65*(4), 353–358.

Avcu, M. T., Zhang, Z., & Chan, D. W. S. (2019). Seizure detection using least EEG channels by deep convolutional neural network. In *ICASSP 2019–2019 IEEE International Conference on Acoustics, Speech and Signal Processing (ICASSP)* (pp. 1120–1124).

Bailey, D. L., Townsend, D. W., Valk, P. E., & Maisy, M. N. (2005). *Positron emission tomography: Basic sciences*. Secaucus, NJ: Springer-Verlag. ISBN 978-1-85233-798-8.

Bairy, G. M., Lih, O. S., Hagiwara, Y., Subha, D. P., Faust, O., et al. (2017). Automated diagnosis of depression electroencephalograph signals using linear prediction Coding

and higher order spectra features. *Journal of Medical Imaging and Health Information*, 7(8), 1857–1862.
Bengio, Y., Courville, A., & Vincent, P. (2013). Representation learning: A review and new perspectives. *IEEE Transactions on Pattern Analysis and Machine Intelligence*, *35*(8), 1798–1828.
Berardelli, A., Wenning, G. K., Antonini, A., Berg, D., Bloem, B. R., et al. (2013). EFNS/MDS-ES recommendations for the diagnosis of Parkinson's disease. *European Journal of Neurology*, *2013*.
Bhat, S., Acharya, U. R., Adeli, H., Bairy, G. M., & Adeli, A. (2014). Automated diagnosis of autism: In search of a mathematical marker. *Reviews in the Neurosciences*, *25*(6), 851–861.
Boto, E., Holmes, N., Leggett, J., Roberts, G., Shah, V., Meyer, S. S., et al. (2018). Moving magnetoencephalography towards real-world applications with a wearable system. *Nature*, *555*(7698), 657–661.
Brammer, M. (2009). The role of neuroimaging in diagnosis and personalized medicine—Current position and likely future directions. *Dialogues in Clinical Neuroscience*, *11*(4), 389–396.
Burns, G. A., Li, X., & Peng, N. (2019). Building deep learning models for evidence classification from the open access biomedical literature. *Database: The Journal of Biological Databases and Curation*, *2019*, baz034.
Campbell, I. G., Kraus, A. M., Burright, C. S., & Feinberg, I. (2016). Restricting time in bed in early adolescence reduces both NREM and REM sleep but does not increase slow wave EEG. *Sleep*, *39*(9), 1663–1670. https://doi.org/10.5665/sleep.6088.
Cao, C., Liu, F., Tan, H., Song, D., Shu, W., Li, W., et al. (2018). Deep learning and its applications in biomedicine. *Genomics, Proteomics & Bioinformatics*, *16*(1), 17–32.
CB Insights Research (2017). Healthcare remains the hottest AI category for deals (Last accessed on 2018 Mar 24). Available from: https://www.cbinsights.com/research/artificial-intelligence-healthcare-startups-investors/.
Chen, D., Liu, S., Kingsbury, P., Sohn, S., Storlie, C. B., Habermann, E. B., et al. (2019). Deep learning and alternative learning strategies for retrospective real-world clinical data. *NPJ Digital Medicine*, *2*, 43.
Chennubhotla, C., Clarke, L. P., Fedorov, A., Foran, D., Harris, G., Helton, E., et al. (2017). An assessment of imaging informatics for precision medicine in cancer. *Yearbook of Medical Informatics*, *26*, 110–119.
Cummins, N., Matcham, F., Klapper, J., & Schuller, B. (2020). *Artificial intelligence to aid the detection of mood disorders*. Elsevier. https://doi.org/10.1016/B978-0-12-817133-2.00010-0.
Darcy, A. M., Louie, A. K., & Roberts, L. W. (2016). Machine learning and the profession of medicine. *JAMA*, *315*, 551–552.
Dash, S. (2014). *Learning using hybrid intelligence techniques, computational intelligence for big data analysis: Frontier advances and applications, studies in adaptation, learning, and optimization series*. *Vol. 9* (pp. 74–94). Heidelberg: Springer-Verlag. https://doi.org/10.1007/978-3-319-16598-1. ISSN 1867–4534, ISBN 978–3-319–16597–4.
Dash, S., Abraham, A., & Rehman, A. (2018). Kernel based chaotic firefly algorithm for diagnosing Parkinson's disease. In A. Madureira, et al. (Eds.), *Vol. 923*. *Advances in intelligent systems and computing* (pp. 178–188). Springer Nature.
Dash, S., Chakraborty, C., Giri, S. K., Pani, S. K., & Frnda, J. (2021). BIFM: Big-data driven intelligent forecasting model for COVID-19. *IEEE Access*, *9*, 97505–97517. https://doi.org/10.1109/ACCESS.2021.3094658.
Dash, S., Patra, B., & Tripathy, B. K. (2012). Study of classification accuracy of microarray data for cancer classification using multivariate and hybrid feature selection method, IOSR. *IOSR Journal of Engineering*, *2*(8), 112–119.

Egbuna, C., Gupta, E., Ezzat, S. M., Jeevanandam, J., Mishra, N., Akram, M., et al. (2020). Aloe species as valuable sources of functional bioactives. In C. Egbuna, & G. Dable Tupas (Eds.), *Functional Foods and Nutraceuticals*. Cham: Springer. CB Insights research 2017.

Fortino, V., Wisgrill, L., Werner, P., Suomela, S., Linder, N., Jalonen, E., et al. (2020). Machine-learning–driven biomarker discovery for the discrimination between allergic and irritant contact dermatitis. *Proceedings of the National Academy of Sciences*, *117*(52), 33474–33485. https://doi.org/10.1073/pnas.2009192117.

Gilani, S. Z., & Mian, A. (2017). *Learning from millions of 3D scans for large-scale 3D face recognition*. http://arxiv.org/abs/1711.05942.

Goodfellow, I., Bengio, Y., & Courville, A. (2016). Deep learning. In P. Kantor (Ed.), *Foundations of statistical natural language processing* (1st ed.). MIT Press.

Harvey, H., & Topol, E. J. (2020). More than meets the AI: Refining image acquisition and resolution. *Lancet*, *396*, 1479.

Hooker, J. M., & Carson, R. E. (2019). Human positron emission tomography neuroimaging. *Annual Review of Biomedical Engineering*, *21*, 551–581.

Jensen, P. B., Jensen, L. J., & Brunak, S. (2012). Mining electronic health records: Towards better research applications and clinical care. *Nature Reviews. Genetics*, *13*(6), 395–405.

Jha, S., & Topol, E. J. (2016). Adapting to artificial intelligence: Radiologists and pathologists as information specialists. *JAMA*, *316*, 2353–2354.

Kececi, H., & Degirmenci, Y. (2008). Quantitative EEG and cognitive evoked potentials in anemia. *Clinical Neurophysiology*, *38*(2), 137–143.

Khan, J., Wei, J. S., Ringner, M., Saal, L. H., Ladanyi, M., Westermann, F., et al. (2001). Classification and diagnostic prediction of cancers using gene expression profiling and artificial neural networks. *Nature Medicine*, 7, 673–679. https://doi.org/10.1038/89044.

Khorram, S., Gideon, J., McInnis, M., & Provost, E. M. (2016). Recognition of depression in bipolar disorder: leveraging cohort and person-specific knowledge. In *Proceedings of INTERSPEECH 2016, 17th annual conference of the international speech communication association* (pp. 1215–1219).

Kohli, M. D., Summers, R. M., & Geis, J. R. (2017). Medical image data and datasets in the era of machine learning-whitepaper from the 2016 C-MIMI meeting dataset session. *Journal of Digital Imaging*, *30*(4), 392–399.

Koumakis, L. (2020). Deep learning models in genomics; are we there yet? *Computational and Structural Biotechnology Journal*, *18*, 1466–1473.

Lambin, P., Rios-Velazquez, E., Leijenaar, R., Carvalho, S., Van Stiphout, R. G., Granton, P., et al. (2012). Radiomics: Extracting more information from medical images using advanced feature analysis. *European Journal of Cancer*, *48*, 441–446.

LeCun, Y., Bengio, Y., & Hinton, G. (2015). Deep learning. *Nature*, *521*(7553), 436–444.

Lecun, Y., Bottou, L., Bengio, Y., & Haffner, P. (1998). Gradient-based learning applied to document recognition. *Proceedings of the IEEE*, *86*, 2278–2324.

Li, H., Lin, Z., Shen, X., Brandt, J., & Hua, G. (2015). A convolutional neural network cascade for face detection. In *Proceedings of IEEE conference on computer vision and pattern recognition. Boston, MA* (pp. 5325–5334).

Logothetis, N. K., Pauls, J., Auguth, M., Trinath, T., & Oeltermann, A. (2001). A neurophysiological investigation of the basis of the BOLD signal in fMRI. *Nature*, *412*(6843), 150–157.

Luo, J., Wu, M., Gopukumar, D., & Zhao, Y. (2016). Big data application in biomedical research and health care: A literature review. *Biomed Inform Insights*, *19*(8), 1–10.

Maqueda, A. I., Loquercio, A., Gallego, G., Garcia, N., & Scaramuzza, D. (2018). *Event-based vision meets deep learning on steering prediction for self-driving cars*. http://arxiv.org/abs/1804.01310.

Mohammad-Rezazadeh, I., Frohlich, J., Loo, S. K., & Jeste, S. S. (2016). Brain connectivity in autism spectrum disorder. *Current Opinion in Neurology*, *29*(2), 137–147.

Montagnon, E., Cerny, M., Cadrin-Chênevert, A., Hamilton, V., Derennes, T., Ilinca, A., et al. (2020). Deep learning workflow in radiology: A primer. *Insights Into Imaging*, *11*(1), 22. https://doi.org/10.1186/s13244-019-0832-5.

Munsell, B. C., Wee, C. Y., et al. (2015). Evaluation of machine learning algorithms for treatment outcome prediction in patients with epilepsy based on structural connectome data. *NeuroImage*, *118*, 219–230.

Nadri, H., Rahimi, B., Timpka, T., & Sedghi, S. (2017). The top 100 articles in the medical informatics: A bibliometric analysis. *Journal of Medical Systems*, *41*(10), 150.

Neill, D. (2013). Using artificial intelligence to improve hospital inpatient care. *IEEE Intelligent Systems*, *28*, 92–95.

Palmini, A. (2006). The concept of the epileptogenic zone: A modern look at Penfield and Jasper's views on the role of interictal spikes. *Epileptic Disorders*, *8*(Suppl 2), S10–S15 (Erratum in: Epileptic Disorders 10(2), 191; PMID: 17012068).

Pampouchidou, A., Simos, P., Marias, K., Meriaudeau, F., Yang, F., Pediaditis, M., et al. (2017). Automatic assessment of depression based on visual cues: A systematic review. *IEEE Transactions on Affective Computing*, *10*(4), 445–470. https://doi.org/10.1109/TAFFC.2017.2724035 (1 October–December 2019).

Papakostas, G. A., Savio, A., Gran, M., & Kaburlasos, V. G. (2015). A lattice computing approach to Alzheimer's disease computer assisted diagnosis based on MRI data. *Neurocomputing*, *150*, 37–42.

Prashanth, R., Dutta-Roy, S., Mandal, P. K., & Ghosh, S. (2016). High-accuracy detection of early parkinson's disease through multimodal features and machine learning. *International Journal of Media and Information Literacy*, *90*, 13–21.

Rajkomar, A., Oren, E., Chen, K., Dai, A. M., Hajaj, N., Hardt, M., et al. (2018). Scalable and accurate deep learning with electronic health records. *NPJ Digital Medicine*, *1*, 18.

Ramanishka, V., Chen, Y.-T., Misu, T., & Saenko, K. (2018). Toward driving scene understanding: a dataset for learning driver behavior and causal reasoning. In *Proceedings of IEEE conference on computer vision and pattern recognition* (pp. 7699–7707). IEEE Computer Society. Institute of Computer Science and Engineering. Salt Lake City, UT.

Rastegar, A. D., Ho, N., Halliday, G. M., & Dzamko, N. (2019). Parkinson's progression prediction using machine learning and serum cytokines. *Npj Parkinson Disease*, *5*(1), 14.

Ravi, D., Wong, C., Deligianni, F., Berthelot, M., Andreu-Perez, J., Lo, B., et al. (2017). Deep learning for health informatics. *IEEE Journal of Biomedical and Health Informatics*, *21*, 4–21.

Ringeval, F., et al. (2017). AVEC 2017: Real-life depression, and affect recognition workshop and challenge. In *Proceedings of the 7th annual workshop on audio/visual emotion challenge* (pp. 3–9).

Ringeval, F., et al. (2018). AVEC 2018 workshop and challenge: Bipolar disorder and cross-cultural affect recognition. In *Proceedings of the 2018 on audio/visual emotion challenge and workshop* (pp. 3–13).

Roli, O. I., Adetunji, C. O., Mishra, P., Adetunji, J. B., & Mishra, R. R. (2020). Caffeine: Nutraceutical and health benefit of caffeine-containing commodities and products. In P. Mishra, R. R. Mishra, & C. O. Adetunji (Eds.), *Innovations in food technology*. Singapore: Springer.

Sejnowski, T., Koch, C., & Churchland, P. S. (1988). Computational neuroscience. *Science*, *241*(4871), 1299–1306.

Sharma, S. D. (2008). Quality of high-energy X-ray radiotherapy beams: Issues of adequacy of routine experimental verification. *Journal of Medical Physics*, *33*(1), 1–2.

Sridhar, C., Bhat, S., Acharya, U. R., Adeli, H., & Bairy, G. M. (2017). Diagnosis of attention deficit hyperactivity disorder using imaging and signal processing techniques. *Computers in Biology and Medicine*, *88*, 93–99.

Subasi, A., Kevric, J., & Abdullah and Canbaz, M. (2019). Epileptic seizure detection using hybrid machine learning methods. *Neural Computing and Applications*, *31*(1), 317–325.

Thirumuruganathan, S., Tang, N., & Ouzzani, M. (2018). *Data curation with deep learning: towards self driving data curation*. Available via https://arxiv.org/abs/1803.01384. (Accessed 12 Aug 2019).

Tibshirani, R., Hastie, T., Narasimhan, B., & Chu, G. (2002). Diagnosis of multiple cancer types by shrunken centroids of gene expression. *PNAS*, *99*, 6567–6572. https://doi.org/10.1073/pnas.082099299.

Weissman, D. D., & Feinstein, R. B. (1971). X-ray beam profiles and oral radiography. *Oral Surgery, Oral Medicine, and Oral Pathology*, *31*(4), 546–555.

Winters, B., Custer, J., Galvagno, S. M., Jr., Colantuoni, E., Kapoor, S. G., Lee, H., et al. (2012). Diagnostic errors in the intensive care unit: A systematic review of autopsy studies. *BMJ Quality and Safety*, *21*, 894–902.

Wong, H.-S., Zhu, H., & Yip, T. T. C. (2009). A neural network-based biomarker association information extraction approach for cancer classification. *Journal of Biomedical Informatics*, *42*(4), 654–666. https://doi.org/10.1016/j.jbi.2008.12.010. ISSN 1532-0464.

Wu, S., Roberts, K., Datta, S., Du, J., Ji, Z., Si, Y., et al. (2020). Deep learning in clinical natural language processing: A methodical review. *Journal of the American Medical Informatics Association*, *27*(3), 457–470.

Wu, Y., Schuster, M., Chen, Z., et al. (2016). *Google's neural machine translation system: bridging the gap between human and machine translation*. http://arxiv.org/abs/1609.08144.

Zhang, J., Wu, Z., Li, F., Xie, C., Ren, T., Chen, J., et al. (2019). A deep learning framework for driving behavior identification on in-vehicle CAN-BUS sensor data. *Sensors (Basel, Switzerland)*, *19*(6), 1356.

Zhang, S., Dinan, E., Urbanek, J., Szlam, A., Kiela, D., & Weston, J. (2018). *Personalizing dialogue agents: I have a dog, do you have pets too?*. http://arxiv.org/abs/1801.07243.

CHAPTER 14

An intelligent diagnostic approach for epileptic seizure detection and classification using machine learning

Charles O. Adetunji[a], Olugbemi T. Olaniyan[b], Olorunsola Adeyomoye[c], Ayobami Dare[d], Mayowa J. Adeniyi[e], and Alex Enoch[f]
[a]Department of Microbiology, Applied Microbiology, Biotechnology and Nanotechnology Laboratory, Edo State University Uzairue, Iyamho, Nigeria
[b]Department of Physiology, Laboratory for Reproductive Biology and Developmental Programming, Rhema University, Aba, Nigeria
[c]Department of Physiology, University of Medical Sciences, Ondo, Nigeria
[d]Department of Physiology, School of Laboratory Medicine and Medical Sciences, College of Health Sciences, Westville Campus, University of KwaZulu-Natal, Durban, South Africa
[e]Department of Physiology, Edo State University Uzairue, Iyamho, Nigeria
[f]Department of Human Physiology, Ahmadu Bello University Zaria, Zaria, Nigeria

Introduction

This could be severe neurological disorder with special features, treatment of recurrent seizures (Chaudhary, Duncan, & Lemieux, 2011; Reynolds, 2009; WHO, 2006, 2005). Epilepsy is linked to electrical activity disturbance inside the brain (Alarcón & Valentín, 2012; Qu & Gotman, 1993), which could be attributed to several factors (Sazgar & Young, 2019). These factors may include scarcity of oxygen during childbirth, low blood sugar levels, and malformations. More than 50 million people worldwide have epilepsy (Alarcón & Valentín, 2012; WHO, 2015). Typical symptoms of epilepsy include loss of momentary consciousness and sudden breakdown or unusual action in the brain that causes an involuntary modification in a patient's behavior and sensation. Moreover, it has been established that seizures can occur at any time without any aura, which might result in significant injuries, such as fractures and burns, or even death (Hannah & Brodie, 1998; Shafer & Sirven, 2014; WHO, 2005).

Numerous scientists have validated the dynamic system of the human brain in relationship to patients suffering from epilepsy (Andrzejak et al., 2001; Babloyantz & Destexhe, 1986; Frank et al., 1990; Iasemidis, Pardalos, Sackellares, & Shiau, 2001; Le Van Quyen et al., 2001; Lehnertz, Arnhold,

Artificial Intelligence for Neurological Disorders
https://doi.org/10.1016/B978-0-323-90277-9.00011-0

Grassberger, & Elger, 2000; Lehnertz & Elger, 1998; Martinerie et al., 1998; Mormann, Lehnertz, David, & Elger, 2000; Theiler, 1995). Moreover, the application of machine learning has been identified as tool that could help in the classification of *seizure* and *non-seizure* (Fergus, Hignett, Hussain, & Al-Jumeily, 2014). This chapter focuses on the utilization of machine learning and different algorithms that provide smart, powerful, real-time, and low-cost approaches for the classification and detection of different pathophysiological characteristics in epileptic seizure such as unusual behavior, loss of awareness, confusion, and memory and attention deficits utilizing vital information derived from ictal and interictal signals from the brain electroencephalogram waves using discrete wavelet transform.

Epileptic seizure

Epileptic seizures occur due to spontaneous recurrent and unprovoked neuronal discharge. It's a neurological disorder that affects 3% of the global population (Sirven, 2015). It's a generalized type of seizure that affects the two halves of the cerebral hemispheres. Although not all seizures are associated with loss of consciousness, epileptic seizure requires medical care and will cause loss of awareness and responsiveness (Stafstrom & Carmant, 2015). Brain tumor and stroke can trigger epileptic seizure. Furthermore, infections like meningitis, AIDS, and viral encephalitis may play significant role (Vezzani et al., 2016). Both men and women can develop epilepsy and it is not related to age. Some of the symptoms include confusion, uncontrollable jerking movements of arms and legs, fear, and anxiety. Epileptic seizure may be genetic if it runs in the family. Evidence have shown that NRXN 2, AVTS 2, NLGNN 1, CNTNAP 2, GRIN 2A, PRRT 2, NIPA 2, and BMP 5 genes could be implicated in the neuronal development of epileptic seizure (Steinlein, 2008). The World Health Organization (WHO) has reported that epilepsy rates are higher in developing countries that have limited access to medical care (Anwar et al., 2020). In addition, traumatic injury, prenatal injury, and dementia are other factors that could increase the risk of epilepsy. Epileptic seizure can pose great danger to any affected individual (Huff & Murr, 2021). The sudden loss of sensation or consciousness can result in skull or other fractures. It can also result in fatal accidents while driving. During pregnancy, both the mother and the fetus may be severely affected. In fact, certain epileptic medications used during pregnancy may increase the risk of developing birth defects.

Seizure localization

Seizure occurs due to abnormal electrical waves emerging from the brain. These could result in unprovoked and recurrent motor activities and if pronounced, may require surgical interventions. For a successful surgery, it is important to locate the exact site of the seizure using a combination of different structural imaging techniques such as magnetic resonance imaging (MRI), electroencephalogram (EEG), and magnetoencephalography (MEG) (Panachakel & Ramakrishnan, 2021). These are non-invasive techniques and are useful in directing therapeutic interventions, however, these techniques may not fully visualize the abnormally firing neurons (Lenkov, Volnova, Pope, & Tsytsarev, 2013). An advanced imaging technique is required to locate foci seizure.

Single-photon emission computed tomography (SPECT) is a modern technique that enables scanning of cross-sectional and visible images of abnormal neurons (Devous, 2005). This tool is also useful in quantifying the changes in cerebral blood flow. In addition, it can diagnose neuronal diseases and give indications before the clinical manifestation of symptoms (NRC, 1996).

Positron emission tomography (PET) is an imaging procedure that uses emitting radiopharmaceuticals to assess glucose metabolism and blood flow (Vaquero & Kinahan, 2015). It uses different radioactive tracers, for example, 18F-FDG for cancer detection, NaF-F18 for bone formation, and oxygen-15 for measurement of blood flow (Berger, 2003). A radioisotope is attached to a drug and ingested; this serves as a tracer that emits gamma rays that form a three-dimensional image that is captured by a gamma camera.

Physiological and pathophysiological signals

Physiological signals are parameters or predictors that, in addition to indicating health and physiological states, are used in diagnosis and prognosis of diseases. They are representative indices relating to normal intracellular, intercellular, system, and multi-system functioning. Deviation from normal intracellular, intercellular, system, and multi-system functioning is associated with pathophysiological signals. Physiological signals are produced as a result of simple or complex intra- and intercellular communications. Within the body, cells communicate with one another through chemical or electric signals.

Chemical signals as physiological signals

Chemical signals involve transmission of chemical messengers from a control cell to another cell, known as a target cell. Types of chemical signals include autocrine, paracrine, juxtacrine, intracrine, neuroendocrine, neurotransmitter, and endocrine signals. In autocrine signals, chemical messengers bind to surface receptors and influence the same cell that secretes them. Chemical messengers such as interleukin-I produced by macrophages result in further activation of the cell (Ray, 2014). Autocrine signals play a vital role in the control of inflammation and pain. They also play a role in cancer activation and metastasis (Wells, 2006) and development of anti-oncogene interventional measures by targeting Wnt pathway (Bafico, Liu, Goldin, Harris, & Aaronson, 2004; Schlange, Matsuda, Lienhard, Huber, & Hynes, 2007). Vasoactive Epithelial Growth Factor (VEGF) signaling has been reported to control the survival of human stem cells via the autocrine pathway (Gerber et al., 2002). A genetic circuit known as the secrete-and-sense circuit regulates the tendency of a cell to communicate with itself and neighboring cells (Doğaner, Yan, & Youk, 2016).

Paracrine signals are physiological signals released to neighboring cells by control cells. Prostaglandins secreted by macula densa, in a paracrine fashion, induce secretion of rennin by the juxtaglomerular cells. Through quorum sensing, a control cell communicates with its neighboring cells (Doğaner et al., 2016). Using a model of microcavity platform, Müller, Wang, Qiao, et al. (2016) showed how autocrine and paracrine signals affect hematopoietic stem cells. Intracrine signals are physiological signals that influence the same cells that secrete them by binding to intracellular receptors. Juxtacrine signals bind to the neighboring cells in contact with one another. Studies on the antioncogenic mechanism of programmed death-ligand 1 (PD-L1) involve the juxtracrine pathway (Yang et al., 2018). Neuroendocrine signals are physiological signals secreted by neural cells that exert their effects on non-neural cells such as endocrine glands. Endocrine signals are physiological signals secreted in a minute quantity from a control cell into the bloodstream where they exert their effects on other distant cells. Endocrine signals are classified as protein/peptides (parathyroid hormones, calcitonin, insulin, glucagon, and many more), steroids (glucocorticoids and mineralocorticoid, estrogen and androgen), and amine (thyroid hormones).

Endocrine disorders as deviations from physiological signals

Endocrine signals are secreted in a minute quantity within a physiological range by a specific control cell (endocrine gland) and into the blood where they are transported to other far-off body cells (target tissues). Abnormal and persistent increase or decrease in the quantity of these signals results in endocrine disorders. Hyperthyroidism, hyperparathyroidism, Cushing syndrome, osteoporosis, and primary aldosteronism occur due to persistent increase in the secretion of endocrine signals. Diabetes mellitus, diabetes insipidus, hypothyroidism, myxedema, cretinism, hypoparathyroidism, and Addison's disease are due to reduction in endocrine signals secreted by endocrine glands off the physiological limit.

Neurotransmitters, just like neuroendocrine cells, are produced and released by nerve cells but effect physiological changes in other neural cells. Neurotransmitters are classified as amino acids (glutamates, aspartates, gamma amino butyric acid, glycine), amine derivatives (serotonin, histamine, dopamine, norepinephrine, epinephrine, melatonin), opioids (enkephalins, morphins, dynorphins), nitric oxide, and acetylcholine.

Neurotransmitter detection using artificial intelligence

As a physiological signal, determination of neurotransmitters in the central nervous system (CNS) provides information about the functional status of the brain and spinal cord and generates clues vital in diagnosis and prognosis of diseases. Conventional methods such as nuclear medicine tomographic image techniques (PET and SPECT), optical sensing techniques (surface-enhanced Raman spectroscopy, fluorescence, Förster resonance energy transfer and photoinduced electron transfer, chemiluminescence, optical fiber biosensing, and colorimetry), electrophysiological techniques, and receptor blockage techniques have been used in the determination of brain neurotransmitters and diagnosis of conditions such as Alzheimer's disease, Parkinson's disease, attention deficit hyperactivity disorders, and many others. However, these methods do not give a detailed and direct estimation of brain neurotransmitter levels. Although popular, electrochemistry techniques such as fast scan voltammetry in combination with carbon-fiber microelectrodes, amperometry, tandem mass spectrometry, high-performance liquid chromatography, and microdialysis are associated with tissue damage, though relatively minimal (Kozai, Jaquins-Gerstl, Vazquez, Michael, & Cui, 2015; Peters, Miner, Michael, & Sesack, 2004). A study by Komoto, Ohshiro, Yoshida, et al. (2020) developed a technique using machine learning analysis and nanogap electrodes to measure the passage of tunneled current through molecules in the striatum and cortical part of a mouse's brain. With this method, serotonin, dopamine, and norepinephrine were identified with future possibility of measuring the distribution of neurotransmitters in the brain with great spatiotemporal resolution.

Electrical signals as physiological signals

Electric signals are propagable or non-propagable changes in resting membrane potential that take place when excitable tissues discharge. Propagable changes in membrane potential are known as action potentials or electric impulses in nerve cells.

Action potentials

Change in membrane potential is called action potential. It consists of resting membrane potential (potential difference between the cell and its environment), depolarization (increase in membrane potential in positive direction), and repolarization (decrease in membrane potential in the negative direction) (Sanes & Reh, 2012; Tamagawa, Funatani, & Ikeda, 2016). Action potential is a medium through which sensory or motor information is encoded and transmitted between the CNS and peripheral nervous system.

The advent of technology-driven techniques such as genetically encoded voltage indicators, high-density microelectrode array recordings, and subcellular patch-clamp recordings has immensely contributed to the generation of information on the function of neurons and pathology of axons (Müller et al., 2015; Radivojevic et al., 2017; Sasaki, Matsuki, & Ikegaya, 2011; Shu, Hasenstaub, Duque, Yu, & Mccormick, 2006). Martins, Erlhagen, and Freitas (2015) proposed an approximately 3375 nm^3 field effect transistor-based neuroelectric nanosensor that can possibly detect action potentials with a spatiotemporal resolution of 0.1 ms, sufficient for wave form characterization at a peak human neural firing rate of 800 Hz.

Application of electrical signals

Discharge of excitable tissues results in creation of an electric field that travels peripherally. These discharges are picked using sensors and processed and analyzed using non-invasive techniques such as electrocardiography, electromyography, electrodermal technique, electrooculography, electroretinography, electroencephalography, and electrocorticography. Through these techniques, voltage analogues picked from body surfaces are transformed into an understandable language, a wave. A wave's vertical axis indicates the strength of the signal, amplitude, or volume, and the horizontal axis represents duration or frequency of the signals.

Nowadays, physiological signals are integrated in vital sign panel and monitors such that whenever alteration from the physiological limit occurs, a mechanical signal (alarm) or warning light is displayed. Using a highly complex algorithm, it is possible to identify pathophysiological signals such as seizures using electroencephalogram. Modern electrocardiographs are programmed to recognize pathophysiological signs such as myocardial infarction, myocardial ischemia, right and left atrial hypertrophies, right and left ventricular hypertrophies, hyperkalemia, hypokalemia, and others. This helps in accurate diagnosis and prognosis with the leaset possible artifacts.

Artificial intelligence and action potential detection

In the machine learning approach developed by Bazan et al. (2020), loss of low threshold spike is overcome because this technique detects additional neuronal spikes that cannot be measured using classical spike detection. A multiresolution Teager energy operator (MTEO) detector is another method that has shown to produce higher detection ratios at a certain fixed detection ratio than the discrete wavelet transform detector and Teager energy operator (TEO) (Choi, Jung, & Kim, 2006; Iqbal, Sheikh, & Karayannis, 2019). There are several techniques that can be deployed in estimating and detecting mammalian neuronal cells. However, the major challenge in using these techniques is the high complexity of data to be analyzed. The employment of an automated artificial intelligence-based methodology to overcome this and other obstacles has proven to be beneficial. Iqbal et al. (2019) developed an artificial intelligence-based approach to Detect Neurons in Different Brain Regions During Development (DeNeRD) for detecting neurons in a variety of imaging planes and modalities.

Electrocorticography and electroencephalography

Electroencephalography

Electroencephalography (EEG) is a non-invasive technique for measuring brain electrical activity from the scalp and converting electrical phenomena into a readable and interpretable output. EEG was first recorded in human being by physiologist Hans Berger in 1924 (Haas, 2003). EEG looks at the brain's spontaneous activities that result from neuronal ionic current. The technique is based on the fact that when synchronous electrical signals are generated by discharging pyramidal cells, an electric field is created. Multiple electrodes placed on the scalp pick the electrical signals and transmit them to the EEG, where the signal is converted to electroencephalographical waves (alpha, beta, gamma, theta, and delta waves). Each of these waves has a peculiar configuration, pattern, duration, and voltage in physiological states. Alterations in the normal configuration, pattern, duration, and voltage occur in pathophysiological conditions.

Electrocardiograph recording and placement

To obtain an accurate and reliable EEG measurement, it is important that the subject abstains from the use of caffeinated foods or drinks, psychoactive drugs, muscle relaxants, and hypnotics at least 12 h prior to the test. The scalp needs to be cleaned to be free of oils, sprays, and cream. Informed consent needs to be obtained from the subjects, temperature of the laboratory room should be kept at around 25 °C, and subjects must be in a reclining position and relaxed (Chernecky & Berger, 2013).

A routine recording typically takes 20 min. However, in research, the duration is variable depending on the experimental design. EEG recording occurs in five phases: eyes closed, eyes opened, with light stimulation, during hyperventilation, and post hyperventilation (Tatum, Dworetzky, & Schomer, 2011). During the eyes closed phase, the patient relaxes and reclines with their eyes closed. The eyes open phase requires the patient to recline with their eyes open and then blink at certain intervals as specified. The light stimulation phase is designed to elicit epileptic-like discharge, especially in older epileptic patients. The hyperventilation phase involves hyperventilating for 3 min (up to 20 breaths per minute) to evoke epileptic discharge in young patients. Hyperventilation is not advised for old people or patients with atherosclerosis. Besides these phases, other phases may include sleep deprivation, sleep, and mental activity like doing mathematical calculations and singing.

The International 10–20 System makes use of 21 electrodes connected to positions that are measured at 10% and 20% of the circumference of the head with the aid of cranial landmarks as reference points. The four cranial landmarks include inion (bony projection on the back of the head), nasion (meeting of the forehead and the bridge of the nose), and the two preauricular points (Chernecky & Berger, 2013). It is also important to note that in

EEG electrodes are identified according to the cranial part where they are placed. For instance, T stands for temporal, O stands for occipital, C stands for central, F stands for frontal, P stands for parietal, A stands for auricular, and Fp stands for fronto-polar.

Electroencephalography and other non-invasive techniques

EEG has attributes that are not found in other non-invasive techniques used for evaluating the functional status of the brain as well as in diagnosis and prognosis. With EEG, covert neuronal events can be recorded. It can be used to measure electrical activities in the brain during sleep and while transiting from sleep to wakefulness. In addition, patients that are not able to make motor response or reflex reaction are also suitable for EEG. Unlike MRI and PET, EEG is less bulky with relatively high temporal resolution (Goldman, Stern, Engel, & Cohen, 2002). Responses to auditory signals are much more possible with EEG because it is silent. Unlike MRI and PET, EEG does not require the use of a magnetic field or radioactivity. It is tolerant to movement and has lower hardware costs (Vespa, Nenov, & Nuwer, 1999).

EEG exhibits a relatively lower spatial resolution when compared with other non-invasive methods (Srinivasan, 2006). Because of its complex operational protocol, EEG is a much more time-consuming procedure. It measures electrical activities at the lower part of the brain poorly.

Applications of electroencephalography

Physiologically, EEG can be used to monitor changes in brain during different life phases. For instance, EEG sleep analysis can be used to deduce brain development timing (Feinberg & Campbell, 2012). With EEG, non-epileptic seizures can be differentiated from epileptic seizures and catatonia can be distinguished from delirium. Other conditions that can be diagnosed by the technique include sleep disorders, epilepsy, brain dysfunction, brain damage due to head injury, encephalitis, and more.

Electrocorticography

Insertion of electrodes directly into the brain through the sectioned skull may be required, especially when evaluating a patient for epilepsy surgery.

This craniotomy-related technique is known as electrocorticography. Unlike EEG, electrocorticography is an invasive electrophysiological technique that measures cortical electrical activity. The technique was first introduced by surgeons Wilser Penfiels and Herbert Jasper in 1950 (Palmini, 2006). Electrocorticography exhibits a greater spatial resolution than EEG due to absence of cranial barrier. Like EEG, it translates electrical activity taking place in the brain into understandable languages with peculiar configurations, patterns, durations, and voltages in physiological states. Alterations in the normal configuration, pattern, duration, and voltage occur in pathophysiological conditions. Electrocorticographic signals are synchronous discharges transmitted to the electrodes directly from the cerebral cortex, implying that the signals must pass through the cerebrospinal fluid and the meninges before reaching the electrocorticograph (Asano, Juhász, Shah, Sood, & Chugani, 2009; Hashiguchi, Morioka, Yoshida, Miyagi, et al., 2007). Like EEG, electrocorticography makes use of electrodes to measure action potential (Ulbert, Halgren, Heit, & Karmos, 2001). As far as defining epileptogenic zones is concerned, electrocorticogram is the gold standard. In addition, it can be used to measure cortical function and in forecasting epileptic surgical resectioning. In brain–computer interface, electrocardiogram is one of the most promising procedures (Shenoy, Miller, Ojemann, & Rao, 2007).

Role of data scientists in epileptic seizure detection

The use of machine learning in health and disease conditions over the years has produced promising results (Mahmud, Kaiser, Hussain, & Vassanelli, 2018). Scientists from different fields including data science and machine learning are working to identify potential strategies for appropriate diagnosis of epileptic seizure. In this regard, application of machine learning has led to the discovery of relevant patterns/presentations obtained from several data (Yong, Ikou, Jiafu, & Zhu, 2011). Many studies have investigated seizure diagnosis with emphasis on features, classification, and accuracy without considering the difficulties encountered by data scientists during the processing of data obtained from several EEG patterns to develop an algorithm to enhance accurate diagnosis. In recent years, there has been a significant increase in the usage of machine learning to classify and extract major EEG signals that assist in detecting seizure, identifying affected brain areas, and related developments (Abbasi & Goldenholz, 2019; Quintero-Rincón, D'giano, & Batatia, 2019). Several characteristics and classifications of EEG datasets, which are the major prerequisite for the detection of seizure, have been reported in literature. Effective utilization of EEG patterns and adequate analysis using computational and sophisticated statistical procedures have been proposed as a model for automatic seizure diagnosis and this has been validated by several signal processing techniques and data science approaches (Siddiqui, Morales-Menendez, Huang, & Hussain, 2020). Therefore, data scientists have played significant roles in the detection of epileptic seizure using machine learning approaches. These roles are numerous and have been classified into four major categories.

Collection of data: This requires collecting data from the pattern of brain signals using different diagnostic tools, primarily EEG. This raw information is monitored by the scientist and grouped into seizure and non-seizure conditions (Herwig, Satrapi, & Schönfeldt-Lecuona, 2003).

Transformation of data: The next step after data collection is transforming the data into a simple format for proper analysis and to provide the required information on seizure detection. Processing and cleaning data, integrating and storing data, and preliminary data research and exploratory data analysis are all key responsibilities at this stage.

Preparation of data: This is the major stage where diagnostic information is extracted from the data collected. Different techniques are used to extract definitive features on EEG patterns (e.g., time, frequency, discrete wavelet, Fourier transformation, time–frequency ratio) for seizure detection. After extracting the diagnostic information, the data becomes more valuable such that it ultimately serves as a potential model and algorithm. Applying data science techniques and performance evaluation: These techniques include machine learning, statistical modeling, and artificial intelligence to achieve more precise seizure detection. Also, these techniques explore relevant information obtained from EEG data. In terms of performance evaluation, the precision of the result achieved is utilized to validate various methods, with 10-fold cross validation being the most prevalent methodology (Arlot & Celisse, 2010). Data scientists focus mostly on data analysis and research needed to identify suitable machine learning techniques to use and model appropriate algorithms and prototypes for testing.

Intelligent diagnostic approaches: Machine learning and deep learning

Machine learning is the application of a computer algorithm that improves itself over time as a result of experience or data. Machine learning algorithms build a model based on the available sample data (training data) to generate predictions without the need for prior programming. Thus, the underlying principle of machine learning is to input training data to the learning algorithm, which later develops a new set of rules depending on the inferences derived from the data inputted (Davenport & Kalakota, 2019). This rule can be applied to solve complicated assignments that could not be previously programmed. Machine learning is widely used in the healthcare system to improve diagnosis (e.g., neurological disorders) based on data derived from previous experience and inputted to the system to generate a computer algorithm. Machine learning has been found in studies to be capable of delivering good results in seizure diagnosis, suggesting that this technology could pave the way for clinical translation of seizure forecasting devices (Adetunji, Adetunji, & Olaniyan, 2021; Adetunji, Ajayi, et al., 2021; Adetunji et al., 2021a, 2021b; Adetunji, Inobeme, et al., 2021; Adetunji & Varma, 2020; Akram et al., 2020; Das, Mishra, & Saraswathy, 2020; Egbuna et al., 2020; Roli, Adetunji, Mishra, Adetunji, & Mishra, 2020). The detection of epileptic seizures has been a long-standing problem, and numerous seizure detection strategies have been developed, including feature-based techniques combined with machine learning, which need domain expertise from users and compromise model performance (Siddiqui et al., 2020).

Deep learning is a subclass of machine learning that relies on neural networks with multiple layers to automate the extraction of characteristics from enormous data without requiring human domain expertise, resulting in an excellent method with broad acceptance across the therapeutic profession (LeCun, Bengio, & Hinton, 2015). Deep learning is a sophisticated computational technology that allows features to be learned from data automatically (LeCun et al., 2015).

Deep learning is typically applied to a type of algorithm known as a deep neural network to complete specific tasks. Because of the abundance of huge data, deep learning has been used to solve a wide range of problems. Much research has looked at deep learning algorithms for automatic seizure identification; however, the results aren't comparable because different deep learning methods were tested on different datasets with varying input types and sizes (including raw EEG images and patterns). Because deep learning methods produce improved results in seizure detection, it is critical to adopt and evaluate these models across the *epi*lepsy community. It is also critical to implement models with a variety of properties that may be employed in a variety of situations, such as offline seizure detection in the lab or in the clinic. An approach with high sensitivity (such as a deep learning model) could be applied to manual inspection, resulting in a significant reduction of rigorous screening and improved accuracy of recognized seizure events.

Selecting appropriate machine learning classifiers and features

Studies have revealed that the selection of features and classifiers in machine learning is very important in the data pre-processing step. The purpose is to eliminate irrelevant, redundant, and outlier attributes from the dataset with the aim of enhancing and improving the algorithm platform performance. Various selection approaches include support vector machines, wrappers, k-nearest neighbor classifier, random forest, linear discriminant analysis, meta-learning, and filters help to identify important simple data characteristics, statistical data characteristics, human activity recognition using smartphones, and information theoretic data characteristics. Feature selection enables simplification of complex data, reduces time consumed, avoids dimensionality, and decreases overfilling by enhancing generalization (Blum & Langley, 1997).

Summary

Neurological diseases are spontaneous neuronal discharges of physiological signals during a pathological state. They affect at least 3% of the global population, thus their diagnosis and treatment require urgent attention. Epilepsy is a neurological disease associated with diverse pathological signals of the brain. Several non-invasive approaches have been developed to reduce this global burden, such as MRI, SPECT, EEG, PET, surface-enhanced Raman spectroscopy, fluorescence, Förster resonance energy transfer and photoinduced electron transfer, chemiluminescence, optical fiber biosensing, fast scan voltammetry, colorimetry, and magnetoencephalography. These approaches can analyze physiological signals representing the cellular pathophysiological features of this epileptic condition through quorum sensing and microcavity platforms. These platforms analyze the neurotransmitters and chemicals secreted by neurons during epileptic electrical impulse transmission, thus generating codes and action potential waves. This brain–computer interface is a promising procedure, thus scientists from different fields including data science and machine learning are working to identify potential strategies for diagnosing epileptic seizure through appropriate algorithm models and prototypes.

Conclusion and future research

This chapter discussed the roles of neurons in the generation of several physiological and pathophysiological signals useful for neurological analysis in health and disease conditions. Several approaches have been developed by scientists to monitor and analyze these signals such as electrocorticography and EEG, particularly in the diagnosis of neurological disorders such as epileptic seizure. Because these techniques are laborious and time-intensive, there is an urgent need to develop automated techniques to assist medical experts in early diagnosis, detection, and classification of epileptic seizure. The development of intelligent diagnostic techniques in epileptic seizure utilizing artificial intelligence and algorithms for classification and detection of symptoms is important for detailed understanding of the physiology and pathophysiology of seizure. Further research on the development of artificial intelligence models across the epilepsy community to facilitate improved screening and accuracy of recognized seizure events is needed.

References

Adetunji, J. B., Adetunji, C. O., & Olaniyan, O. T. (2021). African walnuts: A natural depository of nutritional and bioactive compounds essential for food and nutritional security in Africa. In O. O. Babalola (Ed.), *Food security and safety*. Cham: Springer.

Abbasi, B., & Goldenholz, D. M. (2019). Machine learning applications in epilepsy. *Epilepsia, 60*(10), 2037–2047. https://doi.org/10.1111/epi.16333.

Adetunji, C. O., Ajayi, O. O., Akram, M., Olaniyan, O. T., Chishti, M. A., Inobeme, A., et al. (2021). Medicinal plants used in the treatment of influenza a virus infections. In K. Dua, S. Nammi, D. Chang, D. K. Chellappan, G. Gupta, & T. Collet (Eds.), *Medicinal plants for lung diseases*. Singapore: Springer.

Adetunji, C. O., Akram, M., Olaniyan, O. T., Ajayi, O. O., Inobeme, A., Olaniyan, S., et al. (2021). Targeting SARS-CoV-2 novel corona (COVID-19) virus infection using medicinal plants. In K. Dua, S. Nammi, D. Chang, D. K. Chellappan, G. Gupta, & T. Collet (Eds.), *Medicinal plants for lung diseases*. Singapore: Springer.

Adetunji, C. O., Akram, M., Olaniyan, O. T., Ajayi, O. O., Inobeme, A., Olaniyan, S., et al. (2021a). Medicinal plants used in the treatment of pulmonary hypertension. In K. Dua, S. Nammi, D. Chang, D. K. Chellappan, G. Gupta, & T. Collet (Eds.), *Medicinal plants for lung diseases*. Singapore: Springer. https://doi.org/10.1007/978-981-33-6850-7_14.

Adetunji, C. O., Inobeme, A., Olaniyan, O. T., Ajayi, O. O., Olaniyan, S., & Adetunji, J. B. (2021). Application of Nanodrugs derived from active metabolites of medicinal plants for the treatment of inflammatory and lung diseases: Recent advances. In K. Dua, S. Nammi, D. Chang, D. K. Chellappan, G. Gupta, & T. Collet (Eds.), *Medicinal plants for lung diseases*. Singapore: Springer.

Adetunji, C. O., & Varma, A. (2020). Biotechnological application of trichoderma: a powerful fungal isolate with diverse potentials for the attainment of food safety, management of pest and diseases, healthy planet, and sustainable agriculture. In C. Manoharachary, H. B. Singh, & A. Varma (Eds.), *Vol. 61. Trichoderma: Agricultural applications and beyond. Soil biology*. Cham: Springer.

Akram, M., Jabeen, F., Daniyal, M., Zainab, R., Haq, U. U., Adetunji, C. O., et al. (2020). Genetic engineering of novel products of health significance: Recombinant DNA technology. In C. Egbuna, & T. G. Dable (Eds.), *Functional foods and nutraceuticals*. Cham: Springer. https://doi.org/10.1007/978-3-030-42319-3_26.

Alarcón, G., & Valentín, A. (2012). *Introduction to epilepsy*. Cambridge: Cambridge University Press.

Andrzejak, R. G., Widman, G., Lehnertz, K., Rieke, C., David, P., & Elger, C. E. (2001). Epileptic process as nonlinear deterministic dynamics in stochastic environment: An evaluation on mesial temporal lobe epilepsy. *Epilepsy Research*, *44*, 129–140.

Anwar, H., Khan, Q. U., Nadeem, N., Pervaiz, I., Ali, M., & Cheema, F. F. (2020). Epileptic seizures. *Discoveries (Craiova, Romania)*, *8*(2), e110. https://doi.org/10.15190/d.2020.7 (PMID: 32577498; PMCID: PMC7305811).

Arlot, S., Celisse, A., et al. (2010). A survey of cross-validation procedures for model selection. *Statistics Surveys*, *4*, 40–79.

Asano, E., Juhász, C., Shah, A., Sood, S., & Chugani, H. T. (2009). Role of subdural electrocorticography in prediction of long-term seizure outcome in epilepsy surgery. *Brain*, *132*(Pt 4), 1038–1047. https://doi.org/10.1093/brain/awp025.

Babloyantz, A., & Destexhe, A. (1986). Low-dimensional chaos in an instance of epilepsy. *Proceedings of the National Academy of Sciences of the United States of America*, *83*, 3513.

Bafico, A., Liu, G., Goldin, L., Harris, V., & Aaronson, S. A. (2004). An autocrine mechanism for constitutive Wnt pathway activation in human cancer cells. *Cancer Cell*, *6*(5), 497–506.

Bazan, C., Gaultois, H., Shaikh, A., Gillespie, K., Frederick, S., Amjad, A., ... Belal, N. (2020). A systematic literature review of theinfluence of the university's environmentand support system on the precursors ofsocial entrepreneurial intention of students. *Journal of Innovation and Entrepreneurship*, *9*, 4. https://doi.org/10.1186/s13731-020-0116-9.

Berger, A. (2003). How does it work? Positron emission tomography. *British Medical Journal (Clinical research ed.)*, *326*(7404), 1449.

Blum, A. L., & Langley, P. (1997). Selection of relevant features and examples in machine learning. *Artificial Intelligence*, *97*(1–2), 245–271. https://doi.org/10.1016/s0004-3702(97)00063-5.

Chaudhary, U. J., Duncan, J. S., & Lemieux, L. (2011). A dialogue with historical concepts of epilepsy from the babylonians to hughlings Jackson: Persistent beliefs. *Epilepsy & Behavior*, *21*(2), 109–114.

Chernecky, C. C., & Berger, B. J. (2013). *Laboratory tests and diagnostic procedures* (6th ed., p. *1232*). St. Louis, Mo: Elsevier.

Choi, J. H., Jung, H. K., & Kim, T. (2006). A new action potential detector using the MTEO and its effects on spike sorting systems at low signal-to-noise ratios. *IEEE Transactions on Biomedical Engineering*, *53*(4), 738–746.

Das, A. K., Mishra, S., & Saraswathy, G. S. (2020). Predicting CoVID-19 community mortality risk using machine learning and development of an online prognostic tool. *PeerJ*, *8*, e10083. https://doi.org/10.7717/peerj.10083 (PMID: 33062451; PMCID: PMC7528809).

Davenport, T., & Kalakota, R. (2019). The potential for artificial intelligence in healthcare. *Future Healthcare Journal*, *6*(2), 94–98.

Devous, M. D., Sr. (2005). Single-photon emission computed tomography in neurotherapeutics. *NeuroRx: The Journal of the American Society for Experimental NeuroTherapeutics*, *2*(2), 237–249.

Doğaner, B. A., Yan, L., & Youk, H. (2016). Autocrine signaling and quorum sensing: Extreme ends of a common Spectrum. *Trends in Cell Biology*, *26*(4), 262–271. https://doi.org/10.1016/j.tcb.2015.11.002.

Egbuna, C., Gupta, E., Ezzat, S. M., Jeevanandam, J., Mishra, N., Muhammad Akram, N., et al. (2020). Aloe species as valuable sources of functional bioactives. In C. Egbuna, & G. Dable Tupas (Eds.), *Functional foods and nutraceuticals*. Cham: Springer.

Feinberg, I., & Campbell, I. G. (2012). Longitudinal sleep EEG trajectories indicate complex patterns of adolescent brain maturation. *AJP: Regulatory, Integrative and Comparative Physiology*, *304*(4), R296–R303.

Fergus, P., Hignett, D., Hussain, A. J., & Al-Jumeily, D. (2014). An advanced machine learning approach to generalised epileptic seizure detection. In D. S. Huang, K. Han, & M. Gromiha (Eds.), *Vol. 8590. Intelligent computing in bioinformatics. ICIC 2014. Lecture notes in computer science*. Cham: Springer. https://doi.org/10.1007/978-3-319-09330-7_14.

Frank, G. W., Lookman, T., Nerenberg, M. A. H., Essex, C., Lemieux, J., & Blume, W. (1990). Chaotic time series analyses of epileptic seizures. *Physica D*, *46*, 427.

Gerber, H. P., Malik, A. K., Solar, G. P., Sherman, D., Liang, X. H., Meng, G., … Ferrara, N. (2002). VEGF regulates haematopoietic stem cell survival by an internal autocrine loop mechanism. *Nature*, *417*(6892), 954–958. https://doi.org/10.1038/nature00821. PMID: 12087404.

Goldman, R. I., Stern, J. M., Engel, J., Jr., & Cohen, M. S. (2002). Simultaneous EEG and fMRI of the alpha rhythm. *Neuroreport*, *13*(18), 2487–2492. https://doi.org/10.1097/01.wnr.0000047685.08940.d0 (PMID: 12499854; PMCID: PMC3351136).

Haas, L. F. (2003). Hans Berger (1873-1941), Richard Caton (1842-1926), and electroencephalography. *Journal of Neurology, Neurosurgery, and Psychiatry*, *74*(1), 9.

Hannah, J. A., & Brodie, M. J. (1998). Epilepsy and learning disabilities—A challenge for the next millennium? *Seizure*, *7*(1), 3–13.

Hashiguchi, K., Morioka, T., Yoshida, F., Miyagi, Y., et al. (2007). Correlation between scalp-recorded electroencephalographic and electrocorticographic activities during ictal period. *Seizure*, *16*(3), 238–247. https://doi.org/10.1016/j.seizure.2006.12.010.

Herwig, U., Satrapi, P., & Schönfeldt-Lecuona, C. (2003). Using the international 10-20 EEG system for positioning of transcranial magnetic stimulation. *Brain Topography*, *16*(2), 95–99. https://doi.org/10.1023/b:brat.0000006333.93597.9d. PMID: 14977202.

Huff, J. S., & Murr, N. (2021). Seizure [Updated 2021 Sep 13]. In *StatPearls [Internet]*. Treasure Island (FL): StatPearls Publishing. Available from: https://www.ncbi.nlm.nih.gov/books/NBK430765/. (Accessed January 2022).

Iasemidis, L. D., Pardalos, P., Sackellares, J. C., & Shiau, D.-S. (2001). Quadratic binary programming and dynamical system approach to determine the predictability of epileptic seizures. *Combinatorial Optimization*, *5*(9).

Iqbal, A., Sheikh, A., & Karayannis, T. D. N. R. D. (2019). High-throughput detection of neurons for brain-wide analysis with deep learning. *Scientific Reports*, *9*, 13828.

Komoto, Y., Ohshiro, T., Yoshida, T., et al. (2020). Time-resolved neurotransmitter detection in mouse brain tissue using an artificial intelligence-nanogap. *Scientific Reports*, *10*, 11244.

Kozai, T. D., Jaquins-Gerstl, A. S., Vazquez, A. L., Michael, A. C., & Cui, X. T. (2015). Brain tissue responses to neural implants impact signal sensitivity and intervention strategies. *ACS Chemical Neuroscience*, *6*, 48–67. https://doi.org/10.1021/cn500256e.

Le Van Quyen, M., Martinerie, J., Navarro, V., Boon, P., Have, M. D., Adam, C., et al. (2001). Anticipation of epileptic seizures from standard EEG recordings. *Lancet*, *357*, 183.

LeCun, Y., Bengio, Y., & Hinton, G. (2015). Deep learning. *Nature*, *521*(7553), 436–444. https://doi.org/10.1038/nature14539. PMID: 26017442.

Lehnertz, K., Arnhold, J., Grassberger, P., & Elger, C. E. (2000). *Chaos in brain* (p. 384). Singapore: World Scientific. https://doi.org/10.1142/4347.

Lehnertz, K., & Elger, C. E. (1998). Can epileptic seizures be predicted? Evidence from nonlinear time series analysis of brain electrical activity. *Physical Review Letters*, *80*(22), 5019–5022.

Lenkov, D. N., Volnova, A. B., Pope, A. R., & Tsytsarev, V. (2013). Advantages and limitations of brain imaging methods in the research of absence epilepsy in humans and animal models. *Journal of Neuroscience Methods*, *212*(2), 195–202. https://doi.org/10.1016/j.jneumeth.2012.10.018 (Epub 2012 Nov. 6; PMID: 23137652).

Mahmud, M., Kaiser, M. S., Hussain, A., & Vassanelli, S. (2018). Applications of deep learning and reinforcement learning to biological data. *IEEE Transactions on Neural Networks and Learning Systems*, *29*(6), 2063–2079. https://doi.org/10.1109/TNNLS.2018.2790388. PMID: 29771663.

Martinerie, J., Adam, C., Le Van Quyen, M., Baulac, M., Clemenceau, S., Renault, B., et al. (1998). Epileptic seizures can be anticipated by non-linear analysis. *Nature Medicine*, *4*, 1173.

Martins, N. R. B., Erlhagen, W., & Freitas, R. A., Jr. (2015). Action potential monitoring using neuronanorobots: Neuroelectric nanosensors. *International Journal of Nanoscience and Nanotechnology*, *1*(2), 1–29.

Mormann, F., Lehnertz, K., David, P., & Elger, C. E. (2000). Mean phase coherence as a measure for phase synchronization and its application to the EEG of epilepsy patients. *Physica D*, *144*, 358.

Müller, E., Wang, W., Qiao, W., et al. (2016). Distinguishing autocrine and paracrine signals in hematopoietic stem cell culture using a biofunctional microcavity platform. *Scientific Reports*, *6*, 31951. https://doi.org/10.1038/srep31951.

Müller, J., Ballini, M., Livi, P., Chen, Y., Radivojevic, M., Shadmani, A., et al. (2015). High-resolution CMOS MEA platform to study neurons at subcellular, cellular, and network levels. *Lab on a Chip*, *15*, 2767–2780.

National Research Council (US) and Institute of Medicine (US) Committee on the Mathematics and Physics of Emerging Dynamic Biomedical Imaging. (1996). Mathematics and physics of emerging biomedical imaging. In *Chapter 5, Single Photon Emission Computed Tomography*. Washington (DC): National Academies Press (US). Available from: https://www.ncbi.nlm.nih.gov/books/NBK232492/.

Palmini, A. (2006). The concept of the epileptogenic zone: A modern look at Penfield and Jasper's views on the role of interictal spikes. *Epileptic Disorders*, *8*(Suppl 2), S10–S15.

Panachakel, J. T., & Ramakrishnan, A. G. (2021). Decoding covert speech from EEG-A comprehensive review. *Frontiers in Neuroscience*, *15*, 642251. https://doi.org/10.3389/fnins.2021.642251 (PMID: 33994922; PMCID: PMC8116487).

Peters, J. L., Miner, L. H., Michael, A. C., & Sesack, S. R. (2004). Ultrastructure at carbon fiber microelectrode implantation sites after acute voltammetric measurements in the striatum of anesthetized rats. *Journal of Neuroscience Methods*, *137*, 9–23.

Qu, H., & Gotman, J. (1993). Improvement in seizure detection performance by automatic adaptation to the eeg of each patient. *Electroencephalography and Clinical Neurophysiology*, *86*(2), 79–87.

Quintero-Rincón, A., D'giano, C., & Batatia, H. A. (2019). A quadratic linear-parabolic model-based EEG classification to detect epileptic seizures. *Journal of Biomedical Research*, *34*(3), 205–212. https://doi.org/10.7555/JBR.33.20190012 (PMID: 32561700; PMCID: PMC7324279).

Radivojevic, M., Franke, F., Altermatt, M., Muller, J., Hierlemann, A., & Bakkum, D. J. (2017). Tracking individual action potentials throughout mammalian axonal arbors. *eLife*, *6*, e30198.

Ray, S. (2014). The cell: A molecular approach. *The Yale Journal of Biology and Medicine*, *87*(4), 603–604.

Reynolds, E. H. (2009). Milestones in epilepsy. *Epilepsia*, *50*(3), 338–342. https://doi.org/10.1111/j.1528-1167.2009.02050.x.

Roli, O. I., Adetunji, C. O., Mishra, P., Adetunji, J. B., & Mishra, R. R. (2020). In P. Mishra, & R. R. Mishra (Eds.), *Caffeine: Nutraceutical and health benefit of caffeine-containing commodities and products*.

Sanes, D. H., & Reh, T. A. (2012). *Development of the nervous system* (3rd ed., pp. 211–214). Elsevier Academic Press.

Sasaki, T., Matsuki, N., & Ikegaya, Y. (2011). Action-potential modulation during axonal conduction. *Science*, *331*, 599–601. https://doi.org/10.1126/science.1197598.

Sazgar, M., & Young, M. G. (2019). *Absolute epilepsy and EEG rotation review: Essentials for trainees*. Springer.

Schlange, T., Matsuda, Y., Lienhard, S., Huber, A., & Hynes, N. E. (2007). Autocrine WNT signaling contributes to breast cancer cell proliferation via the canonical WNT pathway and EGFR transactivation. *Breast Cancer Research*, *9*(5), R63.

Shafer, P. O., & Sirven, J. I. (2014). *Epilepsy statistics*. Epilepsy Foundation.

Shenoy, P., Miller, K. J., Ojemann, J. G., & Rao, R. P. N. (2007). Generalized features for electrocorticographic BCIs (PDF). *IEEE Transactions on Biomedical Engineering*, *55*(1), 273–280.

Shu, Y., Hasenstaub, A., Duque, A., Yu, Y., & Mccormick, D. A. (2006). Modulation of intracortical synaptic potentials by presynaptic somatic membrane potential. *Nature*, *441*, 761–765.

Siddiqui, M. K., Morales-Menendez, R., Huang, X., & Hussain, N. (2020). A review of epileptic seizure detection using machine learning classifiers. *Brain Informatics*, 7(1), 1–18.

Sirven, J. I. (2015). Epilepsy: A spectrum disorder. *Cold Spring Harbor Perspectives in Medicine*, *5*(9), a022848.

Srinivasan, R. (2006). Anatomical constraints on source models for high-resolution EEG and MEG derived from MRI. *Technology in Cancer Research & Treatment*, *5*(4), 389–399.

Stafstrom, C. E., & Carmant, L. (2015). Seizures and epilepsy: An overview for neuroscientists. *Cold Spring Harbor Perspectives in Medicine*, *5*(6), a022426. https://doi.org/10.1101/cshperspect.a022426 (PMID: 26033084; PMCID: PMC4448698).

Steinlein, O. K. (2008). Genetics and epilepsy. *Dialogues in Clinical Neuroscience*, *10*(1), 29–38. https://doi.org/10.31887/DCNS.2008.10.1/oksteinlein.

Tamagawa, H., Funatani, M., & Ikeda, K. (2016). Ling's adsorption theory as a mechanism of membrane potential generation observed in both living and nonliving systems. *Membranes*, *6*(1), 11.

Tatum, W. O., Dworetzky, B. A., & Schomer, D. L. (2011). Artifact and recording concepts in EEG. *Journal of Clinical Neurophysiology*, *28*(3), 252–263. https://doi.org/10.1097/WNP.0b013e31821c3c93. PMID: 21633251.

Theiler, J. (1995). On the evidence for low-dimensional chaos in an epileptic electroencephalogram. *Physics Letters A*, *196*, 335.

Ulbert, I., Halgren, E., Heit, G., & Karmos, G. (2001). Multiple microelectrode-recording system for human intracortical applications. *Journal of Neuroscience Methods*, *106*(1), 69–79.

Vaquero, J. J., & Kinahan, P. (2015). Positron emission tomography: Current challenges and opportunities for technological advances in clinical and preclinical imaging systems. *Annual Review of Biomedical Engineering*, *17*, 385–414. https://doi.org/10.1146/annurev-bioeng-071114-040723 (PMID: 26643024; PMCID: PMC5299095.).

Vespa, P. M., Nenov, V., & Nuwer, M. R. (1999). Continuous EEG monitoring in the intensive care unit: Early findings and clinical efficacy. *Journal of Clinical Neurophysiology*, *16*(1), 1–13.

Vezzani, A., Fujinami, R. S., White, H. S., Preux, P. M., Blümcke, I., Sander, J. W., & Löscher, W. (2016). Infections, inflammation and epilepsy. *Acta Neuropathologica*, *131*(2), 211–234. https://doi.org/10.1007/s00401-015-1481-5 (Epub 2015 Sep. 30; PMID: 26423537; PMCID: PMC4867498).

Wells, A. (2006). *Cell motility in cancer invasion and metastasis* (p. XIII). The Netherlands: Springer Science and Business Media. 353 https://doi.org/10.1007/b103440.

WHO. (2015). *Media center epilepsy, (Fact sheet N999)*. http://www.who.int/mediacentre/factsheets/fs999/en/. Accessed 15 July 2019.
World Health Organization. (2005). *Atlas: Epilepsy care in the world*. Geneva: World Health Organization.
World Health Organization. (2006). *Neurological disorders: Public health challenges*. New York: World Health Organization.
Yang, W. H., Cha, J. H., Xia, W., Lee, H. H., Chan, L. C., Wang, Y. N., et al. (2018). Juxtacrine signaling inhibits antitumor immunity by upregulating PD-L1 expression. *Cancer Research*, *78*(14), 3761–3768. https://doi.org/10.1158/0008-5472.CAN-18-0040.
Yong, Y., Ikou, K., Jiafu, T., & Zhu, J. M. (2011). *Data mining. Concepts, methods and applications in management and engineering design London* (pp. 1–312). London: Springer.

CHAPTER 15

Neural signaling and communication using machine learning

Olugbemi T. Olaniyan[a], Charles O. Adetunji[b], Ayobami Dare[c], Olorunsola Adeyomoye[d], Mayowa J. Adeniyi[e], and Alex Enoch[f]
[a]Department of Physiology, Laboratory for Reproductive Biology and Developmental Programming, Rhema University, Aba, Nigeria
[b]Department of Microbiology, Applied Microbiology, Biotechnology and Nanotechnology Laboratory, Edo State University Uzairue, Iyamho, Nigeria
[c]Department of Physiology, School of Laboratory Medicine and Medical Sciences, College of Health Sciences, Westville Campus, University of KwaZulu-Natal, Durban, South Africa
[d]Department of Physiology, University of Medical Sciences, Ondo, Nigeria
[e]Department of Physiology, Edo State University Uzairue, Iyamho, Nigeria
[f]Department of Human Physiology, Ahmadu Bello University Zaria, Zaria, Nigeria

Introduction

Novel advances in wireless terminals, communication systems, and equipment have led to a revolution in human life. Moreover, the higher rate of limitation in the rate of spectrum resource will not be sufficient to meet the request of Internet of Things (IoT) networks (Khan, Rehmani, & Rachedi, 2017) and coming 5G (Hong, Wang, Wang, & Shi, 2014) that have the capacity to meet the significant challenge of the spectrum application and management. There is a need to have a greater priority, especially to spectrum policy, that could enhance the rate of signal recognition algorithms satisfy the demand for spectrum management, and increase the rate of spectrum sensing. Scientists are currently focusing on the application of cognitive spectrum-sharing techniques and dynamic spectrum access to increase the mode of communication effectiveness. The greatest challenges include explicit situations and poor flexibility to the diverse channel situations and device types.

Novel signal recognition mechanisms and recent spectrum sensing schemes could play a significant role in the application of cognitive radio (CR) communication in the biomedical sector (Adetunji & Varma, 2020; Adetunji, Ajayi, et al., 2021; Adetunji et al., 2021a, 2021b; Adetunji, Inobeme, et al., 2021; Adetunji, Adetunji, et al., 2021; Akram et al., 2020; Egbuna et al., 2020; Roli, Adetunji, Mishra, Adetunji, & Mishra, 2020). As such, new spectrum sensing schemes and novel signal recognition

Artificial Intelligence for Neurological Disorders
https://doi.org/10.1016/B978-0-323-90277-9.00010-9

mechanisms are attracting the attention of researchers hoping to pave the way for CR (Ali & Hamouda, 2017). Furthermore, wireless signal recognition, which entails wireless technology recognition and modulation recognition, has many applications in device identification, signal reconnaissance, and antijamming and interception of signals.

Interestingly, machine learning tools have numerous applications in the designing of challenges in communication entails demodulation schemes (Dörner, Cammerer, Hoydis, & Brink, 2018; O'Shea & Hoydis, 2017), estimating channel model parameters (Lee et al., 2017; O'Shea et al., 2017), multiuser detection in code-division multiple-access (CDMA) systems (Aazhang et al., 1992; Işık & Taşpınar, 2007; Mitra & Poor, 1994; Murillo-Fuentes, Caro, & Perez-Cruz, 2005), detection and channel decoding, (Cammerer, Gruber, Hoydis, & Brink, 2017; Dörner et al., 2018; Nachmani et al., 2018; Liang et al., 2018), and decoding of linear codes (Nachmani, Beery, & Burshtein, 2016). Deep neural networks have several applications in bioinformatics (Li & Yu, 2016), such as image classification (He, Zhang, Ren, & Sun, 2016; Krizhevsky, 2012), machine translation (Bahdanau et al., 2014; Cho et al., 2014), and speech recognition (Amodei et al., 2016; Graves & Jaitly, 2014). This chapter focuses on recent advances in neural signaling, signal decoding, and communication, which form the basis of central nervous system functions, interaction and activity and the application of different biometric data derived from the neural cells.

Electrophysiology of brain waves

Brain waves are produced when masses of neurons communicating with each other synchronously discharge. The potential difference between neuron and extracellular fluid is called resting membrane potential. A neuron has a resting membrane potential averaging −85 mv. The upward stroke of the wave represents depolarization, and the downward stroke represents repolarization. Depolarization is due to the influx of sodium ions into the neurons and repolarization is due to the efflux of potassium ions through potassium channels. Types of waves include alpha, beta, delta, gamma, mu, and sensorimotor rhythm waves.

Electrophysiology of alpha waves

The frequency of an alpha wave ranges from 7 to 12 Hz. Alpha waves can be found in the posterior regions of the head on either side, with higher amplitudes on the dominant side. Physiologically, alpha waves are salient during relaxed wakefulness with eyes closed (resting visual cortex). This also occurs during REM sleep. Alpha waves have also been correlated with inhibition control across the brain. These waves are for controlling cognitive processes (Ito, Nikolaev, & van Leeuwen, 2005; von Stein & Sarnthein, 2000). Pathologically, alpha waves are dominant in comatose patients. Although there is controversy concerning neural structures producing alpha rhythms, the thalamus is recognized as the primary alpha pacemaker. The classic posterior alpha rhythm is orchestrated by the pulvinar and lateral geniculate nucleus (Saalmann, Pinsk, Wang, Li, & Kastner, 2012). There is also an assumption that alpha waves stem from the infragranular layer of the cerebral cortex. They can be recorded using electroencephalography and magnetoencephalography.

Electrophysiology of beta waves

The frequency of beta waves is 12–30 Hz. Beta waves are located on both sides of the brain but most prominently at the front. They are low-amplitude waves. Physiologically, beta waves are salient during wakefulness, alert/working states, when a person is active, busy, or anxious, and during active concentration states. Beta waves indicate GABA-induced regulation of the cortical transmission. Research has shown the nexus between beta waves and rewards, such that a high beta wave is associated with the reception of gain when no gain is expected, and a low beta wave is prominent when a well-deserved gain is omitted (Cahn & Polich, 2006).

Electrophysiology of delta waves

The frequency of delta waves is between 1 and 3 Hz. Delta waves are high-amplitude waves located frontally in adults and posteriorly in children. They can also be found in the thalamus. Physiologically, these waves are salient during slow-wave sleep, but only in adults. In babies, delta waves may also occur during specific continuous attention tasks. Pathologically, delta waves occur in metabolic encephalopathy, hydrocephalus, subcortical lesions, diffuse lesions, and deep midline lesions. Evidence avails that the delta wave is sexually dimorphic, with females having more delta waves than males (Ehlers & Kupfer, 1997).

Electrophysiology of theta waves

The frequency of theta waves is 4–7 Hz. They can be found in the cerebral cortex and hippocampus. Physiologically, cortical theta waves are salient in young children during sleepiness, idling, or awakening in adolescents and adults. They are associated with the inhibition of elicited responses, like when someone is actively repressing a response or action. Hippocampal theta increases during meditation. Pathologically, theta waves occur in hydrocephalus, metabolic encephalopathy, focal subcortical lesions, and deep and midline syndromes (Seager, Johnson, Chabot, Asaka, & Berry, 2002).

Electrophysiology of gamma waves

The frequency of gamma waves is 25 to 140 Hz. Gamma waves occur during cognitive processes such as attention, working memory, and so on. They can be recorded using electroencephalography and magnetoencephalography. Aberration in gamma waves occurs in Alzheimer's disease, schizophrenia, and epilepsy (Hughes, 2008; Jia & Kohn, 2011; van Deursen, Vuurman, Verhey, van Kranen-Mastenbroek, & Riedel, 2008).

Electrophysiology of mu waves

The frequency of mu waves is 7–12 Hz. Mu waves occur when the body is at rest. They represent synchronized discharge from several pyramidal neurons involved in locomotion (Amzica, 2010).

Electrophysiology of sensorimotor rhythms

The frequency of sensorimotor rhythms is 13–15 Hz.

Neural signaling and communication

Neural signaling and communication

Communication between neurons occurs at synapses through electrical signals called action potentials.

Electrical signals as physiological signals

Electrical signals are propagable or non-propagable changes in resting membrane potential, which occur when excitable tissues discharge. Propagable changes in membrane potential are known as action potentials or electric impulses in nerve cells.

Action potentials

Changes in membrane potential are called action potentials. They consist of resting membrane potential (potential difference between the cell and its environment), depolarization (increase in membrane potential in a positive direction), and repolarization (decrease in membrane potential in a negative direction) (Tamagawa, Funatani, & Ikeda, 2016; Sanes and Reh, 2012). The action potential is a medium through which sensory or motor information is encoded and transmitted between the central and peripheral nervous system.

The advent of technology-driven techniques such as genetically encoded voltage indicators, high-density microelectrode array recordings, and subcellular patch-clamp recordings has contributed immensely to the generation of information on the function of neurons and pathology of axons (Müller et al., 2015; Radivojevic et al., 2017; Sasaki et al., 2011). (Martins, Erlhagen, and Freitas (2015)) proposed an approximately 3375 nm^3 field-effect transistor-based neuroelectric nanosensor that can detect action potentials with a spatio-temporal resolution of 0.1 ms, sufficient for waveform characterization at a peak human neural firing rate of 800 Hz.

Application of electrical signals

Discharge of excitable tissues results in the creation of an electric field that travels peripherally. These discharges are picked using sensors and processed and analyzed using non-invasive techniques such as electrocardiography, electromyography, electrodermal technique, electrooculography, electroretinography, electroencephalography, and electrocorticography. Through these techniques, voltage analogs picked from body surfaces are transformed into an understandable language, a wave. The vertical axis of a wave indicates the strength, amplitude, or volume of the signal, and the horizontal axis represents the duration or frequency of the signals.

Nowadays, physiological signals are integrated into vital sign panels and monitored such that whenever alteration from the physiological limit occurs, a mechanical signal (alarm) or warning light is displayed. Using a highly complex algorithm, it is possible to identify pathophysiological signals such as seizures using electroencephalography (EEG). Modern electrocardiographs are programmed to recognize pathophysiological signs such as myocardial infarction, myocardial ischemia, right and left atrial hypertrophies, right and left ventricular hypertrophies, hyperkalemia, hypokalemia, and others. This helps in accurate diagnosis and prognosis with the least possible artifacts.

Brain–computer interface (data acquisition)

The brain–computer interface (BCI) is a system that detects brain neural signals and analyzes and converts them to commands that are transmitted to output devices to perform specific tasks. This type of system creates an alternative pathway in contrast to neuromuscular transmission. It helps individuals with neuromuscular disorders such as stroke, spinal cord injury, cerebral palsy, and many others (Shih, Krusienski, & Wolpaw, 2012). The current use of electroencephalography, electrocorticography, and intracortical signals to initiate complex visual, auditory, and cognitive functions cannot be overemphasized. These brain signals control artificial arms, prostheses, cursors, and many other devices to restore normal functions (Nicolas-Alonso & Gomez-Gil, 2012). BCI could be used for medical rehabilitation of neural circuits that are necessary for performing neuromuscular and cognitive tasks. The advances in the use of the mind to control the environment have continued to gain momentum, and many research experts have continued to show interest. Furthermore, brain–computer technology is a rapidly emerging field, and more studies are needed to understand its mechanisms of operation. In the future, individuals who cannot speak, hear, or use their limbs may have the opportunity of using their thoughts to improve communication within their environment (Daly, Haden, Hagins, Papouchis, & Ramirez, 2015). In BCI technology, there are some basic misconceptions. Some believe that electroencephalography and electrocorticography are BCI systems, but they are now. They are only used to record brain signals and do not produce brain signals that control the environment. It is also a misconception that BCI devices read minds; they only allow physically challenged individuals to interact with the environment by using brain signals instead of muscles (Wolpaw, Birbaumer, McFarland, Pfurtscheller, & Vaughan, 2002). Despite BCI being a growing field, there are limitations in the use of implantable and non-invasive devices to augment or provide alternative ways of communicating with the environment. These devices may not have the capacity to perform complex or multiple tasks concurrently. Furthermore, some persons may not have the capacity to control these devices. In a study conducted by Pichiorri et al. (2015), BCI technology was used to assist motor imagery practise during recovery exercise from stroke. Twenty-eight sub-acute stroke patients were randomized into two groups. Both groups were allowed to undergo motor imagery training for one month. The first group was supported by a BCI device throughout the training period. Results showed better motor recovery outcomes in the

BCI group, suggesting that BCI technology has the potential to improve the rehabilitation of motor functions as well. Similarly, many other studies have reported the use of BCI to enhance motor activities and recovery from stroke, spinal cord injury, speech and auditory impairments, and many other neurological defects (Ang, Chua, & Phua, 2015; Cervera et al., 2018; Mattia et al., 2020).

Algorithm classification of brain functions using machine learning

Data acquisition in healthcare systems has posed a big challenge in the past few decades. Even patient health records are often not properly documented and sometimes difficult to trace (Sarkies et al., 2015). However, with advances in machine learning, data such as health records and other medical files are now being processed electronically. Neuroscience is not an exemption, as many medical imaging techniques use machine learning algorithms to diagnose and classify neurological disorders and assess brain functions (Valliani, Ranti, & Oermann, 2019). Machine learning uses the principle of computer science and statistics to improve data gathering and interpretation. Evidence has shown that machine learning can be used to detect seizures from EEG and perform auto-analysis of medical images (Siddiqui, Morales-Menendez, Huang, & Hussain, 2020). In a study conducted by Todorov et al. (2020), tissue clearing was used to image a mouse's whole brain vasculature. Thereafter, a machine learning approach was used to quantify and analyze the brain vasculature. Secondary intracranial collateral vascularization of the cranium was observed in the mouse. In addition, reduced vascularization was observed in the brainstem compared to the cerebrum. Machine learning was used to score the angio-architecture of the mouse brain and determine which part of the brain receives more perfusion than the others. Furthermore, machine learning has the capacity to classify complex patterns of different radiological images from X-rays, computer tomography (CT), functional magnetic resonance imaging (fMRI), single positron emission computer tomography (SPECT), and so on (Zhang, Harman, Ma, & Liu, 2019). Evidence have shown that machine learning algorithms could be used to predict the response of individuals with post-traumatic stress disorder (PTSD) and depressive disorder after exposure to magnetic stimulation (Zandvakili et al., 2019). In a study conducted by Zandvakili et al. (2019), a group of individuals with PTSD and depressive disorders were exposed to transcranial magnetic stimulation. The results obtained via a machine learning algorithm before and after the stimulations showed a 50% reduction in PTSD and depressive disorder scores.

Artificial intelligence and neural signals, communications

Neural networks are systems of software or hardware that operate like neurons in the human brain. Neural networks comprise different technologies like deep learning and machine learning that form part of artificial intelligence (AI). Artificial neural networks (ANNs) are the major techniques of machine learning. ANNs are formed to model neuron functionality, replicating the way humans learn. Neural networks form the input and output layers and a hidden layer containing units that transform input into the output so that the output layer can make use of the value. This technique has been used to develop several multifaceted signals for programmers to retrieve and train the machine to identify.

ANNs are components of AI that enable computers to mimic the human brain's capabilities to process information (Willamette, 2014). In the brain, external signals (such as a change in temperature, odor, or sound) are received by the brain's neurons (dendrons and axons), then transformed and processed in different brain regions to produce corresponding responses (output) that perform a specific task. Similarly, an ANN can be used to simulate the human brain. The required connections can be made using wires that act like neurons to accept external signals (inputs). These inputs are transformed into electrical signals that pass through several layers of the neural network for processing. An ANN comprises numerous nodes (similar to brain neurons) interconnected for adequate communication. These nodes utilize input data to accomplish tasks on trained data, and the outcomes of these tasks are transferred to other nodes (neurons) to generate an output. In ANNs, individual neurons receive and process an input to produce an output; the system of closely connected numerous neurons exhibits intelligent and rich behaviors (Shiffman, 2012). The outputs are typically in the form of algorithms that allow machines to execute complex tasks (e.g., facial recognition, real-time operations, adaptive learning) like the human brain. ANNs are programmed to replicate the role of the neural structure/architecture of the human brain. They are adaptive, flexible, and teachable with several external and internal signals (Willamette, 2014). Neural signals and the generated algorithms have been extensively applied in medicine to diagnose and treat various diseases.

Structurally, ANNs are easy to understand because they are composed of fewer artificial neurons than the human brain, which contains about 100 billion neurons. These neural networks have great potential in understanding neural signals since they help model and understand several biological neural systems and behavior (Poole, 2010). It has been proposed that the major means of understanding the brain's functions and developing intelligence is via models of the brain (Callataÿ, 2014). Since the brain inspires a model of computation that differs from how computers are being programmed, ANNs have become an exciting area of study that seeks to implement this paradigm. However, unlike ANNs, the brain consists of asynchronously distributed concurrent processes, with no major controller, compared to modern-day computers with inert memory and processors (Poole, 2010). The application of ANNs in medicine is still in its infancy, and it is expected that their use will bring significant advancement to the medical field, especially neuroscience, in the coming years. Recently, the neural network formed an essential part of machine learning to extract features from neurological images (CT scans, MRI) and physiological data to develop algorithms for diagnosing diseases (Stergiou & Siganos, 2014). The concept of AI develops from neuroscience, especially in how the human brain enables thinking, much of which is yet to be fully understood. However, advancements in AI have enabled scientists to continue to progress towards understanding the thought processes in the brain. ANNs are vital components of AI that process data and display intelligent behaviors (Agarwal, 2014).

AI is polysemous, utilizing efforts to comprehend human intelligence by constructing a mind within a machine and formulating machinery that can carry out important functions associated with some level of human intelligence (Ma, 2019). Examples of communicative AI include conversational agents, social robots, and automated writing software. These technologies were basically constructed with modifications in the subfield of AI, including natural language processing (NLP) and natural language generation (NLG). At the primary level, NLP and NLG have the common objective of processing human communication appropriately to facilitate proper interpretation of the messages presented in human language by the machine, rather than machine language, and creating messages in human language (Allen, 2003). For instance, voice-based assistants are developed from the application of NLP, while automated writing programs are developed from the application of NLG (Dörr, 2016). AI technologies differ in function as communicators, from the interpersonal interlocutor to a content producer. These technologies can act as voice-based assistants or automated news-writing programs. The common similarity in the functions of these technologies as a communicator is that they all act, in the field of communication, traditionally has been limited to humans. AI, other than facilitating communication, automates communication and the social processes dependent upon it (Gehl & Bakardjieva, 2017; Reeves, 2016). Also, AI-based communication technologies are developed as a platform through which messages are shared, which differs from the historical role of media developed as a means through which people communicate with one another (Gunkel, 2012).

Challenges and opportunities

It is evident that neurons are vital for optimal brain function, and the neural network in AI. ANNs are exceptionally important for real-time applications due to their quick response and parallel architecture. Although the reasoning principle of neural networks has great potential, an expert will only achieve optimal output by combining them with computing. Generally, putting a neural network into a system makes the system more intelligent in decision-making. ANNs can perform complex tasks that improve the brain's processing mechanisms (Jagannath, Polosky, Jagannath, Restuccia, & Melodia, 2019). There is a huge advantage to applying neural networks in medical science. Thus, it is recommended that studies relating to the application of neural networks should be introduced into neuroscience. Furthermore, studies in the field of computational neuroscience should be greatly promoted by providing facilities and teaching personnel in the field (Guzman & Lewis, 2020). However, despite the many opportunities and benefits of neural signals in neuroscience, there are significant challenges that may compromise the effectiveness and acceptability of these technologies. These include hardware dependence resulting from the architecture of the system and requirement of a processor with parallel processing power; unexplained functioning of the system, which is the major problem of neural networks because they do not give any clarity of concepts when probing solutions, thereby causing lack of trust in the network; lack of assurance of proper network structure because the developed structure is generated via trial and error, thus there are no specific guidelines to determine the appropriate structure; the need to translate the data to numerical form before introducing it to the network; and inability to calculate the time required for the network because the network performance is reduced to a certain value of the error on the sample, which could compromise the results (Fink et al., 2020).

Summary

Neural signaling and communication using machine learning establish an interaction between the human brain and a computer device to coordinate or execute a specific task by means of brain signals that are later translated into a device action (Mansoor, Usman, Jamil, & Naeem, 2020). This system relies on the central nervous system rather than muscles or peripheral nerves to establish a direct link between the brain and the external device (Abiri, Borhani, Sellers, et al., 2019).

Although not widely used now, BCI devices have the potential for wide application in the future. They are the only way for people that are physically impaired and unable to use their brain signals to move their muscles to function. BCI technology t can trigger the brain to enhance the remodeling of the patient's central nervous system (Malekmohammadi, Mohammadzade, Chamanzar, et al., 2019). The ability to correctly classify EEG features of a task and translate them into external control commands is a critical component of the BCI communication system (Ramírez-Moreno & Gutiérrez, 2019).

In recent years, deep learning has made outstanding advances in natural language processing, image identification, and speech recognition (Haeb-Umbach, Watanabe, Nakatani, et al., 2019; Tian, Fei, Zheng, et al., 2020). Because of its advantages in self-learning features, deep learning has been an excellent machine learning method in these fields (Kurkin, Hramov, Chholak, et al., 2020; Oh, Vicnesh, Ciaccio, Yuvaraj, & Acharya, 2019). As a result, deep learning is gradually being applied to pattern classification of EEG signals and feature extraction, which improves accuracy (Li, Wang, Yang, et al., 2019; Schirrmeister, Springenberg, Fiederer, et al., 2017).

It has also been found that adaptive classifiers gave the most accurate result among the classification algorithms for EEG-based control systems. Therefore, if an improved model with great accuracy for the non-invasive technique for classifying brain signals is needed, an ANN has better precision and is thus ideal. Because an ANN allows for fast processing, a guidance system could be offered during training to allow the user to improve the classifier's performance. More so, some trade-offs, such as model compatibility with brain signals (Mansoor et al., 2020).

Conclusion and future perspectives

Artificial neural networks are utilized to generate physiological signals that are used by BCIs for the development of neural networks through mapping, augmenting, assisting, and repairing sensory-motor or cognitive functions. EEG-based control systems use algorithm classifiers like tensor classifiers, adaptive classifiers, deep learning, and transfer learning approaches as non-invasive protocols to regulate electrical devices/circuits. This chapter provided a general overview of the recent advances in neural signaling, signal decoding, and communication as the basis of central nervous system functions, interaction, activity, and the application of different biometric data derived from the neural cells. Using a highly complex algorithm, it is possible to identify pathophysiological signals such as seizures using EEG. Thus, it is recommended that studies relating to the application of neural networks and computational physiology be introduced into neuroscience, which will help mitigate various challenges identified in these technologies.

References

Aazhang, B., et al. (1992). Neural networks for multiuser detection in codedivision multiple-access communications. *IEEE Transactions on Communications*, *40*(7), 1212–1222.

Abiri, R., Borhani, S., Sellers, E. W., et al. (2019). A comprehensive review of EEG-based brain-computer interface paradigms. *Journal of Neural Engineering*, *16*(1), 011001.

Adetunji, J. B., Adetunji, C. O., & Olaniyan, O. T. (2021). African walnuts: A natural depository of nutritional and bioactive compounds essential for food and nutritional security in Africa. In O. O. Babalola (Ed.), *Food security and safety*. Cham: Springer.

Adetunji, C. O., Ajayi, O. O., Akram, M., Olaniyan, O. T., Chishti, M. A., Inobeme, A., et al. (2021). Medicinal plants used in the treatment of influenza a virus infections. In K. Dua, S. Nammi, D. Chang, D. K. Chellappan, G. Gupta, & T. Collet (Eds.), *Medicinal plants for lung diseases*. Singapore: Springer.

Adetunji, C. O., Akram, M., Olaniyan, O. T., Ajayi, O. O., Inobeme, A., Olaniyan, S., et al. (2021a). Medicinal plants used in the treatment of pulmonary hypertension. In K. Dua, S. Nammi, D. Chang, D. K. Chellappan, G. Gupta, & T. Collet (Eds.), *Medicinal plants for lung diseases*. Singapore: Springer. https://doi.org/10.1007/978-981-33-6850-7_14.

Adetunji, C. O., Akram, M., Olaniyan, O. T., Ajayi, O. O., Inobeme, A., Olaniyan, S., et al. (2021b). Targeting SARS-CoV-2 novel corona (COVID-19) virus infection using medicinal plants. In K. Dua, S. Nammi, D. Chang, D. K. Chellappan, G. Gupta, & T. Collet (Eds.), *Medicinal plants for lung diseases*. Singapore: Springer.

Adetunji, C. O., Inobeme, A., Olaniyan, O. T., Ajayi, O. O., Olaniyan, S., & Adetunji, J. B. (2021). Application of nanodrugs derived from active metabolites of medicinal plants for the treatment of inflammatory and lung diseases: Recent advances. In K. Dua, S. Nammi, D. Chang, D. K. Chellappan, G. Gupta, & T. Collet (Eds.), *Medicinal plants for lung diseases*. Singapore: Springer.

Adetunji, C. O., & Varma, A. (2020). Biotechnological application of trichoderma: A powerful fungal isolate with diverse potentials for the attainment of food safety, management of pest and diseases, healthy planet, and sustainable agriculture. In C. Manoharachary, H. B. Singh, & A. Varma (Eds.), *Vol. 61. Trichoderma: Agricultural applications and beyond. Soil biology*. Cham: Springer.

Agarwal, T. (2014). *Artificial neural networks and their types*. Retrieved from http://www.elprocus.com/artificial-neural-networks-ann-and-their-types/.

Akram, M., Jabeen, F., Daniyal, M., Zainab, R., Haq, U. U., Adetunji, C. O., et al. (2020). Genetic engineering of novel products of health significance: Recombinant DNA technology. In C. Egbuna, & T. G. Dable (Eds.), *Functional foods and nutraceuticals*. Cham: Springer. https://doi.org/10.1007/978-3-030-42319-3_26.

Ali, A., & Hamouda, W. (2017). Advances on spectrum sensing for cognitive radio networks: Theory and applications. *IEEE Communication Surveys and Tutorials, 19*(2), 1277–1304.

Allen, J. (2003). Natural language processing. In A. Ralston, E. Reilly, & D. Hemmendinger (Eds.), *Encyclopedia of computer science* (4th ed.). Willey: Chichester.

Amodei, D., et al. (2016). Deep speech 2: End-to-end speech recognition in English and mandarin. In *International conference on machine learning* (pp. 173–182).

Amzica, F. (2010). Comment on "The human K-complex represents an isolated cortical down-state". *Science, 330*(6000), 35. https://doi.org/10.1126/science.1182138 (author reply 35).

Ang, K. K., Chua, K. S. G., Phua, K. S., et al. (2015). A randomized controlled trial of EEG-based motor imagery brain-computer interface robotic rehabilitation for stroke. *Clinical EEG and Neuroscience, 46*(4), 310–320. https://doi.org/10.1177/1550059414522229.

Bahdanau, D., et al. (2014). *Neural machine translation by jointly learning to align and translate.* arXiv:1409.0473 [cs, stat.

Cahn, B. R., & Polich, J. (2006). Meditation states and traits: EEG, ERP, and neuroimaging studies. *Psychological Bulletin, 132*(2), 180–211.

Callataÿ, A. (2014). *Natural and artificial intelligence: Misconceptions about brains and neural networks* (p. 694). Psychology.

Cammerer, S., Gruber, T., Hoydis, J., & Brink, S. T. (2017). *Scaling deep learning-based decoding of polar codes via partitioning.* arXiv preprint arXiv:1702.06901.

Cervera, M. A., Soekadar, S. R., Ushiba, J., Millán, J. D. R., Liu, M., Birbaumer, N., & Garipelli, G. (2018). Brain-computer interfaces for post-stroke motor rehabilitation: A meta-analysis. *Annals of Clinical and Translational Neurology, 5*(5), 651–663. https://doi.org/10.1002/acn3.544.

Cho, K., et al. (2014). *Learning phrase representations using RNN encoder-decoder for statistical machine translation.* arXiv preprint arXiv:1406.1078.

Daly, L. A., Haden, S. C., Hagins, M., Papouchis, N., & Ramirez, P. M. (2015). Yoga and emotion regulation in high school students: A randomized controlled trial. *Evidence-Based Complementary and Alternative Medicine, 2015*, 794928. https://doi.org/10.1155/2015/794928.

Dörner, S., Cammerer, S., Hoydis, J., & Brink, S.t. (2018). Deep learning based communication over the air. *IEEE Journal of Selected Topics in Signal Processing, 12*(1), 132–143. https://doi.org/10.1109/JSTSP.2017.2784180.

Dörr, K. N. (2016). Mapping the field of algorithmic journalism. *Digital Journalism, 4*, 700–722.

Egbuna, C., Gupta, E., Ezzat, S. M., Jeevanandam, J., Mishra, N., Akram, M., et al. (2020). Aloe species as valuable sources of functional bioactives. In C. Egbuna, & T. G. Dable (Eds.), *Functional foods and nutraceuticals*. Cham: Springer.

Ehlers, C. L., & Kupfer, D. J. (1997). Sleep: Do young adult men and women age differently? *Journal of Sleep Research, 6*(3), 211–215.

Fink, O., Wang, Q., Svensén, M., Dersin, P., Lee, W.-J., & Ducoffe, M. (2020). Potential, challenges and future directions for deep learning in prognostics and health management applications. *Engineering Applications of Artificial Intelligence, 92*, 103678.

Gehl, R., & Bakardjieva, M. (2017). *Socialbots and their friends: Digital media and the automation of sociality*. New York: Routledge.

Graves, A., & Jaitly, N. (2014). Towards end-to-end speech recognition with recurrent neural networks. In *Proceedings of the 31st international conference on machine learning (ICML-14)* (pp. 1764–1772).

Gunkel, D. (2012). Communication and artificial intelligence: Opportunities and challenges for the 21st century. *Communication. +1*, 1, 1.

Guzman, A., & Lewis, S. (2020). Artificial intelligence and communication: A human–machine communication research agenda. *New Media & Society*, *22*, 70–86.

Haeb-Umbach, R., Watanabe, S., Nakatani, T., et al. (2019). Speech processing for digital home assistants: Combining signal processing with deep-learning techniques. *IEEE Signal Processing Magazine*, *36*(6), 111–124.

He, K., Zhang, X., Ren, S., & Sun, J. (2016). Deep residual learning for image recognition. In *Proceedings of the IEEE conference on computer vision and pattern recognition* (pp. 770–778). IEEE.

Hong, X., Wang, J., Wang, C.-X., & Shi, J. (2014). Cognitive radio in 5G: A perspective on energy-spectral efficiency trade-off. *IEEE Communications Magazine*, *52*(7), 46–53.

Hughes, J. R. (2008 Jul). Gamma, fast, and ultrafast waves of the brain: Their relationships with epilepsy and behavior. *Epilepsy & Behavior*, *13*(1), 25–31. https://doi.org/10.1016/j.yebeh.2008.01.011. Epub 2008 Apr 24 18439878.

Işık, Y., & Taşpınar, N. (2007). Multiuser detection with neural network and pic in CDMA systems for AWGN and Rayleigh fading asynchronous channels. *Wireless Personal Communications*, *43*(4), 1185–1194.

Ito, J., Nikolaev, A. R., & van Leeuwen, C. (2005). Spatial and temporal structure of phase synchronization of spontaneous alpha EEG activity. *Biological Cybernetics*, *92*, 54–60.

Jagannath, J., Polosky, N., Jagannath, A., Restuccia, F., & Melodia, T. (2019). Neural networks for signal intelligence. Machine learning for future wireless communications. *Ad Hoc Networks*, 243–264.

Jia, X., & Kohn, A. (2011). Gamma rhythms in the brain. *PLoS Biology*, *9*(4). https://doi.org/10.1371/journal.pbio, e1001045.

Khan, A., Rehmani, M. H., & Rachedi, A. (2017). Cognitive-radio-based internet of things: Applications, architectures, spectrum related functionalities, and future research directions. *IEEE Wireless Communications Magazine*, *24*(3), 17–25.

Krizhevsky, A., et al. (2012). Imagenet classification with deep convolutional neural networks. In *Advances in neural information processing systems* (pp. 1097–1105).

Kurkin, S., Hramov, A., Chholak, P., et al. (2020). Localizing oscillatory sources in a brain by MEG data during cognitive activity. In *Proceedings of the 2020 4th international conference on computational intelligence and networks (CINE)*. Kolkata, India: IEEE. February 2020.

Lee, C., et al. (2017). Machine learning based channel modeling for molecular MIMO communications. In *IEEE International Workshop on Signal Processing Advances in Wireless Communications (SPAWC)*.

Li, M., Wang, R., Yang, J., et al. (2019). An improved refined composite multivariate multiscale fuzzy entropy method for MI-EEG feature extraction. *Computational Intelligence and Neuroscience*, *2019*, 7529572. 12 pages.

Li, Z., & Yu, Y. (2016). *Protein secondary structure prediction using cascaded convolutional and recurrent neural networks*. arXiv:1604.07176.

Liang, F., et al. (2018). An iterative BP-CNN architecture for channel decoding. *IEEE Journal of Selected Topics in Signal Processing*, *12*, 144–159.

Ma, K. (2019). Artificial unintelligence: How computers misunderstand the world. *The Information Society*, *35*, 314–315.

Malekmohammadi, A., Mohammadzade, H., Chamanzar, A., et al. (2019). An efficient hardware implementation for a motor imagery brain computer interface system. *Scientia Iranica*, *26*, 72–94.

Mansoor, A., Usman, M. W., Jamil, N., & Naeem, M. A. (2020). Deep learning algorithm for brain-computer Interface. *Scientific Programming*, *2020*, 12. https://doi.org/10.1155/2020/576214.

Martins, N. R. B., Erlhagen, W., & Freitas, R. A. (2015). Action potential monitoring using neuronanorobots: Neuroelectric nanosensors. *International Journal of Nanomaterials and Nanostructure*, *1*(2), 20–40.

Mattia, D., Pichiorri, F., Colamarino, E., Masciullo, M., Morone, G., Toppi, J., et al. (2020). The promotoer, a brain-computer interface-assisted intervention to promote upper limb functional motor recovery after stroke: A study protocol for a randomized controlled trial to test early and long-term efficacy and to identify determinants of response. *BMC Neurology*, *20*(1), 254. https://doi.org/10.1186/s12883-020-01826-w.

Mitra, U., & Poor, H. V. (1994). Neural network techniques for adaptive multiuser demodulation. *IEEE Journal on Selected Areas in Communications*, *12*(9), 1460–1470.

O'Shea, T., & Hoydis, J. (2017). An introduction to deep learning for the physical layer. *IEEE Transactions on Cognitive Communications and Networking*, *3*(4), 563–575. https://doi.org/10.1109/TCCN.2017.2758370.

O'Shea, T. J., et al. (2017). *Learning approximate neural estimators for wireless channel state information*. arXiv preprint arXiv:1707.06260.

Müller, J., Ballini, M., Livi, P., Chen, Y., Radivojevic, M., Shadmani, A., et al. (2015). High-resolution CMOS MEA platform to study neurons at subcellular, cellular, and network levels. *Lab on a Chip*, *15*(13), 2767–2780. https://doi.org/10.1039/c5lc00133a.

Murillo-Fuentes, J., Caro, S., & Perez-Cruz, F. (2005). Gaussian processes for multiuser detection in CDMA receivers. In *Advances in neural information processing systems*. Vancouver, CA: NIPS.

Nachmani, E., Marciano, E., Lugosch, L., Gross, W. J., Burshtein, D., & Be'ery, Y. (2018). Deep learning methods for improved decoding of linear codes. *IEEE Journal of Selected Topics in Signal Processing*, *12*(1), 119–131. https://doi.org/10.1109/JSTSP.2017.2788405.

Nicolas-Alonso, L. F., & Gomez-Gil, J. (2012). Brain computer interfaces, a review. *Sensors (Basel)*, *12*(2), 1211–1279. https://doi.org/10.3390/s120201211.

Oh, S. L., Vicnesh, J., Ciaccio, E. J., Yuvaraj, R., & Acharya, U. R. (2019). Deep convolutional neural network model for automated diagnosis of schizophrenia using EEG signals. *Applied Sciences*, *9*(14), 2870.

Pichiorri, F., Morone, G., Petti, M., Toppi, J., Pisotta, I., Molinari, M., et al. (2015). Brain-computer interface boosts motor imagery practice during stroke recovery. *Annals of Neurology*, 77(5), 851–865.

Poole, D. (2010). *Artificial intelligence*. Retrieved from http://artint.info/index.html.

Radivojevic, M., Franke, F., Altermatt, M., Muller, J., Hierlemann, A., & Bakkum, D. J. (2017). Tracking individual action potentials throughout mammalian axonal arbors. *eLife*, *6*. https://doi.org/10.7554/eLife.30198, e30198.

Ramírez-Moreno, M. A., & Gutiérrez, D. (2019). Evaluating a semiautonomous brain-computer interface based on conformal geometric algebra and artificial vision. *Computational Intelligence and Neuroscience*, *2019*, 9374802. 19 pages.

Reeves, J. (2016). Automatic for the people: The automation of communicative labor. *Communication and Critical/Cultural Studies*, *13*, 150–165.

Roli, O. I., Adetunji, C. O., Mishra, P., Adetunji, J. B., & Mishra, R. R. (2020). Caffeine: Nutraceutical and health benefit of caffeine-containing commodities and products. In P. Mishra, R. R. Mishra, & C. O. Adetunji (Eds.), *Innovations in food technology*. Singapore: Springer.

Saalmann, Y. B., Pinsk, M. A., Wang, L., Li, X., & Kastner, S. (2012). The pulvinar regulates information transmission between cortical areas based on attention demands. *Science*, *337*, 753–756.

Sanes, H., Thomas A. Reh, & William A. Harris (2012). *Development of the Nervous System*, Third Edition, 978-0-08-092320-8, Academic Press. https://doi.org/10.1016/C2009-0-02132-1.

Sarkies, M. N., Bowles, K.-A., Skinner, E. H., Mitchell, D., Haas, R., Ho, M., et al. (2015). Data collection methods in health services research: Hospital length of stay and discharge destination. *Applied Clinical Informatics*, *6*(1), 96–109. https://doi.org/10.4338/ACI-2014-10-RA-0097.

Sasaki, K., Suzuki, M., Mieda, M., Tsujino, N., Roth, B., & Sakurai, T. (2011). Pharmacogenetic modulation of orexin neurons alters sleep/wakefulness states in mice. *PLoS One*, *6*(5), e20360. https://doi.org/10.1371/journal.pone.0020360.

Schirrmeister, R. T., Springenberg, J. T., Fiederer, L. D. J., et al. (2017). Deep learning with convolutional neural networks for EEG decoding and visualization. *Human Brain Mapping*, *38*(11), 5391–5420.

Seager, M. A., Johnson, L. D., Chabot, E. S., Asaka, Y., & Berry, S. D. (2002). Oscillatory brain states and learning: Impact of hippocampal theta-contingent training. *Proceedings of the National Academy of Sciences of the United States of America*, *99*(3), 1616–1620.

Shiffman, D. (2012). *The nature of code*. http://natureofcode.com/book/chapter-10-neural-networks/. accessed 12.06.14.

Shih, J. J., Krusienski, D. J., & Wolpaw, J. R. (2012). Brain-computer interfaces in medicine. *Mayo Clinic Proceedings*, *87*(3), 268–279. https://doi.org/10.1016/j.mayocp.2011.12.008.

Siddiqui, M. K., Morales-Menendez, R., Huang, X., & Hussain, N. (2020). A review of epileptic seizure detection using machine learning classifiers. *Brain Informatics*, 7(1), 5. https://doi.org/10.1186/s40708-020-00105-1.

Stergiou, C., & Siganos, D. (2014). *Artificial neural network in medicine*. Retrieved from http://www-students.doc.ic.ac.uk/~cbp/article2.html.

Tamagawa, H., Funatani, M., & Ikeda, K. (2016). Ling's adsorption theory as a mechanism of membrane potential generation observed in both living and nonliving systems. *Membranes (Basel)*, *6*(1). https://doi.org/10.3390/membranes6010011.

Tian, C., Fei, L., Zheng, W., et al. (2020). Deep learning on image denoising: An overview. *Neural Networks*, *131*, 251–275.

Todorov, M. I., Paetzold, J. C., Schoppe, O., Tetteh, G., Shit, S., Efremov, V., et al. (2020). Machine learning analysis of whole mouse brain vasculature. *Nature Methods*, *17*(4), 442–449.

Valliani, A. A.-A., Ranti, D., & Oermann, E. K. (2019). Deep learning and neurology: A systematic review. *Neurology and Therapy*, *8*(2), 351–365. https://doi.org/10.1007/s40120-019-00153-8.

van Deursen, J. A., Vuurman, E. F., Verhey, F. R., van Kranen-Mastenbroek, V. H., & Riedel, W. J. (2008). Increased EEG gamma band activity in Alzheimer's disease and mild cognitive impairment. *Journal of Neural Transmission (Vienna)*, *115*(9), 1301–1311. https://doi.org/10.1007/s00702-008-0083-y.

von Stein, A., & Sarnthein, J. (2000). Different frequencies for different scales of cortical integration: From local gamma to long range alpha/theta synchronization. *International Journal of Psychophysiology*, *38*, 301–313.

Willamette. (2014). *Computation in the brain: The brain as an information processing System*. Retrieved from http://www.willamete.edu/~gorr/classes/as449/intro.html.

Wolpaw, J. R., Birbaumer, N., McFarland, D. J., Pfurtscheller, G., & Vaughan, T. M. (2002). Brain–computer interfaces for communication and control. *Clinical Neurophysiology*, *113*(6), 767–791. https://doi.org/10.1016/S1388-2457(02)00057-3.

Zandvakili, A., Philip, N. S., Jones, S. R., Tyrka, A. R., Greenberg, B. D., & Carpenter, L. L. (2019). Use of machine learning in predicting clinical response to transcranial magnetic stimulation in comorbid posttraumatic stress disorder and major depression: A resting state electroencephalography study. *Journal of Affective Disorders*, *252*, 47–54.

Zhang, J., Harman, M., Ma, L., & Liu, Y. (2019). Machine learning testing: Survey, landscapes and horizons. *IEEE Transactions on Software Engineering*, *48*, 1–36.

Nachmani, E., Beery, Y., & Burshtein, D. (2016). Learning to decode linear codes using deep learning. In *54th Annual allerton conference on communication, control, and computing (Allerton 2016)* (pp. 341–346).

CHAPTER 16

Classification of neurodegenerative disorders using machine learning techniques

Charles O. Adetunji[a], Olugbemi T. Olaniyan[b], Olorunsola Adeyomoye[c], Ayobami Dare[d], Mayowa J. Adeniyi[e], and Alex Enoch[f]

[a]Department of Microbiology, Applied Microbiology, Biotechnology and Nanotechnology Laboratory, Edo State University Uzairue, Iyamho, Nigeria
[b]Department of Physiology, Laboratory for Reproductive Biology and Developmental Programming, Rhema University, Aba, Nigeria
[c]Department of Physiology, University of Medical Sciences, Ondo, Nigeria
[d]Department of Physiology, School of Laboratory Medicine and Medical Sciences, College of Health Sciences, Westville Campus, University of KwaZulu-Natal, Durban, South Africa
[e]Department of Physiology, Edo State University Uzairue, Iyamho, Nigeria
[f]Department of Human Physiology, Ahmadu Bello University Zaria, Zaria, Nigeria

Introduction

The three most common types of diseases responsible for neurological disorders are schizophrenia, Alzheimer's disease (AD), and Parkinson's disease (PD). These diseases are characterized by the disruption of regular brain functions and operations (Danielyan & Nasrallah, 2009; Islam & Zhang, 2018; Tolosa, Wenning, & Poewe, 2006). The diseases cause diverse types of health challenges that affect many people worldwide. Hence it has become necessary to search for sustainable techniques that can help in the early detection of these diseases (Fontana et al., 2017; Leparulo et al., 2020). Typical examples of recent neuroimaging techniques include deep learning-based analysis, magnetic resonance imaging (MRI), positron emission tomography (PET), and computed tomography (CT) for easy and quick identification and diagnosis of neurological diseases as well as for adequate deployment of treatment (Islam & Zhang, 2018; Mahmud & Vassanelli, 2016, 2019; Mathew, Vivek, & Anurenjan, 2018; Poldrack et al., 2019; Shatte, Hutchinson, & Teague, 2019; Salvatore et al., 2014).

Machine learning is a typical and sustainable technique applied in mining of biological data (Mahmud, Kaiser, & Hussain, 2020; Mahmud, Kaiser, Hussain, & Vassanelli, 2018), strategic game playing (Silver et al., 2016),

Artificial Intelligence for Neurological Disorders
https://doi.org/10.1016/B978-0-323-90277-9.00028-6

image analysis (Ali, Kaiser, & Mahmud, 2019), natural language processing (Rabby, Azad, Mahmud, Zamli, & Rahman, 2020; Watkins, Fabietti, & Mahmud, 2020), anomaly detection (Fabietti et al., 2020; Yahaya, Lotfi, & Mahmud, 2019), financial forecasting (Orojo, Tepper, McGinnity, & Mahmud, 2019), and disease detection (Miah, Prima, Seema, Mahmud, & Kaiser, 2020; Noor, Zenia, Kaiser, Mahmud, & Al Mamun, 2019a, 2019b). The analysis of neurological disorders using multimodal approaches and the detection of these disorders used deep learning-based approaches including data fusion play a crucial role in personalized services (Chand et al., 2020; Islam & Zhang, 2020; Kaiser et al., 2018; Lei et al., 2020; Liu et al., 2015). Therefore, this chapter focuses on recent advances in the classification of neurodegenerative diseases utilizing machine learning algorithms.

Patient datasets

Machine learning helps a computer learn from data to develop an algorithm to solve a complicated task, for example, diagnosing a disease (Adetunji, Adetunji, and Olaniyan, 2021; Adetunji, Ajayi, et al., 2021; Adetunji, Akram, et al., 2021a; Adetunji, Akram, et al., 2021b; Adetunji, Inobeme, et al., 2021; Adetunji & Varma, 2020; Akram et al., 2020; Das et al., 2020; Egbuna et al., 2020; Roli, Adetunji, Mishra, Adetunji, & Mishra, 2020). Data handling and processing are vital to the effective use of machine learning. Patient datasets must be well handled by a data scientist who collects, transforms, prepares, and inputs the data into a computer to develop a classification model as well as evaluate the performance of the model for proper application of machine learning in the classification of neurodegenerative diseases. This activity will culminate in seemingly stress-free diagnosis and classification of diseases (Golriz Khatami, Mubeen, and Hofmann-Apitius, 2020). A patient dataset in neurodegenerative disease includes numerous amounts of data collected from various patients that will be used to train a computer to develop a model to classify disorders without prior programming. The heterogenous nature of neurodegenerative disease requires adequate quantification of various biomarkers of all modalities such as imaging (e.g., CT, PET, fMRI) and nonimaging (plasma and cerebrospinal fluid sample data) techniques that are subjected to computational analysis independently or combined with other sophisticated analysis to generate data that can translate into clinical applications (Whitwell et al., 2017). Patient datasets are obtained from adequate medical examinations and processed to form a model approach for identifying biomarkers and anatomical and physiological changes in the brain, especially in the hippocampus and neurons. Volumetric (hippocampal volume and medial temporal lobe atrophy), morphometric (frontal and temporal lobe atrophy), and vascular (extensive white matter changes with lacunar and cortical infarcts) characteristics are among the numerous data extracted from neuroimaging, with a multiclass classifier via disease state index methodology to provide continuous probability indices for each disease in improving decision-making during diagnosis (Frisoni et al., 1999; Kantarci et al., 2012; Klöppel et al., 2008).

These datasets provide pathological and genetic characteristics of the disease to assist in proper diagnosis, classification, and management (Rohini & Surendran, 2019). In clinical settings, these datasets enable early diagnosis of neurodegenerative disease and are relatively general, thereby requiring subjective assessment and expertise by the clinician. However, the application of machine learning will promote automatic image assessment and computerized decision support to objectively extract numerous data and investigate the similarity between patient data and typical (developed algorithm) data for neurodegenerative disease. This will improve decision-making during diagnosis (Koikkalainen et al., 2016).

Related medical examinations

The pathogenesis of neurodegenerative disorders is usually multifaceted and there is no single test for their diagnosis. Thus, different diagnostic investigations and brain imaging procedures have been used to assist clinicians during diagnosis. We discuss some of these imaging techniques in the sections that follow.

Brain imaging.

Brain imaging methods are noninvasive medical procedures for analyzing brain structures.

1. Computed tomography (CT): In this method, an X-ray machine captures a sequence of images along the axial position of the subject to produce a CT scan. It is widely utilized in the diagnosis of AD, where a progressive loss of cognitive function is linked to a change in brain size. CT scans of patients with AD show significant expansion of the third and lateral ventricles compared to scans of control patients (Tagaris, Kollias, Stafylopatis, Tagaris, & Kollias, 2018).
2. Positron Emission Tomography (PET): This is an imaging procedure that evaluate metabolic process in the body and are often used to diagnose and monitor the progression of neurodegenerative disorders. For

instance, PET is used to detect glucose hypo-metabolism in the temporoparietal region, a condition that varies directly with the severity of the disease (Duffy, Boyle, & Vasdev, 2019).
3. Magnetic Resonance Imaging (MRI): This is a type of medical imaging that uses magnetic and radio waves to record the inside structure of organs. MRI scans can capture images from different anatomical positions and produce several sequences containing multiple slices to reveal the three-dimensional mapping of the area of interest in the brain. They can identify cerebral atrophy (a major anatomical damage associated with neurodegenerative diseases) as well as the temporal lobe, lentiform nucleus, caudate nucleus, and the temporoparietal cortex (Noor et al., 2019a, 2019b).
4. Single-Photon Emission Computed Tomography (SPECT): This imaging procedure uses gamma rays to show the interior of the body via injection of gamma-emitting radioisotopes into the systemic circulation. These isotopes are specific to different regions of the body and thus subsequently produce different images. A dopamine transporter (DaT) scan, a form of SPECT that measures dopamine levels in the brain, has been used to diagnose PD (Pahuja, Nagabhushan, & Prasad, 2020).

Clinical tests

These tests include various rating scales, interviews, surveys, and laboratory tests that evaluate a patient's performance.

1. Unified Parkinson's Disease Rating Scale (UPDRS): This is a commonly used test to evaluate the overall clinical state (nonmotor and motor experience in daily life, motor examination, and motor complications) of patients with PD (Goetz, Nutt, & Stebbins, 2008; Goetz et al., 2008).
2. Mini-Mental State Examination (MMSE): This is a questionnaire used to evaluate the severity of cognitive impairment with respect to spatial orientation, memory recall, and attention calculation. It is commonly used in AD and PD (Tombaugh & McIntyre, 1992).
3. Unified Dyskinesia Rating Scale (UDysRS): This scale is commonly used to evaluate involuntary movement in neurodegenerative diseases (Goetz, Nutt, & Stebbins, 2008; Goetz et al., 2008).
4. Freezing of Gait (FOG) questionnaire: This test assesses the patient's FOG via questions on gait in daily life, FOG rate and intensity, festinating gait rate in connection to falls, and the frequency and severity of falls (Giladi et al., 2000).

5. Core Assessment Program for Surgical Interventional Therapies in Parkinson's Disease (CAPSIT-PD): This is a protocol that uses timed tests to assess motor syndromes (Defer, Widner, Marié, Rémy, & Levivier, 1999).

Biomarkers

These are early indicators such as neurochemicals used to monitor the risk, development, and progression of neurodegenerative diseases. A commonly used biomarker is the presence of amyloid beta in body fluid (including cerebrospinal fluid or plasma), which is useful in differentiating between several neurodegenerative diseases.

Clinical tests and biomarkers

Neurodegenerative disorders are diagnosed using arrays of different imaging techniques to identify damaged neurons or lesions in the brain. In a bid to further elucidate the dysfunction of abnormal neurons, more sophisticated imaging techniques are being developed and adopted to understand the physiology and pathophysiologic mechanisms of the brain or cognitive function. CT is the use of a monitored X-rays source to produce beam of rays directed to a patient to identify or elucidate areas of abnormalities (Garvey & Hanlon, 2002). The patient is placed in a prone position on a bed that moves through the gantry. The CT scanner contains an X-ray detection that absorbs the X-ray from the patient to produce slices of images that can be stacked together to produce a three-dimensional image of the tissue on the computing system (Patel & De Jesus, 2021). In certain cases, a contrast agent may be injected to allow soft tissues or organ to be easily visualized. CT can be used to detect or visualize the lungs and heart. The risk of using CT is minimal and the radiation from it cannot ionize body cells. However, in children, the CT machine should be adjusted to limit the exposure to radiation (Goske, 2014). In addition, individuals with kidney disease should avoid using contrast agents, which can compromise kidney functions (Kaller & An, 2021). MRI is a diagnostic technique that uses a magnetic field to generate images of body organs (Grover et al., 2015). It is an alternative to X-ray and exposure to ionizing radiation. It provides detailed information of body structures. Diffuse optical imaging (DOI) uses infrared or florescence to create a three-dimensional image of a structure. Its role in diagnosis of neurodegenerative disease has been widely established (Huppert, Franceschini, & Boas, 2009). It is also useful in monitoring cancer progression

(Hadjipanayis, Jiang, Roberts, & Yang, 2011). DOI detects both oxy and deoxygenated hemoglobin levels in the blood. Body tissue is a strong diffusing medium and has the potential to allow light propagation from diffuse optical medium to create images (Durduran, Choe, Baker, & Yodh, 2010). Magnetoencephalography (MEG) is designed to map out brain activities. It measures the magnetic field produced by electric current in the brain. It evaluates and monitors cognitive processes. Furthermore, it assesses the areas of pathology in the brain to carry out surgical interventions (Singh, 2014). Electroencephalography (EEG) is noninvasive method of measuring brain electrical activities using connecting electrodes placed on the scalp. It can diagnose epilepsy, sleep disorders, coma, encephalopathy, and brain death (Smith, 2005). In addition, it can diagnose tumor, stroke, and other brain disorders (Sachs, 1986). PET uses radioactive tracers for imaging and assessing metabolic processes, including blood flow, body composition, and absorption (Vaquero & Kinahan, 2015). A radioactive tracer is often ingested as tracer agent that gives off rays that are captured to form three-dimensional images. A more advanced imaging approach is SPECT. Like PET, SPECT is used for monitoring blood flow and also uses gamma rays and radioactive tracers (Mariani et al., 2010). Functional magnetic resonance imaging (fMRI) and event-related optical signals are two additional imaging modalities that can be utilized to measure cognitive function and other neurodegenerative illnesses.

Classification of neurodegenerative diseases

The classification of neurodegenerative disorders is critical for identifying the genes involved in disease development and investigating neurochemical alterations, neuron loss, and other forms of brain dysfunction (Relja, 2004). AD, PD, and schizophrenia disease are the three most common neurodegenerative illnesses. AD is a neurological illness that primarily affects individuals older than 65 years (Qiu, Kivipelto, & von Strauss, 2009). It causes cognitive impairment with severe effects on language, attention, comprehension, memory, and the ability to reason (Paulsen, 2011). Although treatments are available to manage this condition, there is currently no cure (Ajitkumar & De Jesus, 2021; Yiannopoulou & Papageorgiou, 2020). It is characterized by lack of motivation, disorganization, memory impairment, difficulty in solving problems, and problems with multitasking (Yiannopoulou & Papageorgiou, 2020). Depression, family history, smoking, cerebrovascular accident, and head injury are possible risk factors for the disease (Silva et al., 2019). PD is a neurological illness that causes resting tremor, rigidity, bradykinesia, and postural instability, among other symptoms (Berardelli, Rothwell, Thompson, & Hallett, 2001). Evidence has shown that smoking, pesticides, and caffeine are contributing factors to PD (DeMaagd & Philip, 2015). As in AD, PD occurs in elderly people, and the prevalence of the disease has been projected to increase in the future (Massano & Bhatia, 2012). Usually, there is loss of substantia nigra dopaminergic neurons, thus treatment involves dopamine administration even though other neurotransmitters are also involved in pathogenesis of the disease (Eriksen, Wszolek, & Petrucelli, 2005). Huntington disease is an autosomal disorder characterized by impaired movement and cognitive disturbances. It occurs due to short repeat of cytosine, adenine, and guanine (CAG) in chromosome 4p16.3 (Tanner et al., 2018). It is common in individuals between the ages of 30 and 50 years. Genetic modifiers, CAG repeat, and CAG instability are risk factors for the disease (Massey & Jones, 2018). Other neurodegenerative diseases include Prion's disease, Prick's disease, spinocerebellar ataxia, amyloid lateral sclerosis (ALS), and others.

Machine learning techniques as computer-assisted diagnostic systems

Machine learning has been used in the diagnosis of several diseases. Examples of machine learning algorithms for classification of diseases include linear discriminant analysis, quadratic discriminant analysis, multilayer perceptions, support vector machine, and pixel/voxel values in images. Specifically, linear discriminant analysis and quadratic discriminant analysis have been used for the classification of lung cancers (Rumelhart, Hinton, & Williams, 1986; Vapnik, 1995). A naïve Bayesian classifier is used in distinguishing cardiovascular images with accuracy of up to 96.59%. Support vector machine is used in ultrasound imaging (Prabusankarlal, Thirumoorthy, & Manavalan, 2017). The multiclass AdaBoost algorithm has been utilized in distinguishing carcinomas, fibroadenomas, and cysts.

Multimodal analysis

Multimodal analysis can be defined as qualitative data interpretation and depiction, especially in projects where nonverbal and verbal forms of information are combined. It can analyze nonverbal communication, verbal communication, 3D materials, images, and videos. Social semiotics, sensory anthropology, and interactional sociology are the three major theoretical influences on multimodal analysis.

Conclusion and future perspectives

This chapter presented an overview of recent advances in the classification of neurodegenerative diseases such as Huntington's disease, PD, progressive brain dysfunction, vascular dementia, cognitive impairment, AD, and ALS utilizing machine learning algorithms Several developments in neuroimaging data acquisition platforms like diffusion magnetic resonance and data mining from electroencephalogram are channeled towards classification of different neurodegenerative diseases and treatment options. The challenges encountered in the utilization of artificial intelligence for proper diagnosis, analysis, and classification of neurodegenerative disorders, such as increased dimensionality, nonlinearity, and nonstationarity, should be the focus of future research.

References

Adetunji, J. B., Adetunji, C. O., & Olaniyan, O. T. (2021). African walnuts: A natural depository of nutritional and bioactive compounds essential for food and nutritional security in Africa. In O. O. Babalola (Ed.), *Food security and safety*. Cham: Springer.

Adetunji, C. O., Ajayi, O. O., Akram, M., Olaniyan, O. T., Chishti, M. A., Inobeme, A., et al. (2021). Medicinal plants used in the treatment of influenza a virus infections. In K. Dua, S. Nammi, D. Chang, D. K. Chellappan, G. Gupta, & T. Collet (Eds.), *Medicinal plants for lung diseases*. Singapore: Springer.

Adetunji, C. O., Akram, M., Olaniyan, O. T., Ajayi, O. O., Inobeme, A., Olaniyan, S., et al. (2021a). Medicinal plants used in the treatment of pulmonary hypertension. In K. Dua, S. Nammi, D. Chang, D. K. Chellappan, G. Gupta, & T. Collet (Eds.), *Medicinal plants for lung diseases*. Singapore: Springer. https://doi.org/10.1007/978-981-33-6850-7_14.

Adetunji, C. O., Akram, M., Olaniyan, O. T., Ajayi, O. O., Inobeme, A., Olaniyan, S., et al. (2021b). Targeting SARS-CoV-2 novel corona (COVID-19) virus infection using medicinal plants. In K. Dua, S. Nammi, D. Chang, D. K. Chellappan, G. Gupta, & T. Collet (Eds.), *Medicinal plants for lung diseases*. Singapore: Springer.

Adetunji, C. O., Inobeme, A., Olaniyan, O. T., Ajayi, O. O., Olaniyan, S., & Adetunji, J. B. (2021). Application of nanodrugs derived from active metabolites of medicinal plants for the treatment of inflammatory and lung diseases: Recent advances. In K. Dua, S. Nammi, D. Chang, D. K. Chellappan, G. Gupta, & T. Collet (Eds.), *Medicinal plants for lung diseases*. Singapore: Springer.

Adetunji, C. O., & Varma, A. (2020). Biotechnological application of trichoderma: A powerful fungal isolate with diverse potentials for the attainment of food safety, management of pest and diseases, healthy planet, and sustainable agriculture. In C. Manoharachary, H. B. Singh, & A. Varma (Eds.), *Vol. 61. Trichoderma: Agricultural applications and beyond. Soil biology*. Cham: Springer.

Ajitkumar, A., & De Jesus, O. (2021). Huntington disease. [Updated 2021 Feb 7]. In *StatPearls [Internet]*. Treasure Island (FL): StatPearls Publishing. Available from: https://www.ncbi.nlm.nih.gov/books/NBK559166/.

Akram, M., Jabeen, F., Daniyal, M., Zainab, R., Haq, U. U., Adetunji, C. O., et al. (2020). Genetic engineering of novel products of health significance: Recombinant DNA technology. In C. Egbuna, & T. G. Dable (Eds.), *Functional foods and nutraceuticals*. Cham: Springer. https://doi.org/10.1007/978-3-030-42319-3_26.

Ali, H. M., Kaiser, M. S., & Mahmud, M. (2019). Application of convolutional neural network in segmenting brain regions from mri data. In *International conference on brain informatics* (pp. 136–146). Springer. https://doi.org/10.1007/978-3-030-37078-7_14.

Berardelli, A., Rothwell, J. C., Thompson, P. D., & Hallett, M. (2001). Pathophysiology of bradykinesia in Parkinson's disease. *Brain: A Journal of Neurology*, *124*(Pt 11), 2131–2146.

Chand, G. B., Dwyer, D. B., Erus, G., Sotiras, A., Varol, E., Srinivasan, D., et al. (2020). Two distinct neuroanatomical subtypes of schizophrenia revealed using machine learning. *Brain*, *143*(3), 1027–1038.

Danielyan, A., & Nasrallah, H. A. (2009). Neurological disorders in schizophrenia. *Psychiatric Clinics*, *32*(4), 719–757.

Das, S., Sharangi, A. B., Egbuna, C., Jeevanandam, J., Ezzat, S. M., Adetunji, C. O., et al. (2020). Health benefits of isoflavones found exclusively of plants of the Fabaceae Family. In C. Egbuna, & T. G. Dable (Eds.), *Functional foods and nutraceuticals*. Cham: Springer. https://doi.org/10.1007/978-3-030-42319-3_22.

Defer, G.-L., Widner, H., Marié, R. M., Rémy, P., & Levivier, M. (1999). Core assessment program for surgical interventional therapies in Parkinson's disease (CAPSIT-PD). *Movement Disorders*, *14*(4), 572–584.

DeMaagd, G., & Philip, A. (2015). Parkinson's disease and its management: Part 1: Disease entity, risk factors, pathophysiology, clinical presentation, and diagnosis. *P & T: A Peer-Reviewed Journal for Formulary Management*, *40*(8), 504–532.

Duffy, I. R., Boyle, A. J., & Vasdev, N. (2019). Improving PET imaging acquisition and analysis with machine learning: A narrative review with focus on Alzheimer's disease and oncology. *Molecular Imaging*, *18*, 1536012119869070.

Durduran, T., Choe, R., Baker, W. B., & Yodh, A. G. (2010). Diffuse optics for tissue monitoring and tomography. *Reports on Progress in Physics*, *73*(7), 076701.

Egbuna, C., Gupta, E., Ezzat, S. M., Jeevanandam, J., Mishra, N., Akram, M., et al. (2020). Aloe species as valuable sources of functional bioactives. In C. Egbuna, & T. G. Dable (Eds.), *Functional foods and nutraceuticals*. Cham: Springer.

Eriksen, J. L., Wszolek, Z., & Petrucelli, L. (2005). Molecular pathogenesis of Parkinson disease. *Archives of Neurology*, *62*(3), 353–357.

Fabietti, M., Mahmud, M., Lotfi, A., Averna, A., Guggenmo, D., Nudo, R., et al. (2020). Neural network-based artifact detection in local field potentials recorded from chronically implanted neural probes. In *Proc. IJCNN* (pp. 1–8).

Fontana, R., Agostini, M., Murana, E., Mahmud, M., Scremin, E., Rubega, M., et al. (2017). Early hippocampal hyperexcitability in ps2a pp mice: Role of mutant ps2 and app. *Neurobiology of Aging*, *50*, 64–76. https://doi.org/10.1016/j.neurobiolaging.2016.10.027.

Frisoni, G. L., Beltramello, M., Geroldi, A., Bianchetti, C., Soininen, A., & Trabucchi, H. (1999). Hippocampal and entorhinal cortex atrophy in frontotemporal dementia and Alzheimer's disease. *Neurology*, *52*(1), 91–100.

Garvey, C. J., & Hanlon, R. (2002). Computed tomography in clinical practice. *British Medical Journal (Clinical research ed.)*, *324*(7345), 1077–1080.

Giladi, N., Shabtai, H., Simon, E., Biran, S., Tal, J., & Korczyn, A. (2000). Construction of freezing of gait questionnaire for patients with Parkinsonism. *Parkinsonism & Related Disorders*, *6*(3), 165–170.

Goetz, C., Nutt, G., & Stebbins, J. G. (2008). The unified dyskinesia rating scale: Presentation and clinimetric profile. *Movement Disorders*, *23*(16), 2398–2403.

Goetz, C. G., et al. (2008). Movement disorder society-sponsored revision of the unified Parkinson's disease rating scale (MDS-UPDRS): Scale presentation and clinimetric testing results. *Movement Disorders*, *23*(15), 2129–2170.

Golriz Khatami, S., Mubeen, S., & Hofmann-Apitius, M. (2020). Data science in neurodegenerative disease: Its capabilities, limitations, and perspectives. *Current Opinion in Neurology*, *33*(2), 249–254.

Goske, M. J. (2014). Doctor, is a CT scan safe for my child? *The British Journal of Radiology*, *87*(1034), 20130517.

Grover, V. P., Tognarelli, J. M., Crossey, M. M., Cox, I. J., Taylor-Robinson, S. D., & McPhail, M. J. (2015). Magnetic resonance imaging: Principles and techniques: Lessons for clinicians. *Journal of Clinical and Experimental Hepatology*, *5*(3), 246–255.

Hadjipanayis, C. G., Jiang, H., Roberts, D. W., & Yang, L. (2011). Current and future clinical applications for optical imaging of cancer: From intraoperative surgical guidance to cancer screening. *Seminars in Oncology*, *38*(1), 109–118.

Huppert, T. J., Franceschini, M. A., & Boas, D. A. (2009). Noninvasive imaging of cerebral activation with diffuse optical tomography. In R. D. Frostig (Ed.), *In vivo optical imaging of brain function* (2nd ed.). Boca Raton (FL): CRC Press/Taylor & Francis. Chapter 14. Available from: https://www.ncbi.nlm.nih.gov/books/NBK20225/.

Islam, J., & Zhang, Y. (2018). Brain mri analysis for Alzheimer's disease diagnosis using an ensemble system of deep convolutional neural networks. *Brain Informatics*, *5*, 1–14. https://doi.org/10.1186/s40708-018-0080-3.

Islam, J., & Zhang, Y. (2020). GAN-based synthetic brain PET image generation. *Brain Informatics*, 7(1), 3. https://doi.org/10.1186/s40708-020-00104-2.

Kaiser, M. S., Lwin, K. T., Mahmud, M., Hajializadeh, D., Chaipimonplin, T., Sarhan, A., et al. (2018). Advances in crowd analysis for urban applications through urban event detection. *IEEE Transactions on Intelligent Transportation Systems*, *19*(10), 3092–3112.

Kaller, M. O., & An, J. (2021). Contrast agent toxicity. [Updated 2020 May 25]. In *StatPearls [Internet]*. Treasure Island (FL): StatPearls Publishing. Available from: https://www.ncbi.nlm.nih.gov/books/NBK537159/.

Kantarci, K. L., Boeve, V. J., Weigand, B. F., Senjem, F. D., Przybelski, M. L., Dickson, M. A., et al. (2012). Multimodality imaging characteristics of dementia with Lewy bodies Neurobiol. *Aging*, *33*(9), 2091–2105.

Klöppel, C. M., Stonnington, C., Chu, B., Draganski, R. I., Scahill, J. D., Rohrer, N. C., et al. (2008). Automatic classification of MR scans in Alzheimer's disease. *Brain*, *131*(3), 681–689.

Koikkalainen, J., Rhodius-Meester, H., Tolonen, A., Barkhof, F., Tijms, B., Lemstra, A. W., et al. (2016). Differential diagnosis of neurodegenerative diseases using structural MRI data. *NeuroImage: Clinical*, *11*(2016), 435–449.

Lei, B., et al. (2020). Adaptive sparse learning using multi-template for neurodegenerative disease diagnosis. *Medical Image Analysis*, 101632.

Leparulo, A., Mahmud, M., Scremin, E., Pozzan, T., Vassanelli, S., & Fasolato, C. (2020). Dampened slow oscillation connectivity anticipates amyloid deposition in the ps2a pp mouse model of Alzheimer's disease. *Cell*, *9*(1), 54. https://doi.org/10.3390/cells9010054.

Liu, S., Cai, W., Liu, S., Zhang, F., Fulham, M., Feng, D., et al. (2015). Multimodal neuroimaging computing: A review of the applications in neuropsychiatric disorders. *Brain Informatics*, *2*(3), 167.

Mahmud, M., Kaiser, M. S., & Hussain, A. (2020). *Deep learning in mining biological data* (pp. 1–36). CoRR. arXiv:abs/2003.00108 https://arxiv.org/abs/2003.00108.

Mahmud, M., Kaiser, M. S., Hussain, A., & Vassanelli, S. (2018). Applications of deep learning and reinforcement learning to biological data. *IEEE Transactions on Neural Networks and Learning Systems*, *29*(6), 2063–2079. https://doi.org/10.1109/TNNLS.2018.2790388.

Mahmud, M., & Vassanelli, S. (2016). Processing and analysis of multichannel extracellular neuronal signals: State-of-the-art and challenges. *Frontiers in Neuroscience*. https://doi.org/10.3389/fnins.2016.00248.

Mahmud, M., & Vassanelli, S. (2019). Open-source tools for processing and analysis of in vitro extracellular neuronal signals. In M. Chiappalone, V. Pasquale, & M. Frega (Eds.), *In vitro neuronal networks: From culturing methods to neuro-technological applications* (pp. 233–250). Cham: Springer. https://doi.org/10.1007/978-3-030-11135-9_10.

Mariani, G., Bruselli, L., Kuwert, T., Kim, E. E., Flotats, A., Israel, O., et al. (2010). A review on the clinical uses of SPECT/CT. *European Journal of Nuclear Medicine and Molecular Imaging*, *37*(10), 1959–1985.

Massano, J., & Bhatia, K. P. (2012). Clinical approach to Parkinson's disease: Features, diagnosis, and principles of management. *Cold Spring Harbor Perspectives in Medicine*, *2*(6), a008870.

Massey, T. H., & Jones, L. (2018). The central role of DNA damage and repair in CAG repeat diseases. *Disease Models & Mechanisms*, *11*(1), dmm031930.

Mathew, N. A., Vivek, R. S., & Anurenjan, P. R. (2018). Early diagnosis of Alzheimer's disease from MRI images using PNN. In *Proc. IC4* (pp. 161–164).

Miah, Y., Prima, C. N. E., Seema, S. J., Mahmud, M., & Kaiser, M. S. (2020). Performance comparison of machine learning techniques in identifying dementia from open access clinical datasets. In *Proc. ICACIn* (pp. 69–78).

Noor, M. B. T., Zenia, N. Z., Kaiser, M. S., Mahmud, M., & Al Mamun, S. (2019a). Detecting neurodegenerative disease from MRI: a brief review on a deep learning perspective. In P. Liang, V. Goel, & C. Shan (Eds.), *Vol. 11976. Brain informatics. BI 2019. Lecture notes in computer science*. Cham: Springer.

Noor, M. B. T., Zenia, N. Z., Kaiser, M. S., Mahmud, M., & Al Mamun, S. (2019b). Detecting neurodegenerative disease from MRI: A brief review on a deep learning perspective. In *International Conference on Brain Informatics* (pp. 115–125). https://doi.org/10.1007/978-3-030-37078-7_12.

Orojo, O., Tepper, J., McGinnity, T. M., & Mahmud, M. (2019). A multi-recurrent network for crude oil price prediction. In *Proceedings of the 2019 IEEE symposium series on computational intelligence (SSCI)* (pp. 2953–2958). https://doi.org/10.1109/SSCI44817.2019.9002841.

Pahuja, G., Nagabhushan, T. N., & Prasad, B. (2020). Early detection of Parkinson's disease by using SPECT imaging and biomarkers. *Journal of Intelligent Systems*, *29*(1), 1329–1344.

Patel, P. R., & De Jesus, O. (2021). CT scan. [Updated 2021 Jan 11]. In *StatPearls [Internet]*. Treasure Island (FL): StatPearls Publishing. Available from: https://www.ncbi.nlm.nih.gov/books/NBK567796/.

Paulsen, J. S. (2011). Cognitive impairment in Huntington disease: Diagnosis and treatment. *Current Neurology and Neuroscience Reports*, *11*(5), 474–483.

Poldrack, R., et al. (2019). Computational and informatic advances for reproducible data analysis in neuroimaging. *Annual Review of Biomedical Data Science*, *2*, 119–138.

Prabusankarlal, K. M., Thirumoorthy, P., & Manavalan, R. (2017). Classification of breast masses in ultrasound images using self-adaptive differential evolution extreme learning machine and rough set feature selection. *Journal of medical imaging (Bellingham, Wash.)*, *4*(2). https://doi.org/10.1117/1.JMI.4.2.024507, 024507.

Qiu, C., Kivipelto, M., & von Strauss, E. (2009). Epidemiology of Alzheimer's disease: Occurrence, determinants, and strategies toward intervention. *Dialogues in Clinical Neuroscience*, *11*(2), 111–128.

Rabby, G., Azad, S., Mahmud, M., Zamli, K. Z., & Rahman, M. M. (2020). Teket: A tree-based unsupervised keyphrase extraction technique. *Cognitive Computation*. https://doi.org/10.1007/s12559-019-09706-3.

Relja, M. (2004). Pathophysiology and classification of neurodegenerative diseases. *EJIFCC, 15*(3), 97–99.

Rohini, & Surendran. (2019). Classification of neurodegenerative disease stages using ensemble machine learning classifiers. In *Vol. 165. International conference on recent trends in advanced computing* (pp. 66–73). https://doi.org/10.1016/j.procs.2020.01.071.

Roli, O. I., Adetunji, C. O., Mishra, P., Adetunji, J. B., & Mishra, R. R. (2020). Caffeine: Nutraceutical and health benefit of caffeine-containing commodities and products. In P. Mishra, R. R. Mishra, & C. O. Adetunji (Eds.), *Innovations in food technology*. Singapore: Springer.

Rumelhart, D. E., Hinton, G. E., & Williams, R. J. (1986). Learningrepresentations by back-propagating errors. *Nature, 323*, 533–536.

Sachs, B. (1986). Zur gegenwärtigen Rolle der Elektroenzephalographie für die Hirntumordiagnostik [Current role of electroencephalography in the diagnosis of brain tumors]. *Zentralblatt für Neurochirurgie, 47*(4), 292–295.

Salvatore, C., et al. (2014). Machine learning on brain MRI data for differential diagnosis of Parkinson's disease and progressive supranuclear palsy. *Journal of Neuroscience Methods, 222*, 230–237.

Shatte, A., Hutchinson, D., & Teague, S. (2019). Machine learning in mental health: A scoping review of methods and applications. *Psychological Medicine, 49*, 1–23.

Silva, M., Loures, C., Alves, L., de Souza, L. C., Borges, K., & Carvalho, M. (2019). Alzheimer's disease: Risk factors and potentially protective measures. *Journal of Biomedical Science, 26*(1), 33.

Silver, D., Huang, A., Maddison, C. J., Guez, A., Sifre, L., Van Den Driessche, G., et al. (2016). Mastering the game of go with deep neural networks and tree search. *Nature, 529*(7587), 484.

Singh, S. P. (2014). Magnetoencephalography: Basic principles. *Annals of Indian Academy of Neurology, 17*(Suppl 1), S107–S112.

Smith, S. J. (2005). EEG in the diagnosis, classification, and management of patients with epilepsy. *Journal of Neurology, Neurosurgery, and Psychiatry, 76 Suppl 2*(Suppl 2), ii2–ii7.

Tagaris, A., Kollias, D., Stafylopatis, A., Tagaris, G., & Kollias, S. (2018). Machine learning for neurodegenerative disorder diagnosis—Survey of practices and launch of benchmark dataset. *International Journal on Artificial Intelligence Tools, 27*(03), 1850011.

Tanner, C., Marder, K., Eberly, S., Biglan, K., Oakes, D., Shoulson, I., et al. (2018). Selected health and lifestyle factors, cytosine-adenine-guanine status, and phenoconversion in Huntington's disease. *Movement Disorders: Official Journal of the Movement Disorder Society, 33*(3), 472–478.

Tolosa, E., Wenning, G., & Poewe, W. (2006). The diagnosis of Parkinson's disease. *Lancet Neurology, 5*(1), 75–86.

Tombaugh, T. N., & McIntyre, N. J. (1992). The mini-mental state examination: A comprehensive review. *Journal of the American Geriatrics Society, 40*(9), 922–935.

Vapnik, V. N. (1995). *The nature of statistical learning theory* (pp. 1–188). Berlin: Springer-Verlag.

Vaquero, J. J., & Kinahan, P. (2015). Positron emission tomography: Current challenges and opportunities for technological advances in clinical and preclinical imaging systems. *Annual Review of Biomedical Engineering, 17*, 385–414.

Watkins, J., Fabietti, M., & Mahmud, M. (2020). Sense: A student performance quantifier using sentiment analysis. In *Proc. IJCNN* (pp. 1–6).

Whitwell, J. L., Höglinger, G. U., Antonini, A., Bordelon, Y., Boxer, A. L., Colosimo, C., et al. (2017). Radiological biomarkers for diagnosis in PSP: Where are we and where do

we need to be? Movement Disorder Society-endorsed PSP Study Group. *Movement Disorders*, *32*(7), 955–971.
Yahaya, S. W., Lotfi, A., & Mahmud, M. (2019). A consensus novelty detection ensemble approach for anomaly detection in activities of daily living. *Applied Soft Computing*, *83*, 105613. https://doi.org/10.1016/j.asoc.2019.105613.
Yiannopoulou, K. G., & Papageorgiou, S. G. (2020). Current and future treatments in Alzheimer disease: An update. *Journal of Central Nervous System Disease*, *12*, 1179573520907397.

CHAPTER 17

New trends in deep learning for neuroimaging analysis and disease prediction

Olugbemi T. Olaniyan[a], Charles O. Adetunji[b], Ayobami Dare[c], Olorunsola Adeyomoye[d], Mayowa J. Adeniyi[e], and Alex Enoch[f]

[a]Department of Physiology, Laboratory for Reproductive Biology and Developmental Programming, Rhema University, Aba, Nigeria
[b]Department of Microbiology, Applied Microbiology, Biotechnology and Nanotechnology Laboratory, Edo State University Uzairue, Iyamho, Nigeria
[c]Department of Physiology, School of Laboratory Medicine and Medical Sciences, College of Health Sciences, Westville Campus, University of KwaZulu-Natal, Durban, South Africa
[d]Department of Physiology, University of Medical Sciences, Ondo, Nigeria
[e]Department of Physiology, Edo State University Uzairue, Iyamho, Nigeria
[f]Department of Human Physiology, Ahmadu Bello University Zaria, Zaria, Nigeria

Introduction

In the last two decades, several advancements that depend on mass-univariate analytical techniques such as statistical parametric mapping have been reported in the neuroimaging investigation of neurological and psychiatric patients. The technique is usually applied for comparing of patients with analysis of some diseases free persons having neurofunctional and neuroanatomical variations at group level. This has also led to numerous advancements in understanding neurological disorders and the neurobiology of psychiatric disorders. However, mass-univariate analytical methodology has several challenges. This might be linked to the facts that statistical inference is obtained from multiple independent comparisons which depends on the hypothesis that diverse brain regions act individually but the application of network-level alteration in function and structure is better than focal alternations (Biswal et al., 2010; Fox et al., 2005; Mulders, van Eijndhoven, Schene, Beckmann, & Tendolkar, 2015; Sheffield & Barch, 2016).

The application of machine learning has been identified as a sustainable technology that can help in preventing diverse challenges encountered in the neuroimaging community. It plays a significant role in the development of algorithms that assist in identifying patterns and trends in prevailing data and using this data to make forecasts on new information (Adetunji, Adetunji, & Olaniyan, 2021; Adetunji, Ajayi, et al., 2021; Adetunji, Akram,

Artificial Intelligence for Neurological Disorders
https://doi.org/10.1016/B978-0-323-90277-9.00012-2

et al., 2021; Adetunji, Inobeme, et al., 2021; Adetunji & Varma, 2020; Akram et al., 2020; Egbuna et al., 2020; Roli, Adetunji, Mishra, Adetunji, & Mishra, 2020). This can only be accomplished with the mathematical optimization of computational statistics (Hastie, Tibshirani, & Friedman, 2001). Moreover, the application of machine learning could help in understanding multivariate statistics which could help in analysing inter-correlation among voxels, which could assist in first constraint of mass-univariate analytical method. In addition, machine learning could help in making statistical inferences at a single subject level that have wider application in the diagnosis and prognosis of disease. Hence, machine learning overcomes some of the drawbacks of mass-univariate analytical procedures (Arbabshirani, Plis, Sui, & Calhoun, 2016).

The application of machine learning in obtaining neuroimaging data from neurological and psychiatric patients has been documented by several scientists (Arbabshirani et al., 2016; Bengio, 2009; Calhoun & Sui, 2016; LeCun, Bengio, & Hinton, 2015; Pereira, Mitchell, & Botvinick, 2009; Plis et al., 2014a, 2014b; Vapnik, 1995; Wolfers, Buitelaar, Beckmann, Franke, & Marquand, 2015). Studies have shown that learning is accomplished through several iterative processes of adjusting the interconnections that exist among artificial neurons within a network (Bengio, 2009; Plis et al., 2014a, 2014b). Hence, this chapter provides an overview of different approaches in deep learning for the prediction of major neurological diseases such as mild cognitive impairment, Alzheimer's disease, brain tumors, depressive disorders, traumatic brain injury, schizophrenia, Parkinson's disease, autism spectrum disorder, attention-deficit hyperactivity disorder, epilepsy, stroke, multiple sclerosis. The chapter also discusses current challenges in using deep learning for assessing brain disorders via neuroimaging data.

Deep learning techniques

Deep learning is a specialized form of machine learning that enables a computer to develop an algorithm for learning multiple layers of representation to model complex interactions among data. Studies have revealed that deep learning/artificial intelligence is attracting significant attention in numerous fields. In deep learning, a computer algorithm learns to execute some complex tasks with high accuracy in various conditions (Dash, Acharya, Mittal, Abraham, & Keleman, 2020; Deng & Yu, 2014). Algorithms are developed by using a large amount of labeled data and a neural network structure containing several layers. It should be noted that deep learning differs from machine learning. In deep learning, relevant features and modeling steps are automatically extracted from the data (neuroimages) to perform "end-to-end learning," that is, when the system is loaded with raw data and a specific function to execute (e.g., diagnosis), it automatically learns how to perform this function. In machine learning, features and classifiers to analyze data (neuroimages) are extracted manually. A profound learning network increases with increasing data size, unlike machine learning, which has limited network size and whose performance reaches a plateau when extra data are added to the system (Sejnowski, 2020; Vu et al., 2018).

Deep learning algorithms are developed by using large labeled datasets and different network structures that learn features directly from the data without the need for manual feature extraction (Shrestha & Mahmood, 2019). These networks include classic neural networks convoluted neural networks, recurrent neural networks, generative adversarial networks, self-organizing maps, Boltzmann machines, deep reinforcement learning, autoencoders, backpropagation, and gradient descent. Deep learning has contributed significantly to several applications in science for adaptive testing, biological image classification, computer vision, cancer detection, natural language processing, object detection, facial recognition, handwriting recognition, speech recognition, and much more (Dargan, Kumar, Ayyagari, & Kumar, 2020).

Neuroimaging and data science

Neuroimaging is a form of technology that provides a collection of visual illustrations of the anatomical structures, physiological functions, and metabolic capacities of the nervous system (central or peripheral). In the brain, neuroimaging is broadly categorized into structural imaging (which assesses the brain and detects gross damages) and physiologic imaging (which is concerned with detailed and precise diagnosis of metabolic diseases). Incorporating multimodal imaging procedures may help identify useful biomarkers for clinical and experimental purposes (Pinto et al., 2001; Sharma & Weintraub, 2017). Neuroimaging has expanded over the years, and several neuroimaging techniques are currently employed to assess brain anatomy. We discuss some of these techniques in the sections that follow.

Computed axial tomography (CT or CAT): This is a non-invasive technique that uses the combinations of several X-ray measurements obtained from several angles to develop tomographic images of the body, thus allowing the user to examine visceral organs (Amyot et al., 2015).

Magnetic resonance imaging (MRI): This method produces two- and three-dimensional pictures of the brain and neighboring organs with the help of radio waves and magnetic fields. The MRI is made up of three main components: the primary magnet (which creates a magnetic field of optimal capacity to develop images), gradient magnets (which focus the MRI on the region-of-interest in the body by accurately modifying the magnetic field), and a coil (which emits a radiofrequency that causes the nuclei to align their spin orientation to the opposite orientation) (Irimia et al., 2012).

Diffusion-weighted (DWI) and diffusion tensor imaging (DTI): These are advanced MRI imaging techniques for measuring the difference in the rate and directionality of water molecules in diseased or normal brain tissues (Huisman, 2010).

Magnetic resonance spectroscopy (MRS): This type of metabolic, molecular imaging uses similar biophysical principles as MRI. However, the major difference in this technique is that it solely depends on the magnetic properties of water molecules or lipids, which helps to differentiate structures within the brain (Di Gialleonardo, Wilson, & Keshari, 2016).

Functional magnetic resonance imaging (fMRI): This is a special type of MRI scan that depends on increased blood flow in areas of local vasculature in the brain and spinal cord associated with neuronal activity (Kim & Bandettini, 2006).

Positron emission tomography (PET): This is a nuclear medical imaging procedure for producing images of active processes in the body. This technique involves administering a radio-labeled substance into the blood and analyzing the emissions produced (Ricker, Arenth, & Wagner, 2013).

Single-photon emission tomography (SPECT): This is a nuclear imaging procedure that assesses regional cerebral blood flow and indirectly measures metabolic activity in the brain. SPECT, like PET, utilizes gamma ray-emitting radioisotopes detected by a gamma camera. The camera transfers the data to a computer that processes the information and produces a two- and three-dimensional images (Lee & Park, 2020).

Magnetoencephalography (MEG): This is a noninvasive imaging procedure that quantifies the magnetic fields generated by electrical activity in brain neurons. MEG is like electroencephalography (EEG), but the major difference is that MEG analyzes magnetic fields and EEG analyzes electrical potentials (Singh, 2014).

Diffuse optical imaging (DOI): This procedure uses infrared rays to produce images of the body by measuring variations in hemodynamics. Variations in optical absorption of water and oxygenated and deoxygenated hemoglobin are measured to visualize internal body structures (Boas, Dale, & Franceschini, 2004).

Recent trends in neuroimaging have witnessed the application of deep learning in the analysis of brain images to improve the diagnosis of neurological disorders and have been reported to be superior compared to classic imaging techniques (Plis et al., 2014a, 2014b; Zaharchuk, Gong, Wintermark, Rubin, & Langlotz, 2018). The application of deep learning techniques in neuroimaging can help to reduce the risk associated with image acquisition via traditional methods, accelerate image acquisition, reduce artifacts, and improve quality and analysis, thus improving identification, classification, diagnosis, and even management of brain diseases (Dash, Chakraborty, Giri, Pani, & Frnda, 2021; Zhu et al., 2019).

Data science is an interdisciplinary field that utilizes scientific methods and processes to acquire knowledge from data and apply such knowledge across various domains (Dash, Thulasiram, & Thulasiram, 2019; Dhar, 2013). The importance of data science in deep learning techniques suggests that there is a direct application of data science in neuroimaging. A large amount of data is generated from brain images, which are then processed, transformed, and used to develop algorithms used in deep learning approaches to improve diagnosis of neurological disorders. Thus, data science is undoubtedly playing a major role in the new trend of neuroimaging analysis.

Cognitive impairment

Cognition is the process of acquiring knowledge and comprehension. Thinking, remembering, judging, knowing, and solving problems are some examples of cognitive processes. Any lesion or damage to areas of the brain can cause cognitive impairment, which may profoundly affect an individual's health and general wellbeing (Zang, Zhang, & Castellanos, 2010; Lopez, 2013). Cognitive impairment can result in cognitive decline, which can interfere with quality of life (Stites, Harkins, Rubright, & Karlawish, 2018). Some of the risk factors for cognitive impairment include age, alcohol consumption, smoking, obesity, depression, genetic factors, and many others (Chen, Cheng, Lin, Lee, & Chou, 2018). Cognitive impairment is common in the elderly, affecting memory, attention, and other cognitive functions (Murman, 2015). Age-related cognitive decline may be influenced by lifestyle patterns, for example, through diet and physical exercise. Neurodegenerative diseases like Alzheimer's disease and Parkinson's disease

(Dash, Abraham, & Rehman, 2018) are common in individuals older than 65 years and could further contribute to loss of cognitive functions (Wilson, Leurgans, Boyle, Schneider, & Bennett, 2010).

Furthermore, evidence has shown that global alcohol consumption has escalated over the years and can cause a decline in cognition, especially in older adults (Koch et al., 2019). However, several studies have reported that moderate alcohol intake enhances cognitive function and reduces the risk of dementia (Duong, Patel, & Chang, 2017). Smoking has also been associated with cognitive decline, but because of the decrease in the life expectancy of smokers, there is a paucity of information on its impact on cognitive function, and the extent to which smoking increases cognitive decline remains unclear. Obesity is a public health problem and its prevalence has continued to increase over the years due to exposure to environmental and genetic factors (Hruby & Hu, 2015). Studies have linked mid-life obesity with a decline in cognitive function in late life (Nguyen, Killcross, & Jenkins, 2014). Depression is a mood disorder characterized by a persistent feeling of sadness and loss of interest. The cognitive deficits associated with depression include lack of attention, short-term and working memory, visual and auditory processing, problem-solving, and decline in motor activities (Morimoto, Kanellopoulos, & Alexopoulos, 2014).

Images, text, sounds, waves, biomarkers, and physiological signals

With the exponential increase in information technology, there has been a massive surge in visual data (Jen, Kerndt, & Korvek, 2021). Multiple images from videos and photos are taken and shared online every moment. The healthcare system is affected by these advances in imaging. Imaging techniques have caused a surge in medical knowledge to increase effectiveness of disease diagnosis (Krupinski, 2010). X-ray radiography, MRI medical ultrasonography, CT, and PET are some of the imaging techniques used in medical science. The use of deep learning to enhance medical diagnosis using sound is gaining momentum. For instance, heart sound classification has been a significant challenge, but a band filter can remove noise from heartbeat signals (Raza et al., 2019). Because of the increase in the amount of biological data being generated, effective analytical methods and interpretation are necessary. The use of deep learning to process these large volumes of data has emerged. Studies have proposed an additional integrated layer to a deep learning structure to analyze biological data and discover biomarkers for many diseases (Xu, Xue, & Zhang, 2019). Structural neuroimaging in conjunction with deep learning could provide stable biomarkers for many neurological conditions. Deep learning can explore neuroimaging data to generate a superior representation for characterizing a neurological disorder. The use of biomarkers for investigating neurodegenerative disease has increased and has contributed to the understanding of the pathogenesis of disease conditions (Balyen & Peto, 2019). Medical instruments can generate data from sound, graphs, images, text, and signals (individually and collectively), giving rise to more complex data that can be analyzed and interpreted by deep learning models (Egho et al., 2012). For example, electronic health records contain various types of patient data. Therefore, text classification is crucial in retrieving and organizing medical records. Text classification often utilizes different rules in traditionally locating a health record. However, with deep learning, text classification and location of health records are now easier, faster, and effective (Xiao, Choi, & Sun, 2018).

Artificial intelligence and disease diagnosis and prediction

Artificial intelligence has grown tremendously since the late 1980s and has great potential to succeed the almighty human in almost all spheres, including medical diagnosis and forecasting. Some time ago, an algorithm that confirms the identification of encrypted retinal images to recognize individuals with cardiovascular risks was developed (Bicciato, 2004; Gulshan et al., 2016; Adetunji et al., 2020). Apart from the cardiovascular system, there are many other systems where artificial intelligence is being implemented.

Radiology is a branch of medicine that uses radioactivity in the diagnosis and management of disease. Through the use of machine learning, especially in medical imaging such as radiomics, disease diagnosis has become less difficult (Suzuki, 2017). Today, there are digital slide scanners that digitalize histology slides to form images with high resolution. Automation of complex tasks in pathology can reduce the need for hospital appointments as well as decrease instances of incorrect diagnoses and erroneous prescriptions (Suddhasuchi et al., 2020; CADTH, 2018; Dash, Chakraborty, Giri, & Pani, 2021).

Cognitive computing in oncology has reduced healthcare professionals' workload, allowing them to arrive at diagnosis more quickly (Kantarjian & Yu, 2015). In neurology, artificial intelligence algorithms have been created to forecast disease development (CADTH, 2018) with massive potential in availing information on anatomy, physiology, and brain behaviors (Yu &

Chen, 2005). Algorithms also exist that produce early warning signs of stroke (Zhang, Xie, Ye, & Pang, 2013); the earlier stroke is detected, the better the treatment outcomes will be.

Current challenges of heterogeneous multisite datasets and opportunities

Amassing data from different sources and integrating them are very challenging and time-intensive tasks. This is true for data that exhibit different types and formats. For instance, it takes a lot of time to make extracted data of different formats compatible with the desired pattern. Another challenge is the quality of the extracted data. Usually, poor data quality and incorrect information result in poor integration.

Summary

Deep learning is a specialized form of machine learning that enables a computer to develop an algorithm for learning multiple layers of representation to model complex interactions among data. Deep learning algorithms are developed by using large labeled datasets and different network structures that learn features directly from the data without the need for manual feature extraction. Various deep learning networks include classic neural networks, convoluted neural networks, recurrent neural networks, generative adversarial networks, self-organizing maps, Boltzmann machines, deep reinforcement learning, autoencoders, backpropagation, and gradient descent. Incorporating multimodal imaging procedures may help identify useful biomarkers for clinical and experimental purposes. Recent trends in neuroimaging have witnessed the application of deep learning in the analysis of brain images to improve the diagnosis of neurological disorders and have been reported to be superior compared to classic imaging techniques. There is great potential to use deep learning to automate complex tasks in neurology, thus reducing the need for hospital appointments and instances of incorrect diagnoses and erroneous prescriptions.

Conclusion and future directions

This chapter highlighted various important applications of deep learning techniques for diagnosing and predicting disease conditions from neuroimaging. The advantage of big data and artificial intelligence approaches and innovations in the treatment and diagnosis of various neurological diseases are currently being explored by scientists in numerous fields. The utilization of machine learning to generate large datasets from images, text, sounds, graphs, and signals is an emerging trend in neuroscience and disease management. Deep learning can be used to predict diseases like cancer, cardiovascular disease, and metabolic dysfunctions by obtaining and evaluating patient electronic health records, biological system information, physiological signals, biomarkers, and biomedical images to diagnose cognitive functions across different brain sub-regions. Hence, the current challenges of utilizing deep learning in assessing brain disorders in neuroimaging data can be mitigated through the advanced development and incorporation of computational physiology and neural networks into neuroscience teaching and research.

References

Adetunji, C. O., & Varma, A. (2020). Biotechnological application of trichoderma: A powerful fungal isolate with diverse potentials for the attainment of food safety, management of pest and diseases, healthy planet, and sustainable agriculture. In C. Manoharachary, H. B. Singh, & A. Varma (Eds.), *Vol. 61. Trichoderma: Agricultural applications and beyond. Soil biology*. Cham: Springer.

Adetunji, C. O., Tijjani, H., Olisah, M. C., Patrick-Iwuanyanwu, K. C., Adetunji, J. B., Ifemeje, J. C., Akram, M., Moboladji, B. M., & Onyeike, P. C. (2020). Health benefits of Isoflavones found exclusively of plants of the Fabaceae Family. In C. Egbuna, & T. G. Dable (Eds.), *Functional foods and nutraceuticals*. Cham: Springer. https://doi.org/10.1007/978-3-030-42319-3_22.

Adetunji, C. O., Akram, M., Olaniyan, O. T., Ajayi, O. O., Inobeme, A., Olaniyan, S., Hameed, L., & Adetunji, J. B. (2021). Medicinal plants used in the treatment of pulmonary hypertension. In K. Dua, S. Nammi, D. Chang, D. K. Chellappan, G. Gupta, & T. Collet (Eds.), *Medicinal plants for lung diseases*. Singapore: Springer. https://doi.org/10.1007/978-981-33-6850-7_14.

Adetunji, C. O., Inobeme, A., Olaniyan, O. T., Ajayi, O. O., Olaniyan, S., & Adetunji, J. B. (2021). Application of nanodrugs derived from active metabolites of medicinal plants for the treatment of inflammatory and lung diseases: Recent advances. In K. Dua, S. Nammi, D. Chang, D. K. Chellappan, G. Gupta, & T. Collet (Eds.), *Medicinal plants for lung diseases*. Singapore: Springer.

Adetunji, C. O., Ajayi, O. O., Akram, M., Olaniyan, O. T., Chishti, M. A., Inobeme, A., Olaniyan, S., Adetunji, J. B., Olaniyan, M., & Awotunde, S. O. (2021). Medicinal plants used in the treatment of influenza A virus infections. In K. Dua, S. Nammi, D. Chang, D. K. Chellappan, G. Gupta, & T. Collet (Eds.), *Medicinal plants for lung diseases*. Singapore: Springer.

Adetunji, J. B., Adetunji, C. O., & Olaniyan, O. T. (2021). African walnuts: A natural depository of nutritional and bioactive compounds essential for food and nutritional security in Africa. In O. O. Babalola (Ed.), *Food security and safety*. Cham: Springer.

Akram, M., Jabeen, F., Daniyal, M., Zainab, R., Haq, U. U., Adetunji, C. O., Egbuna, C., Ephraim-Emmanuel, B. C., Patrick-Iwuanyanwu, K. C., & Ogbo, A. B. (2020). Genetic engineering of novel products of health significance: Recombinant DNA technology. In

C. Egbuna, & T. G. Dable (Eds.), *Functional foods and nutraceuticals*. Cham: Springer. https://doi.org/10.1007/978-3-030-42319-3_26.

Amyot, F., Arciniegas, D. B., Brazaitis, M. P., Curley, K. C., Diaz-Arrastia, R., Gandjbakhche, A., Herscovitch, P., Hinds, S. R., 2nd, Manley, G. T., Pacifico, A., Razumovsky, A., Riley, J., Salzer, W., Shih, R., Smirniotopoulos, J. G., & Stocker, D. (2015). A review of the effectiveness of neuroimaging modalities for the detection of traumatic brain injury. *Journal of Neurotrauma, 32*, 1693–1721.

Arbabshirani, M. R., Plis, S., Sui, J., & Calhoun, V. D. (2016). Single subject prediction of brain disorders in neuroimaging: Promises and pitfalls. *NeuroImage*, 137–165.

Balyen, L., & Peto, T. (2019). Promising artificial intelligence-machine learning-deep learning algorithms in ophthalmology. *Asia-Pacific Journal of Ophthalmology, 8*(3), 264–272.

Bengio, Y. (2009). Learning deep architectures for AI found. *Foundations and Trends® in Machine Learning, 2*(2009), 1–127.

Bicciato, S. (2004). Artificial neural network technologies to identify biomarkers for therapeutic intervention. *Current Opinion in Molecular Therapeutics, 6*(6), 616–623. 15663326.

Biswal, B. B., Mennes, M., Zuo, X. N., Gohel, S., Kelly, C., Smith, S. M., Beckmann, C. F., Adelstein, J. S., Buckner, R. L., Colcombe, S., Dogonowski, A. M., Ernst, M., Fair, D., Hampson, M., Hoptman, M. J., Hyde, J. S., Kiviniemi, V. J., Kotter, R., Li, S. J., … Milham, M.P. (2010). Toward discovery science of human brain function. *Proceedings of the National Academy of Sciences of the United States of America, 107*, 4734–4739.

Boas, D. A., Dale, A. M., & Franceschini, M. A. (2004). Diffuse optical imaging of brain activation: Approaches to optimizing image sensitivity, resolution, and accuracy. *NeuroImage, 23*(Supp.l 1), S275–S288.

CADTH, 2018 An overview of clinical applications of artificial intelligence. Ottawa, September (CADTH issues in emerging health technologies); issue 174.

Calhoun, V. D., & Sui, J. (2016). Multimodal fusion of brain imaging data: A key to finding the missing link(s) in complex mental illness. *Biological Psychiatry: Cognitive Neuroscience and Neuroimaging, 1*, 230–244.

Chen, P. H., Cheng, S. J., Lin, H. C., Lee, C. Y., & Chou, C. H. (2018). Risk factors for the progression of mild cognitive impairment in different types of neurodegenerative disorders. *Behavioural Neurology, 2018*, 6929732.

Dargan, S., Kumar, M., Ayyagari, M. R., & Kumar, G. (2020). A survey of deep learning and its applications: A new paradigm to machine learning. *Archives of Computational Methods in Engineering, 27*, 1071–1092.

Dash, S., Abraham, A., & Rehman, A. (2018). Kernel based chaotic firefly algorithm for diagnosing Parkinson's disease. In *18th international conference on hybrid intelligent systems (HIS 2018), advances in intelligent systems and computing, Porto, Portugal* (pp. 176–188).

Dash, S., Acharya, B. R., Mittal, M., Abraham, A., & Keleman, A. (2020). Deep learning techniques for biomedical and health informatics. In *Studies in big data series* Springer. ISSN:2197–6511, ISBN:978–3–030-33966-1.

Dash, S., Chakraborty, C., Giri, S. K., & Pani, S. K. (2021). Intelligent computing on time-series data analysis and prediction of covid-19 pandemics. *Pattern Recognition Letters, 151*, 69–75. https://doi.org/10.1016/j.patrec.2021.07.027. Elsevier.

Dash, S., Chakraborty, C., Giri, S. K., Pani, S. K., & Frnda, J. (2021). BIFM: Big-data driven intelligent forecasting model for COVID-19. *IEEE Access, 9*, 97505–97517. https://doi.org/10.1109/ACCESS.2021.3094658.

Dash, S., Thulasiram, R., & Thulasiram, P. (2019). A modified firefly based meta-search algorithm for feature selection: A predictive model for medical data. *International Journal of Swarm Intelligence, 10*(2), 2. IGI Global, USA.

Deng, L., & Yu, D. (2014). Deep learning: Methods and applications. *Foundations and Trends® in Signal Processing*, 7, 197–387.

Dhar, V. (2013). Data science and prediction. *Communications of the ACM, 56*, 64–73.

Di Gialleonardo, D. V., Wilson, D. M., & Keshari, K. R. (2016). The potential of metabolic imaging. *Seminars in Nuclear Medicine*, *46*, 28–39.

Duong, S., Patel, T., & Chang, F. (2017). Dementia: What pharmacists need to know. *Canadian Pharmacists Journal: CPJ = Revue des pharmaciens du Canada: RPC*, *150*(2), 118–129.

Egbuna, C., Gupta, E., Ezzat, S. M., Jeevanandam, J., Mishra, N., Akram, M., Sudharani, N., Adetunji, C. O., Singh, P., Ifemeje, J. C., Deepak, M., Bhavana, A., Walag, A. M. P., Ansari, R., Adetunji, J. B., Laila, U., Olisah, M. C., & Onyekere, P. F. (2020). Aloe species as valuable sources of functional bioactives. In C. Egbuna, & T. G. Dable (Eds.), *Functional foods and nutraceuticals*. Cham: Springer.

Egho, E., Raïssi, C., Ienco, D., Jay, N., Napoli, A., Poncelet, P., Quantin, C., & Teisseire, M. (2012). Healthcare trajectory mining by combining multidimensional component and itemsets. In *International workshop on new frontiers in mining complex patterns* (pp. 109–123). Springer.

Fox, M. D., Snyder, A. Z., Vincent, J. L., Corbetta, M., Van Essen, D. C., & Raichle, M. E. (2005). The human brain is intrinsically organized into dynamic, anticorrelated functional networks. *Proceedings of the National Academy of Sciences of the United States of America*, *102*, 9673–9678.

Gulshan, V., Peng, L., Coram, M., Stumpe, M., Wu, D., Narayanaswamy, A., et al. (2016). Development and validation of a deep learning algorithm for detection of diabetic retinopathy in retinal fundus photographs. *Journal of the American Medical Association*, *316*(22), 240.

Hastie, T., Tibshirani, R., & Friedman, J. (2001). *The elements of statistical learning: Data mining, inference and prediction. 2001*. New York, NY: Springer-Verlag.

Hruby, A., & Hu, F. B. (2015). The epidemiology of obesity: A big picture. *PharmacoEconomics*, *33*(7), 673–689.

Huisman, T. A. G. M. (2010). Diffusion-weighted and diffusion tensor imaging of the brain, made easy. *Cancer Imaging: The Official Publication of the International Cancer Imaging Society*, *10*(Spec. A), S163–S171.

Irimia, A., Wang, B., Aylward, S. R., Prastawa, M. W., Pace, D. F., Gerig, G., Hovda, D. A., Kikinis, R., Vespa, P. M., & Van Horn, J. D. (2012). Neuroimaging of structural pathology and connectomics in traumatic brain injury: Toward personalized outcome prediction. *NeuroImage: Clinical*, *1*, 1–17.

Jen, M. Y., Kerndt, C. C., & Korvek, S. J. (2021). Health information technology [Updated 2020 Sep 16]. In *StatPearls [Internet]*. Treasure Island (FL): StatPearls Publishing. Available from: https://www.ncbi.nlm.nih.gov/books/NBK470186/.

Kantarjian, H., & Yu, P. P. (2015). Artificial intelligence, big data, and cancer. *JAMA Oncology*, *1*(5), 573–574.

Kim, S.-G., & Bandettini, P. (2006). Principles of functional MRI. In S. Faro, & F. Mohamed (Eds.), *Functional MRI: Basic principles and applications*. New York: McGraw-Hill Professional.

Koch, M., Fitzpatrick, A. L., Rapp, S. R., Nahin, R. L., Williamson, J. D., Lopez, O. L., DeKosky, S. T., Kuller, L. H., Mackey, R. H., Mukamal, K. J., Jensen, M. K., & Sink, K. M. (2019). Alcohol consumption and risk of dementia and cognitive decline among older adults with or without mild cognitive impairment. *JAMA Network Open*, *2*(9), e1910319.

Krupinski, E. A. (2010). Current perspectives in medical image perception. *Attention, Perception & Psychophysics*, *72*(5), 1205–1217.

LeCun, Y., Bengio, Y., & Hinton, G. (2015). Deep learning. *Nature*, *521*, 436–444.

Lee, S. J., & Park, H. J. (2020). Single photon emission computed tomography (SPECT) or positron emission tomography (PET) imaging for radiotherapy planning in patients with lung cancer: A meta-analysis. *Scientific Reports*, *10*, 14864.

Lopez, O. L. (2013). Mild cognitive impairment. *Continuum (Minneapolis, Minn.)*, *19*(2 dementia), 411–424.

Morimoto, S. S., Kanellopoulos, D., & Alexopoulos, G. S. (2014). Cognitive impairment in depressed older adults: Implications for prognosis and treatment. *Psychiatric Annals*, *44*(3), 138–142.

Mulders, P. C., van Eijndhoven, P. F., Schene, A. H., Beckmann, C. F., & Tendolkar, I. (2015). Resting-state functional connectivity in major depressive disorder: A review. *Neuroscience and Biobehavioral Reviews*, *56*, 330–344.

Murman, D. L. (2015). The impact of age on cognition. *Seminars in Hearing*, *36*(3), 111–121.

Nguyen, J. C., Killcross, A. S., & Jenkins, T. A. (2014). Obesity and cognitive decline: Role of inflammation and vascular changes. *Frontiers in Neuroscience*, *8*, 375.

Pereira, F., Mitchell, T., & Botvinick, M. (2009). Machine learning classifiers and fMRI: A tutorial overview. *NeuroImage*, *45*, S199–S209.

Pinto, H., Han, J., Pei, J., Wang, K., Chen, Q., & Dayal, U. (2001). Multi-dimensional sequential pattern mining. In *Proceedings of the tenth international conference on Information and knowledge management* (pp. 81–88). ACM.

Plis, S. M., Hjelm, D. R., Salakhutdinov, R., Allen, E. A., Bockholt, H. J., Long, J. D., Johnson, H. J., Paulsen, J. S., Turner, J. A., & Calhoun, V. D. (2014a). Deep learning for neuroimaging: A validation study. *Frontiers in Neuroscience*, *8*.

Plis, S. M., Hjelm, D. R., Salakhutdinov, R., Allen, E. A., Bockholt, H. J., Long, J. D., Johnson, H. J., Paulsen, J. S., Turner, J., & Calhoun, V. D. (2014b). Deep learning for neuroimaging: A validation study front. *Neuroscience*, *8*, 1–11.

Raza, A., Mehmood, A., Ullah, S., Ahmad, M., Choi, G. S., & On, B. W. (2019). Heartbeat sound signal classification using deep learning. *Sensors (Basel, Switzerland)*, *19*(21), 4819.

Ricker, J., Arenth, P., & Wagner, A. (2013). Functional neuroimaging. In *Brain injury medicine* (2nd ed., pp. 218–229). New York: Demos.

Roli, O. I., Adetunji, C. O., Mishra, P., Adetunji, J. B., & Mishra, R. R. (2020). Caffeine: Nutraceutical and health benefit of caffeine-containing commodities and products. In P. Mishra, R. R. Mishra, & C. O. Adetunji (Eds.), *Innovations in food technology*. Singapore: Springer.

Sejnowski, T. J. (2020). The unreasonable effectiveness of deep learning in artificial intelligence. *Proceedings of the National Academy of Sciences*, *117*, 30033–30038.

Sharma, A., & Weintraub, A. (2017). Neuroimaging. In J. Kreutzer, J. Deluca, & B. Caplan (Eds.), *Encyclopedia of clinical neuropsychology*. Cham: Springer International Publishing.

Sheffield, J. M., & Barch, D. M. (2016). Cognition and resting-state functional connectivity in schizophrenia. *Neuroscience and Biobehavioral Reviews*, *61*, 108–120.

Shrestha, A., & Mahmood, A. (2019). Review of deep learning algorithms and architectures. *IEEE Access*, 7, 53040–53065.

Singh, S. P. (2014). Magnetoencephalography: Basic principles. *Annals of Indian Academy of Neurology*, *17*, S107–S112.

Stites, S. D., Harkins, K., Rubright, J. D., & Karlawish, J. (2018). Relationships between cognitive complaints and quality of life in older adults with mild cognitive impairment, mild Alzheimer disease dementia, and normal cognition. *Alzheimer Disease and Associated Disorders*, *32*(4), 276–283.

Suddhasuchi, D., Sharangi, A. B., Egbuna, C., Jeevanandam, J., Ezzat, S. M., Adetunji, C. O., Tijjani, H., Olisah, M. C., Patrick-Iwuanyanwu, K. C., Adetunji, J. B., Ifemeje, J. C., Akram, M., Moboladji, B. M., & Onyeike, P. C. (2020). *Health benefits of isoflavones found exclusively of plants of the Fabaceae Family*.

Suzuki, K. (2017). Overview of deep learning in medical imaging. *Radiological Physics and Technology*, *10*(3), 257–273.

Vapnik, V. N. (1995). *The nature of statistical learning theory. 1995*. Springer.

Vu, M. T., Adali, T., Ba, D., Buzsaki, G., Carlson, D., Heller, K., et al. (2018). A shared vision for machine learning in neuroscience. *The Journal of Neuroscience*, *38*(7), 1601–1607.

Wilson, R. S., Leurgans, S. E., Boyle, P. A., Schneider, J. A., & Bennett, D. A. (2010). Neurodegenerative basis of age-related cognitive decline. *Neurology*, *75*(12), 1070–1078.

Wolfers, T., Buitelaar, J. K., Beckmann, C. F., Franke, B., & Marquand, A. F. (2015). From estimating activation locality to predicting disorder: A review of pattern recognition for neuroimaging-based psychiatric diagnostics. *Neuroscience and Biobehavioral Reviews*, *57*, 328–349.

Xiao, C., Choi, E., & Sun, J. (2018). Opportunities and challenges in developing deep learning models using electronic health records data: A systematic review. *Journal of the American Medical Informatics Association*, *25*(10), 1419–1428.

Xu, J., Xue, K., & Zhang, K. (2019). Current status and future trends of clinical diagnoses via image-based deep learning. *Theranostics*, *9*(25), 7556–7565.

Yu, C.-C., & Chen, Y.-L. (2005). Mining sequential patterns from multidimensional sequence data. *IEEE Transactions on Knowledge and Data Engineering*, *17*(1), 136–140.

Zaharchuk, G., Gong, E., Wintermark, M., Rubin, D., & Langlotz, C. P. (2018). Deep learning in neuroradiology. *AJNR. American Journal of Neuroradiology*, *39*, 1776–1784.

Zang, H. Y., Zhang, F. X., & Castellanos, M. P. M. (2010). Toward discovery science of human brain function. *Proceedings of the National Academy of Sciences*, *107*, 4734–4739.

Zhang, Q., Xie, Y., Ye, P., & Pang, C. (2013). Acute ischaemic stroke prediction from physiological time series patterns. *The Australasian Medical Journal*, *6*(5), 280–286.

Zhu, G., Jiang, B., Tong, L., Xie, Y., Zaharchuk, G., & Wintermark, M. (2019). Applications of deep learning to neuro-imaging techniques. *Frontiers in Neurology*, *10*.

CHAPTER 18

Prevention and diagnosis of neurodegenerative diseases using machine learning models

Olugbemi T. Olaniyan[a], Charles O. Adetunji[b], Ayobami Dare[c], Olorunsola Adeyomoye[d], Mayowa J. Adeniyi[e], and Alex Enoch[f]
[a]Department of Physiology, Laboratory for Reproductive Biology and Developmental Programming, Rhema University, Aba, Nigeria
[b]Department of Microbiology, Applied Microbiology, Biotechnology and Nanotechnology Laboratory, Edo State University Uzairue, Iyamho, Nigeria
[c]Department of Physiology, School of Laboratory Medicine and Medical Sciences, College of Health Sciences, Westville Campus, University of KwaZulu-Natal, Durban, South Africa
[d]Department of Physiology, University of Medical Sciences, Ondo, Nigeria
[e]Department of Physiology, Edo State University Uzairue, Iyamho, Nigeria
[f]Department of Human Physiology, Ahmadu Bello University Zaria, Zaria, Nigeria

Introduction

Neurodegenerative diseases such as Alzheimer's disease, Parkinson's disease, Huntington disease, multiple sclerosis, and others are debilitating disorders (Gitler, Dhillon, & Shorter, 2017). Most neurodegenerative diseases are age-related and are the leading cause of morbidity and cognitive impairment in older individuals (Wyss-Coray, 2016). Early prevention of these diseases and identification of their risk factors can help solve the challenges of disease burden (Adetunji, Adetunji, & Olaniyan, 2021; Adetunji, Ajayi, et al., 2021; Adetunji et al., 2021a, 2021b; Adetunji, Inobeme, et al., 2021; Adetunji & Varma, 2020; Akram et al., 2020; Das et al., 2020; Egbuna et al., 2020; Roli, Adetunji, Mishra, Adetunji, & Mishra, 2020). Poor nutrition, stress, depression, smoking, and alcohol consumption are some of the risk factors for neurodegenerative diseases (Adetunji et al., 2020; Brown, Lockwood, & Sonawane, 2005). Although the diagnosis of these diseases may be difficult, early detection is pertinent and could even provide clues for understanding their pathogenesis. Several diagnostic tools have been employed to assess cognitive impairment in neurodegenerative disease using already established criteria. Neurofunctional imaging techniques allow for greater specificity in clinical assessments, even though these tests have different variability (Veldsman & Egorova, 2017). With the rapid increase in the older

Artificial Intelligence for Neurological Disorders
https://doi.org/10.1016/B978-0-323-90277-9.00009-2

population and the rise in neurodegenerative disease cases, there is a need to provide alternative methods of diagnosis to reduce the burden of disease on the populace and healthcare providers.

Machine learning (ML) is a branch of artificial intelligence (AI) that involves using a variety of algorithms to perform predictions using a wide range of datasets. Evidence has shown that ML can extract information with greater accuracy than health professionals can, making it a useful and powerful tool (Golriz Khatami, Mubeen, & Hofmann-Apitius, 2020). Furthermore, it can process large volumes of data with a high capacity for algorithm design (Nichols, Herbert Chan, & Baker, 2019). Therefore, ML can help researchers, clinicians, and patients address some of the difficulties posed by neurodegenerative disease. It can help in early detection and prevention of disease, interpretation of medical images, and development of therapeutic strategies.

Neurodegenerative diseases

Neurodegenerative diseases are debilitating disorders characterized by the loss or death of many neurons (Dugger & Dickson, 2017). The most common of these diseases is Alzheimer's disease (Selkoe & Lansbury Jr., 1999). Others include Parkinson's disease, multiple sclerosis, and so on. Alzheimer's disease is the most common type of dementia affecting individuals older than 65 years (Guerreiro & Bras, 2015). It interferes with both behavioral and cognitive functions in the affected individual, including attention, comprehension, language, reasoning, memory, and judgment. Symptoms include short-term memory loss, difficulty in solving problems, lack of motivation, and more. Alzheimer's disease has been reported to affect 24 million people globally, and this number is expected to increase in the future with a surge in the global population (Qiu, Kivipelto, & von Strauss, 2009).

Parkinson's disease is another type of neurodegenerative disorder associated with motor disorders, including bradykinesia, resting tremor, and rigidity. Although the etiology of the disease is poorly understood, evidence has shown that environmental factors like smoking, caffeine consumption, and pesticide exposure could be risk factors (Emamzadeh & Surguchov, 2018). The deficiency in motor activity could be due to the loss of dopaminergic neurons of the substantia nigra. Other factors like mitochondria dysfunction and nerve inflammation could play a significant role (DeMaagd & Philip, 2015). Parkinson's disease is also associated with depression and anxiety, cognitive impairment, and autonomic dysfunction.

Multiple sclerosis is an inflammatory disease of the central nervous system characterized by physical and cognitive impairment. Although the pathogenesis of the disease is unclear, reports have shown that it occurs due to neuronal cell death, nerve demyelination, and neuronal dysfunction (Loma & Heyman, 2011). The cause of the disease is multifactorial. Smoking, infection, and vitamin C deficiency are possible risk factors. The therapeutic approach utilized for managing multiple sclerosis is based on the use of antiinflammatory and immunomodulatory agents (Dargahi et al., 2017). Other examples of neurodegenerative disease include prion disease, motor neuron disease, spinocerebellar ataxia, amyloidosis, tauopathies, α-synucleinopathies, and proteinopathies.

Artificial intelligence (AI) and machine learning (ML)

AI and ML are new technologies used to create an intelligent system that can perform complicated tasks like the human brain does. Recently, they have been widely applied in the healthcare system to improve the diagnosis and treatment of several disorders, including neurological diseases. The two terms are often used interchangeably. However, there are clear differences between them. AI is a broader term that involves creating an intelligent system that can simulate human thinking capability, whereas ML is a form of AI that allows the system to learn from data to perform a specific task without prior programming for such task (Botha, 2017).

An AI system does not require prior programming; instead, the system uses an algorithm that can work with its intelligence to solve complex tasks like humans do. These algorithms can be ML algorithms (e.g., reinforcement learning algorithms) and deep learning neural networks. AI has wider application and is concerned with maximizing the chances of success and comprehensively dealing with structured, semistructured, and unstructured data (Hutson, 2019). ML promotes efficient use of data to highlight disease mechanisms and assists with early detection, prognosis, patient stratification, and new treatment modalities. Many researchers have obtained rich, high-dimensional datasets from healthy individuals and patients suffering from neurodegenerative disorders with these goals in mind.

Supervised ML algorithms are presently the most applied approaches to neurodegenerative disease-related data that use labeled datasets acquired manually by an expert. After labeling a dataset, the ML algorithm builds a model of the relationship between the input features (e.g., atrophy on the region-of-interest on the image) and the label (e.g., diagnostic classification).

The algorithm can then apply this model to new, unlabeled datasets to predict the label based on the new input features (Rohini & Surendran, 2019). In the diagnosis of neurodegenerative disease, a physician often interprets test results, which may cause a delay in diagnosis; such delays can be minimized by incorporating ML techniques to analyze data acquired from the hospital. At the same time, these data can be used to determine patient prognosis by comparing disease progression at any given time with historical data from patients sharing the same endotype or phenotype (Myszczynska et al., 2020).

Furthermore, ML is strictly evidence-based and can analyze problems in an unbiased manner. The approach helps make objective diagnoses from medical images and frequently exceeds the performance of physicians with respect to duration, precision, and accuracy (Summers et al., 2017; Zhu et al., 2019). Computer-aided diagnosis (CAD), driven mainly by ML methods, can further enhance the interpretation of neuroimaging data and help recognize subtle anomalies on images that radiologists may not detect. For instance, convoluted neural networks (CNNs) can categorize images by recognizing and mapping a large number of features (He, Zhang, Ren, & Sun, 2016) and have been used to predict the diagnosis of Alzheimer's disease (Gray, Aljabar, Heckemann, Hammers, & Rueckert, 2013) and to study cognitive aging (Choi, Kang, & Lee, 2018) from magnetic resonance imaging (MRI) and positron emission tomography (PET) scans. Thus, ML approaches can improve efficiency in the management of neurodegenerative diseases.

AI and clinical practice

It has been proposed that the field that will be most affected by AI is health care due to the large volumes of data generated in clinical practice, from macrolevel physiology and behavior to laboratory and imaging studies. The coming age in medicine will be associated with "big data," which will require sophisticated techniques to be analyzed and interpreted and to develop clinically relevant algorithms. Therefore, there is a paradigm shift to new techniques in AI, especially ML, to handle this data and develop models that will assist in clinical practice (Obermeyer & Emanuel, 2016). AI-powered medical technologies have been well received by the general population partly because they promote the 4P model of clinical practice (i.e., prevent, personalized, participate, and predict) and encourage patient autonomy, which may not be possible via physical contact with a physician

(Orth et al., 2019). For example, smartphones are being used to complete and distribute electronic personal health records, assess and screen important body functions via biosensors, and assist in reaching optimal therapeutic compliance (Morawski et al., 2018; Topol, 2019). The development of intelligent medical technologies is improving several areas in clinical practice. Numerous AI-based algorithms have been approved recently by the Food and Drug Administration (FDA) and could therefore be implemented soon. Different areas of clinical practice in which AI has been effectively applied include cardiology, where AI techniques are used to detect atrial fibrillation and cardiovascular risk (Dorado-Díaz, Sampedro-Gómez, Vicente-Palacios, & Sánchez, 2019; Turakhia et al., 2019); pulmonary medicine, where AI is used to interpret pulmonary function tests (Topalovic et al., 2019); endocrinology, where AI has been used to monitor blood glucose (Christiansen et al., 2017); gastroenterology, where AI algorithms and other deep learning techniques have been used to analyze endoscopy and ultrasound images as well as other gastrointestinal diseases (Ichimasa et al., 2018; Yang & Bang, 2019); neurology, where AI-powered mobile applications have been applied to monitor neurological disorders such as epilepsy, gait, and tremor (Bruno et al., 2018; Dorsey, Glidden, Holloway, Birbeck, & Schwamm, 2018); oncology, where AI-based algorithms have been used to diagnosis cancer with high accuracy (Campanella et al., 2019); and radiology (Liu et al., 2019).

Neurodegenerative diseases and physiological signals

Neurodegenerative diseases are diseases characterized by a progressive deterioration in the normal structure and function of neurons over time. Neural degeneration can exist at any level of organization as far as the nervous system is concerned; it can occur at subcellular, cellular, tissue, and system levels. Unfortunately, neurodegenerative diseases, like other degenerative diseases, are incurable (Priller et al., 2006), with age being one of the risk factors. Diseases identified as neurodegenerative include Alzheimer's disease, Parkinson's disease, Huntington's disease, multiple sclerosis, and many others. In these diseases, deterioration occurs with neurotransmitter productions and subcellular structures. Degenerative diseases tend to exhibit different pathophysiological features and mechanisms. Research activities are under greater intensity to identify the nexus among neurodegenerative diseases with respect to presentations, disease progression, and pathophysiological mechanisms to improve management (De Vos, Grierson, Ackerley, & Miller, 2008).

Alzheimer's disease is the most prevalent neurodegenerative disease. In Alzheimer's disease, deterioration of neurons occurs in the cerebrum leading to attenuation of the cingulate gyrus, hippocampus, parietal lobe, temporal lobe, and frontal lobe of the brain. Alzheimer's disease enunciates the dangers associated with attenuation of the amyloid precursor protein, a protein involved in the repair, growth, and survival of neurons (Priller et al., 2006; Turner, O'Connor, Tate, & Abraham, 2003). Common pathophysiological features of the disease include the presence of beta-amyloid. Beta-amyloid is produced when amyloid precursor protein is broken down in the presence of beta-secretase and alpha-secretase (Hooper, 2005). Beta-amyloid protein condenses, forming what is known as amyloid plaques (Ohnishi & Takano, 2004; Tiraboschi, Hansen, Thal, & Corey-Bloom, 2004).

Parkinson's disease elucidates the function of a neurotransmitter secreted by the substantia nigra known as dopamine and disruption of the nigrostriatal dopaminergic pathway. It is the second most common neurodegenerative disease and is characterized by hypokinetic and hyperkinetic features such as rigidity, tremor, and so on. The specific mechanisms responsible for the deterioration of dopaminergic neurons of the pars compacta in the substantia nigra are unclear. It is possible that dysfunction of a transport protein called RAB1 results in the accumulation of alpha-synuclein-ubiquitin (Varkey et al., 2010).

In Huntington's disease, there is a malfunction of the GABA-secreting neurons of the pars reticulata in the substantia nigra (Purves et al., 2001). It is a rare neurodegenerative disorder characterized by hyperkinetic features such as chorea (Crossman, 2000). It is also associated with neurogenesis (Barnat et al., 2020).

Multiple sclerosis presents a scenario in which the body's immune proteins launch an attack against the covering of the neuronal axons causing injury due to inflammation (Kaufman & Milstein, 2013; Stys & Tsutsui, 2019) reduction in conduction of information from the central nervous system. Due to the absence of myelin sheath in the central nervous system, the brain and spinal cord are the soft hubs for the attack.

Neurodegenerative disease data acquisition

Neurodegenerative diseases are deteriorations in physiological functions affecting the neurons over time. Alzheimer's disease accounts for between 55% and 75% of dementia cases worldwide (Lane, Hardy, & Schott, 2018).

Unfortunately, assessment of Alzheimer's disease remains subjective. There are different ways through which neurocognitive data are acquired. We discuss some of these methods in the sections that follow.

Magnetic resonance imaging (MRI): MRI is a noninvasive technique that uses electromagnetic radiation to produce images of the brain and spinal cord (Bozzali, Serra, & Cercignani, 2016). Neurodegenerative disease markers are produced with this method. Changes in cellular level can be monitored using modified MRI techniques such as magnetization transfer, susceptibility imaging, diffusion MRI, and others (Deistung, Schweser, & Reichenbach, 2017; Laule, Vavasour, Kolind, et al., 2007; Wolff & Balaban, 1989).

Positron emission tomography (PET): PET uses a metabolite-bound radioactive tracer to explore the brain and spinal cord, monitor the profile of beta-amyloid (Jagust, 2010), and diagnose neurodegenerative diseases (Teune, Bartels, & Leenders, 2013).

Bioassays: A bioassay is an invasive procedure that measures amounts of neurofibrillary tangles or misfolded proteins in the cerebrospinal fluid (Blennow, Dubois, Fagan, et al., 2015; Blennow & Hampel, 2003). A problem with this procedure is reproducibility of results.

Computer-based models for neurodegenerative diseases: The DeepSymNet architecture developed by Pena et al. (2019) recognizes Alzheimer's disease progression and longitudinal changes without depending on previous brain regions of interest. This method was compared with FreeSurfer longitudinal pipelines and voxel-based methods using the Alzheimer's Disease Neuroimaging Initiative (ADNI) database. Peña-Nogales et al. (2018) also designed a package called longitudinal connectome data as a marker for neurodegenerative progression.

Data-oriented disease progression models (DPMs): These techniques determine long-term temporal patterns of disease progression directly from cross-sectional or short-term longitudinal datasets on the assumption of some degree of commonality of progression over a patient cohort. Bilgel, Prince, Wong, Resnick, and Jedynak (2016) developed a multivariate nonlinear mixed-effects model for determining the trajectories of voxel-wise neuroimaging biomarkers from longitudinal data that accounts for such differences across individuals. Donohue et al. (2014) designed a semiparametric model and iterative estimation procedure that simultaneously estimates long-term growth curves and pathological timing.

Event-based model (EBM): This has been utilized in the study of Huntington's disease, familial Alzheimer's disease (Fonteijn, Modat, Clarkson, et al., 2012; Oxtoby, Young, Lorenzi, et al., 2016), and sporadic Alzheimer's disease (Oxtoby et al., 2016). Other models include discriminative models such as supervised ML approaches that use labeled data to differentiate between people who have disease from those who do not and between those who will clinically progress and those who will not (Klöppel, Stonnington, Chu, et al., 2008; Young, Modat, Cardoso, et al., 2013).

Challenges in data handling

Healthcare services such as preventing and treating diseases, injuries, and other physical and mental disorders require a large amount of data, including patient personal information, insurance records, prescriptions, healthcare provider descriptions, electronic medical records, X-ray images, and all types of scanned images (Kaddi & Patil, 2020; Pashazadeh & Navimipour, 2018). However, there is a lack of effective algorithms for converting billions of data across all Internet-connected devices into their appropriate format (Hong et al., 2018).

Analysis, modeling for classifications, and forecasting of tasks will always be severely limited by the irregularity of the data and the missing values that result (Weerakody, Wong, Wang, & Ela, 2021). Unbalanced data presents a significant challenge to data mining and ML algorithms. Usually, the unbalanced dataset comprises majority and minority classes, with the majority class taking the lead over the minority class. The assumption of so many algorithms, in general, is based on a balanced class distribution. If these learning algorithms predict unbalanced data, the accuracy will increase for most classes, which may ultimately lead to poor performance (Rekha, Tyagi, & Reddy, 2019).

Most traditional pattern classifiers assume that their input data is well organized in all data instances in terms of analogous class distributions, the presence of a full set of observed features, and balanced class size. On the other hand, some datasets exhibit a variety of irregularities that occur frequently enough to confuse a classifier, thereby reducing its ability to learn from the data. Real-world data is frequently marred by several irregularities of distribution and feature bias, including classification with missing features and class imbalance (Das, Datta, & Chaudhuri, 2018).

Several research papers emphasize that preprocessing data before generating a classifier will result in a solution that is better when compared to solutions of other methods. Consequently, this will result in either addition of new information or removal of information that is not useful to obtain balanced data. More so, it is pertinent to note that artificially rebalancing the class distributions will have no substantial effect on the performance of the base classifier. Furthermore, it has been noticed that quite a lot of research focuses on the problem of binary class imbalance while ignoring that of multiclass imbalance. This could be because the binary class problem is almost always the result of most domain-specific applications, and the complexity of solving a multiclass problem is much greater than that of a binary class imbalance problem. The future trend is expected to be the development of new techniques for dealing with multiclass imbalance problems.

Many published studies used cutting-edge ML algorithms such as support vector machine (SVM), decision tree (DT), K-nearest neighbors (KNN), and naive Bayes (NB) to curb the problem of class imbalance. There is also a perceived increase in research on ensemble techniques to solve this same challenge. Applying two or more techniques, known as the hybrid approach, provides a better solution to the class imbalance problem. Another very good technique that can be employed in classifying imbalanced data is the feature selection method. However, note that the nature of the problem heavily influences a feature selection algorithm (Abd Elrahman & Abraham, 2015; Abolkarlou, Niknafs, & Ebrahimpour, 2014).

Furthermore, class overlapping, rather than just mere class imbalance, is the main cause of the class imbalance problem resulting in poor classifier performance. Consequently, this has led to many studies adopting different sampling techniques during the preprocessing phase to solve the overlapping class problem. However, these sampling techniques may not be enough to solve the problem of overlapping. Some studies revealed that redundant features cause overlapping class problems, and future research will focus on different feature selection techniques for dealing with class overlapping problems.

The major challenge in classification is developing an efficient way of processing massive amounts of data that are being generated daily. The problem of class imbalance can also have an impact on big data. In addition to the massive increase in data volume, the nature of the problem poses a challenge to existing learning algorithms. Data handling has its characteristics in health care, such as incompleteness, heterogeneity, timeliness, longevity, ownership, and privacy. These features present many challenges for data storage, mining, and sharing to advance health-related research.

Summary

Medical imaging is a technology used to visual represent the inner parts of the human body to assist clinicians in identifying, evaluating, and managing diseases earlier and more efficiently (Brody, 2013). Typical examples of medical imaging include PET, X-ray, mammography, MRI, and computed tomography (CT) (Heidenreich, Desgrandschamps, & Terrier, 2002). These methods play a vital role by providing functional and anatomical details about organs for adequate and swift identification and diagnosis of disease. They also have a practical role in translational research. Furthermore, their application in clinical environment is mainly applied for description of numerous medical images that needs to be carried out by human experts such and clinicians and radiologists. One of the biggest challenges of imaging is that it is time consuming and results can be affected by biases and human error. Thus, CAD systems are now being used to enhance the effectiveness of medical imaging.

ML is applied in CAD to evaluate medical images to obtain information. Most of the time, medical images are multifaceted, and the feature selection step is still performed by a human specialist on the basis of their domain-specific information (Erickson, Korfiatis, Akkus, & Kline, 2017; Li, Meng, Shi, Initiative, et al., 2019; Pandya, Shah, & Jardosh, 2019; Wernick, Yang, Brankov, Yourganov, & Strother, 2010; Wu et al., 2013). Therefore, the application of deep learning could help in the quick interpretation of medical images through traditional ML algorithms (Goodfellow, Bengio, Courville, & Bengio, 2016; LeCun, Bengio, & Hinton, 2015; Lian, Liu, Zhang, & Shen, 2018; Schmidhuber, 2015).

Conclusion and future perspectives

The prevalence of neurodegenerative diseases is increasing worldwide. Thus, it is vital to adopt and utilize ML techniques to analyze medical and health data to predict disease progression to improve diagnosis, classification, and treatment of neurodegenerative diseases. Cutting-edge computational approaches and models in AI can improve the general understanding, diagnosis, and prevention of neurodegenerative disease. This chapter presented an in-depth analysis of the importance of ML techniques for diagnosing and preventing neurodegenerative diseases.

References

Abd Elrahman, S. M., & Abraham, A. (2015). Class imbalance problem using a hybrid ensemble approach. *International Journal of Hybrid and Intelligent System, 12*, 219–227. https://doi.org/10.3233/HIS-160217.

Abolkarlou, N. A., Niknafs, A. A., & Ebrahimpour, M. K. (2014). Ensemble imbalance classification: Using data preprocessing, clustering algorithm and genetic algorithm. In *Proceedings of the 4th international eConference on computer and knowledge engineering, Oct. 29-30.*

Adetunji, J. B., Adetunji, C. O., & Olaniyan, O. T. (2021). African walnuts: A natural depository of nutritional and bioactive compounds essential for food and nutritional security in Africa. In O. O. Babalola (Ed.), *Food security and safety*. Cham: Springer.

Adetunji, C. O., Ajayi, O. O., Akram, M., Olaniyan, O. T., Chishti, M. A., Inobeme, A., et al. (2021). Medicinal plants used in the treatment of influenza A virus infections. In K. Dua, S. Nammi, D. Chang, D. K. Chellappan, G. Gupta, & T. Collet (Eds.), *Medicinal plants for lung diseases*. Singapore: Springer.

Adetunji, C. O., Akram, M., Olaniyan, O. T., Ajayi, O. O., Inobeme, A., Olaniyan, S., et al. (2021a). Medicinal plants used in the treatment of pulmonary hypertension. In K. Dua, S. Nammi, D. Chang, D. K. Chellappan, G. Gupta, & T. Collet (Eds.), *Medicinal plants for lung diseases*. Singapore: Springer. https://doi.org/10.1007/978-981-33-6850-7_14.

Adetunji, C. O., Akram, M., Olaniyan, O. T., Ajayi, O. O., Inobeme, A., Olaniyan, S., et al. (2021b). Targeting SARS-CoV-2 novel Corona (COVID-19) virus infection using medicinal plants. In K. Dua, S. Nammi, D. Chang, D. K. Chellappan, G. Gupta, & T. Collet (Eds.), *Medicinal plants for lung diseases*. Singapore: Springer.

Adetunji, C. O., Inobeme, A., Olaniyan, O. T., Ajayi, O. O., Olaniyan, S., & Adetunji, J. B. (2021). Application of nanodrugs derived from active metabolites of medicinal plants for the treatment of inflammatory and lung diseases: Recent advances. In K. Dua, S. Nammi, D. Chang, D. K. Chellappan, G. Gupta, & T. Collet (Eds.), *Medicinal plants for lung diseases*. Singapore: Springer.

Adetunji, C. O., Tijjani, H., Olisah, M. C., Patrick-Iwuanyanwu, K. C., Adetunji, J. B., Ifemeje, J. C., et al. (2020). Health benefits of isoflavones found exclusively of plants of the Fabaceae Family. In C. Egbuna, & T. G. Dable (Eds.), *Functional foods and nutraceuticals*. Cham: Springer. https://doi.org/10.1007/978-3-030-42319-3_22.

Adetunji, C. O., & Varma, A. (2020). Biotechnological application of trichoderma: A powerful fungal isolate with diverse potentials for the attainment of food safety, management of pest and diseases, healthy planet, and sustainable agriculture. In C. Manoharachary, H. B. Singh, & A. Varma (Eds.), *Vol. 61. Trichoderma: Agricultural applications and beyond. Soil biology*. Cham: Springer.

Akram, M., Jabeen, F., Daniyal, M., Zainab, R., Haq, U. U., Adetunji, C. O., et al. (2020). Genetic engineering of novel products of health significance: Recombinant DNA technology. In C. Egbuna, & T. G. Dable (Eds.), *Functional foods and nutraceuticals*. Cham: Springer. https://doi.org/10.1007/978-3-030-42319-3_26.

Barnat, M., Capizzi, M., Aparicio, E., Boluda, S., Wennagel, D., Kacher, R., et al. (2020). Huntington's disease alters human neurodevelopment. *Science*, *369*(6505), 787–793.

Bilgel, M., Prince, J. L., Wong, D. F., Resnick, S. M., & Jedynak, B. M. (2016). A multivariate nonlinear mixed effects model for longitudinal image analysis: Application to amyloid imaging. *NeuroImage*, *134*, 658–670.

Blennow, K., Dubois, B., Fagan, A. M., et al. (2015). Clinical utility of cerebrospinal fluid biomarkers in the diagnosis of early Alzheimer's disease. *Alzheimer's & Dementia*, *11*, 58–69.

Blennow, K., & Hampel, H. (2003). CSF markers for incipient Alzheimer's disease. *Lancet Neurology*, *2*, 605–613.

Botha, A. P. (2017). Machine innovation—A future reality. In *International association for management of technology conference proceedings* (pp. 1–16).

Bozzali, M., Serra, L., & Cercignani, M. (2016). Quantitative MRI to understand Alzheimer's disease pathophysiology. *Current Opinion in Neurology*, *29*, 437–444.

Brody, H. (2013). Medical imaging. *Nature*, *502*, S81. https://doi.org/10.1038/502S81a.

Brown, R. C., Lockwood, A. H., & Sonawane, B. R. (2005). Neurodegenerative diseases: An overview of environmental risk factors. *Environmental Health Perspectives, 113*(9), 1250–1256. https://doi.org/10.1289/ehp.7567.

Bruno, E., Simblett, S., Lang, A., Biondi, A., Odoi, C., Schulze-Bonhage, A., et al. (2018). Wearable technology in epilepsy: The views of patients, caregivers, and healthcare professionals. *Epilepsy & Behavior, 85*, 141–149.

Campanella, G., Hanna, M. G., Geneslaw, L., Miraflor, A., Werneck Krauss Silva, V., Busam, K. J., et al. (2019). Clinical-grade computational pathology using weakly supervised deep learning on whole slide images. *Nature Medicine, 25*, 1301–1309.

Choi, H., Kang, H., & Lee, D. S. (2018). Predicting aging of brain metabolic topography using variational autoencoder. *Frontiers in Aging Neuroscience, 10*, 212.

Christiansen, M. P., Garg, S. K., Brazg, R., Bode, B. W., Bailey, T. S., Slover, R. H., et al. (2017). Accuracy of a fourth-generation subcutaneous continuous glucose sensor. *Diabetes Technology & Therapeutics, 19*, 446–456.

Crossman, A. R. (2000). Functional anatomy of movement disorders. *Journal of Anatomy, 196(Pt 4), 519–525.*

Dargahi, N., Katsara, M., Tselios, T., Androutsou, M. E., de Courten, M., Matsoukas, J., et al. (2017). Multiple sclerosis: Immunopathology and treatment update. *Brain Sciences*, 7(7), 78.

Das, S., Datta, S., & Chaudhuri, B. B. (2018). Handling data irregularities in classification: Foundations, trends, and future challenges. *Pattern Recognition, 81*, 674–693. https://doi.org/10.1016/j.patcog.2018.03.008.

Das, S., Sharangi, A. B., Egbuna, C., Jeevanandam, J., Ezzat, S. M., & Adetunji, C. O. (2020). *Health benefits of isoflavones found exclusively of plants of the Fabaceae Family* (pp. 473–508). Springer International Publishing. https://doi.org/10.1007/978-3-030-42319-3_22.

De Vos, K. J., Grierson, A. J., Ackerley, S., & Miller, C. C. (2008). Role of axonal transport in neurodegenerative diseases. *Annual Review of Neuroscience, 31*, 151–173. https://doi.org/10.1146/annurev.neuro.31.061307.090711.

Deistung, A., Schweser, F., & Reichenbach, J. R. (2017). Overview of quantitative susceptibility mapping. *NMR in Biomedicine, 30*, e3569.

DeMaagd, G., & Philip, A. (2015). Parkinson's disease and its management: Part 1: Disease entity, risk factors, pathophysiology, clinical presentation, and diagnosis. *P & T: A Peer-Reviewed Journal for Formulary Management, 40*(8), 504–532.

Donohue, M. C., Jacqmin-Gadda, H., Le Goff, M., Thomas, R. G., Raman, R., Gamst, A. C., et al. (2014). Estimating long-term multivariate progression from short-term data. *Alzheimer's & Dementia, 10*, S400–S410.

Dorado-Díaz, P. I., Sampedro-Gómez, J., Vicente-Palacios, V., & Sánchez, P. L. (2019). Applications of artificial intelligence in cardiology. The future is already here. *Revista Española de Cardiología (English edition), 72*, 1065–1075.

Dorsey, E. R., Glidden, A. M., Holloway, M. R., Birbeck, G. L., & Schwamm, L. H. (2018). Teleneurology and mobile technologies: The future of neurological care. *Nature Reviews. Neurology, 14*, 285–297.

Dugger, B. N., & Dickson, D. W. (2017). Pathology of neurodegenerative diseases. *Cold Spring Harbor Perspectives in Biology, 9*(7), a028035.

Egbuna, C., Gupta, E., Ezzat, S. M., Jeevanandam, J., Mishra, N., Akram, M., et al. (2020). Aloe species as valuable sources of functional bioactives. In C. Egbuna, & T. G. Dable (Eds.), *Functional foods and nutraceuticals*. Cham: Springer.

Emamzadeh, F. N., & Surguchov, A. (2018). Parkinson's disease: Biomarkers, treatment, and risk factors. *Frontiers in Neuroscience, 12*, 612.

Erickson, B. J., Korfiatis, P., Akkus, Z., & Kline, T. L. (2017). Machine learning for medical imaging. *Radiographics, 37*, 505–515. https://doi.org/10.1148/rg.2017160130.

Fonteijn, H. M., Modat, M., Clarkson, M. J., et al. (2012). An event-based model for disease progression and its application in familial Alzheimer's disease and Huntington's disease. *NeuroImage, 60*, 1880–1889.

Gitler, A. D., Dhillon, P., & Shorter, J. (2017). Neurodegenerative disease: Models, mechanisms, and a new hope. *Disease Models & Mechanisms, 10*(5), 499–502.

Golriz Khatami, S., Mubeen, S., & Hofmann-Apitius, M. (2020). Data science in neurodegenerative disease: Its capabilities, limitations, and perspectives. *Current Opinion in Neurology, 33*(2), 249–254.

Goodfellow, I., Bengio, Y., Courville, A., & Bengio, Y. (2016). *Deep learning. Vol. 1*. Cambridge, MA: MIT Press.

Gray, K. R., Aljabar, P., Heckemann, R. A., Hammers, A., & Rueckert, D. (2013). Random forest-based similarity measures for multi-modal classification of Alzheimer's disease. *NeuroImage, 65*, 167–175.

Guerreiro, R., & Bras, J. (2015). The age factor in Alzheimer's disease. *Genome Medicine*, 7, 106.

He, K., Zhang, X., Ren, S., & Sun, J. (2016). Deep residual learning for image recognition. In *2016 IEEE conference on computer vision and pattern recognition (CVPR), 27–30 June 2016* (pp. 770–778).

Heidenreich, A., Desgrandschamps, F., & Terrier, F. (2002). Modern approach of diagnosis and management of acute flank pain: Review of all imaging modalities. *European Urology, 41*, 351–362. https://doi.org/10.1016/S0302-2838(02)00064-7.

Hong, L., Luo, M., Wang, R., Lu, P., Lu, W., & Lu, L. (2018). Big data in health care: Applications and challenges. *Data and Information Management*. https://doi.org/10.2478/dim-2018-0014.

Hooper, N. M. (2005). Roles of proteolysis and lipid rafts in the processing of the amyloid precursor protein and prion protein. *Biochemical Society Transactions, 33*(Pt 2), 335–338.

Hutson, M. (2019). *No coding required: Companies make it easier than ever for scientists to use artificial intelligence*. Retrieved from: https://www.sciencemag.org/news/2019/07/no-coding-required-companies-make-it-easier-ever-scientists-use-artificial-intelligence.

Ichimasa, K., Kudo, S. E., Mori, Y., Misawa, M., Matsudaira, S., Kouyama, Y., et al. (2018). Artificial intelligence may help in predicting the need for additional surgery after endoscopic resection of T1 colorectal cancer. *Endoscopy, 50*, 230–240.

Jagust, W. J. (2010). Amyloid imaging: Coming to a PET scanner near you. *Annals of Neurology, 68*, 277–278.

Kaddi, S. S., & Patil, M. M. (2020). Challenges and trends in clinical data analytics. *Journal of Science and Technology, 5*(4), 348–360.

Kaufman, D. M., & Milstein, M. J. (2013). Multiple sclerosis. In D. M. Kaufman, & M. J. Milstein (Eds.), *Kaufman's clinical neurology for psychiatrists* (7th ed., pp. 329–349). Philadelphia: W.B. Saunders, ISBN:978-0-7234-3748-2 (chapter 15).

Klöppel, S., Stonnington, C. M., Chu, C., et al. (2008). Automatic classification of MR scans in Alzheimer's disease. *Brain, 131*, 681–689.

Lane, C. A., Hardy, J., & Schott, J. M. (2018). Alzheimer's disease. *European Journal of Neurology, 25*, 59–70. https://doi.org/10.1111/ene.13439.

Laule, C., Vavasour, I. M., Kolind, S. H., et al. (2007). Magnetic resonance imaging of myelin. *Neurotherapeutics, 4*, 460–484.

LeCun, Y., Bengio, Y., & Hinton, G. (2015). Deep learning. *Nature, 521*, 436. https://doi.org/10.1038/nature14539.

Li, Y., Meng, F., Shi, J., Initiative, A. D. N., et al. (2019). Learning using privileged information improves neuroimaging-based CAD of Alzheimer's disease: A comparative study. *Medical & Biological Engineering & Computing, 57*, 1605–1616. https://doi.org/10.1007/s11517-019-01974-3.

Lian, C., Liu, M., Zhang, J., & Shen, D. (2018). Hierarchical fully convolutional network for joint atrophy localization and Alzheimer's disease diagnosis using structural MRI. *IEEE Transactions on Pattern Analysis and Machine Intelligence*, *42*, 880–893. https://doi.org/10.1109/TPAMI.2018.2889096.

Liu, X., Faes, L., Kale, A. U., Wagner, S. K., Fu, D. J., Bruynseels, A., et al. (2019). A comparison of deep learning performance against health-care professionals in detecting diseases from medical imaging: A systematic review and meta-analysis. *The Lancet Digital Health*, *1*, e271–e297.

Loma, I., & Heyman, R. (2011). Multiple sclerosis: Pathogenesis and treatment. *Current Neuropharmacology*, *9*(3), 409–416.

Morawski, K., Ghazinouri, R., Krumme, A., Lauffenburger, J. C., Lu, Z., Durfee, E., et al. (2018). Association of a smartphone application with medication adherence and blood pressure control: The MedISAFE-BP randomized clinical trial. *JAMA Internal Medicine*, *178*, 802–809.

Myszczynska, M. A., Ojamies, P. N., Lacoste, A. M. B., Neil, D., Saffari, A., Mead, R., et al. (2020). Applications of machine learning to diagnosis and treatment of neurodegenerative diseases. *Nature Reviews. Neurology*, *16*, 440–456.

Nichols, J. A., Herbert Chan, H. W., & Baker, M. (2019). Machine learning: Applications of artificial intelligence to imaging and diagnosis. *Biophysical Reviews*, *11*(1), 111–118.

Obermeyer, Z., & Emanuel, E. J. (2016). Predicting the future—Big data, machine learning, and clinical medicine. *The New England Journal of Medicine*, *375*, 1216–1219.

Ohnishi, S., & Takano, K. (2004). Amyloid fibrils from the viewpoint of protein folding. *Cellular and Molecular Life Sciences*, *61*(5), 511–524.

Orth, M., Averina, M., Chatzipanagiotou, S., Faure, G., Haushofer, A., Kusec, V., et al. (2019). Opinion: Redefining the role of the physician in laboratory medicine in the context of emerging technologies, personalised medicine and patient autonomy ('4P medicine'). *Journal of Clinical Pathology*, *72*, 191–197.

Oxtoby, N. P., Young, A. L., Lorenzi, M., et al. (2016). Model-based comparison of autosomal-dominant and late-onset Alzheimer's disease progression in the DIAN and ADNI studies. *Alzheimer's & Dementia*, *12*, 668–669.

Pandya, M. D., Shah, P. D., & Jardosh, S. (2019). Medical image diagnosis for disease detection: A deep learning approach. In *U-healthcare monitoring systems* (pp. 37–60). Elsevier. https://doi.org/10.1016/B978-0-12-815370-3.00003-7.

Pashazadeh, A., & Navimipour, N. J. (2018). Big data handling mechanisms in the healthcare applications: A comprehensive and systematic literature review. *Journal of Biomedical Informatics*, *82*, 47–62.

Pena, D., Barman, A., Suescun, J., Jiang, X., Schiess, M. C., Giancardo, L., et al. (2019). Quantifying neurodegenerative progression with DeepSymNet, an end-to-end data-driven approach. *Frontiers in Neuroscience*, *13*, 1053. https://doi.org/10.3389/fnins.2019.01053.

Peña-Nogales, Ó., Ellmore, T. M., de Luis-García, R., Suescun, J., Schiess, M. C., & Giancardo, L. (2018). Longitudinal connectomes as a candidate progression marker for prodromal Parkinson's disease. *Frontiers in Neuroscience*, *12*, 967. https://doi.org/10.3389/fnins.2018.00967.

Priller, C., Bauer, T., Mitteregger, G., Krebs, B., Kretzschmar, H. A., & Herms, J. (2006). Synapse formation and function is modulated by the amyloid precursor protein. *The Journal of Neuroscience*, *26*(27), 7212–7221.

Purves, D., Augustine, G. A., Fitzpatrick, D., Hall, W., LaMantia, A. S., McNamara, J. O., et al. (2001). Modulation of movement by the basal ganglia—Box A. Huntington's disease. In D. Purves (Ed.), *Neuroscience* (2nd ed.). Sunderland, MA: Sinauer Associates, ISBN:978-0-87893-742-4.

Qiu, C., Kivipelto, M., & von Strauss, E. (2009). Epidemiology of Alzheimer's disease: Occurrence, determinants, and strategies toward intervention. *Dialogues in Clinical Neuroscience*, *11*(2), 111–128.

Rekha, G., Tyagi, A. K., & Reddy, V. K. (2019). A wide scale classification of class imbalance problem and its solutions: A systematic literature review. *Journal of Computer Science*, *15*(7), 886–929.

Rohini, M., & Surendran, D. (2019). Classification of neurodegenerative disease stages using ensemble machine learning classifiers. *Procedia Computer Science*, *165*, 66–73.

Roli, O. I., Adetunji, C. O., Mishra, P., Adetunji, J. B., & Mishra, R. R. (2020). Caffeine: Nutraceutical and health benefit of caffeine-containing commodities and products. In P. Mishra, R. R. Mishra, & C. O. Adetunji (Eds.), *Innovations in food technology*. Singapore: Springer.

Schmidhuber, J. (2015). Deep learning in neural networks: An overview. *Neural Networks*, *61*, 85–117. https://doi.org/10.1016/j.neunet.2014.09.003.

Selkoe, D. J., & Lansbury, P. J., Jr. (1999). Alzheimer's disease is the most common neurodegenerative disorder. In G. J. Siegel, B. W. Agranoff, R. W. Albers, et al. (Eds.), *Basic neurochemistry: Molecular, cellular and medical aspects* (6th ed.). Philadelphia: Lippincott-Raven. Available from: https://www.ncbi.nlm.nih.gov/books/NBK27944/.

Stys, P. K., & Tsutsui, S. (2019). Recent advances in understanding multiple sclerosis. *F1000Research*, *8*, 2100.

Summers, M. J., Madl, T., Vercelli, A. E., Aumayr, G., Bleier, D. M., & Ciferri, L. (2017). Deep machine learning application to the detection of preclinical neurodegenerative diseases of aging. *DigitCult—Scientific Journal on Digital Cultures*, *2(2)*. https://doi.org/10.4399/97888255088952. *ICT & Neurodegenerative Diseases [special issue]DO -.*

Teune, L. K., Bartels, A. L., & Leenders, K. L. (2013). FDG- PET imaging in neurodegenerative brain diseases. In *Functional brain mapping and the endeavor to understand the working brain* InTech.

Tiraboschi, P., Hansen, L. A., Thal, L. J., & Corey-Bloom, J. (2004). The importance of neuritic plaques and tangles to the development and evolution of AD. *Neurology*, *62*(11), 1984–1989.

Topalovic, M., Das, N., Burgel, P. R., Daenen, M., Derom, E., Haenebalcke, C., et al. (2019). Artificial intelligence outperforms pulmonologists in the interpretation ofpulmonary function tests. *The European Respiratory Journal*, *53*.

Topol, E. J. (2019). A decade of digital medicine innovation. *Science Translational Medicine*, *11*.

Turakhia, M. P., Desai, M., Hedlin, H., Rajmane, A., Talati, N., Ferris, T., et al. (2019). Rationale and design of a large-scale, app-based study to identify cardiac arrhythmias using a smartwatch: The apple heart study. *American Heart Journal*, *207*, 66–75.

Turner, P. R., O'Connor, K., Tate, W. P., & Abraham, W. C. (2003). Roles of amyloid precursor protein and its fragments in regulating neural activity, plasticity and memory. *Progress in Neurobiology*, *70*(1), 1–32.

Varkey, J., Isas, J. M., Mizuno, N., Jensen, M. B., Bhatia, V. K., Jao, C. C., et al. (2010). Membrane curvature induction and tubulation are common features of synucleins and apolipoproteins. *The Journal of Biological Chemistry*, *285*(42), 32486–32493.

Veldsman, M., & Egorova, N. (2017). Advances in neuroimaging for neurodegenerative disease. *Advances in Neurobiology*, *15*, 451–478.

Weerakody, P. B., Wong, K. W., Wang, G., & Ela, W. (2021). A review of irregular time series data handling with gated recurrent neural networks. *Neurocomputing*, *441*, 161–178. https://doi.org/10.1016/j.neucom.2021.02.046.

Wernick, M. N., Yang, Y., Brankov, J. G., Yourganov, G., & Strother, S. C. (2010). Machine learning in medical imaging. *IEEE Signal Processing Magazine*, *27*, 25–38. https://doi.org/10.1109/MSP.2010.936730.

Wolff, S. D., & Balaban, R. S. (1989). Magnetization transfer contrast (MTC) and tissue water proton relaxationin vivo. *Magnetic Resonance in Medicine, 10*, 135–144.

Wu, G., Kim, M., Wang, Q., Gao, Y., Liao, S., & Shen, D. (2013). Unsupervised deep feature learning for deformable registration of MR brain images. In *International conference on medical image computing and computer-assisted intervention* (pp. 649–656). Shenzhen: Springer. https://doi.org/10.1007/978-3-642-40763-5_80.

Wyss-Coray, T. (2016). Ageing, neurodegeneration and brain rejuvenation. *Nature, 539*(7628), 180–186.

Yang, Y. J., & Bang, C. S. (2019). Application of artificial intelligence in gastroenterology. *World Journal of Gastroenterology, 25*, 1666–1683.

Young, J., Modat, M., Cardoso, M. J., et al. (2013). Accurate multimodal probabilistic prediction of conversion to Alzheimer's disease in patients with mild cognitive impairment. *NeuroImage: Clinical, 2*, 735–745.

Zhu, G., Jiang, B., Tong, L., Xie, Y., Zaharchuk, G., & Wintermark, M. (2019). Applications of deep learning to neuro-imaging techniques. *Frontiers in Neurology, 10*.

CHAPTER 19

Artificial intelligence-based early detection of neurological disease using noninvasive method based on speech analysis

Olugbemi T. Olaniyan[a], Charles O. Adetunji[b], Ayobami Dare[c], Olorunsola Adeyomoye[d], Mayowa J. Adeniyi[e], and Alex Enoch[f]
[a]Department of Physiology, Laboratory for Reproductive Biology and Developmental Programming, Rhema University, Aba, Nigeria
[b]Department of Microbiology, Applied Microbiology, Biotechnology and Nanotechnology Laboratory, Edo State University Uzairue, Iyamho, Nigeria
[c]Department of Physiology, School of Laboratory Medicine and Medical Sciences, College of Health Sciences, Westville Campus, University of KwaZulu-Natal, Durban, South Africa
[d]Department of Physiology, University of Medical Sciences, Ondo, Nigeria
[e]Department of Physiology, Edo State University Uzairue, Iyamho, Nigeria
[f]Department of Human Physiology, Ahmadu Bello University Zaria, Zaria, Nigeria

Introduction

Neurological disorders are diseases of the central nervous and peripheral systems. Typical symptoms of neurological disease include loss of consciousness, muscle weakness, poor coordination, paralysis, and seizures (WHO, 2017). Interestingly, more than 600 diseases have been associated with the nervous system, such as traumatic brain injuries, brain tumor, stroke, dementia, Parkinson's disease, neuroinfections, multiple sclerosis, headache disorders, epilepsy, and more. Furthermore, it has been established that some microbial-induced infections such as parasitic infections (malaria and Chagas), bacterial infections (*Mycobacterium tuberculosis*, *Neisseria meningitides*), viral infections (enteroviruses, HIV, West Nile virus, Zika), and fungal infections (*Cryptococcus* and *Aspergillus*) can affect the nervous system (Acharya, Hagiwara, & Adeli, 2018; Acharya, Oh, Hagiwara, Tan, & Adeli, 2018; Acharya et al., 2018; Bairy et al., 2017; Bhat, Acharya, Adeli, Bairy, & Adeli, 2014a, 2014b; Sridhar, Bhat, Acharya, Adeli, & Bairy, 2017; WHO, 2017). This might be linked to the fact that such neurological symptoms could be due to immune inflammation or infection. Hundreds of millions of people worldwide are affected by neurological disorders (Acharya, Hagiwara, & Adeli, 2018; Acharya, Oh, Hagiwara, Tan, & Adeli, 2018; Acharya, Oh, Hagiwara,

Artificial Intelligence for Neurological Disorders
https://doi.org/10.1016/B978-0-323-90277-9.00008-0

Tan, Adeli, & Subha, 2018; Adetunji, Adetunji, & Olaniyan, 2021; Adetunji, Ajayi, et al., 2021; Adetunji et al., 2021a, 2021b; Adetunji & Varma, 2020; Adetunji, Inobeme, et al., 2021; Akram et al., 2020; Das et al., 2020; Egbuna et al., 2020; Jahmunah et al., 2019; Roli, Adetunji, Mishra, Adetunji, & Mishra, 2020; Sridhar et al., 2017). For example, stroke affects about 6 million people in middle- and low-income countries, resulting in high rates of mortality in these areas (Ortega-Zamorano, Jerez, Gómez, & Franco, 2017; WHO, 2017; Yıldırım, Baloglu, & Acharya, 2018). Furthermore, about 47.5 million people suffer from dementia (Acharya et al., 2019; Fang et al., 2018; Khedher et al., 2017; López-Sanz et al., 2017; Valenzuela, Jiang, Carrillo, & Rojas, 2018; WHO, 2017) and 50 million people suffer from epilepsy (WHO, 2017) worldwide. Anomalous or abnormal neurological conditions are usually identified by neuropathological investigation. However, most anomalous neurological conditions are not related with neurological disorders (Acharya et al., 2019)

Numerous scientific breakthroughs have been documented for identification of neurological diseases using clinical biomarkers (Alzheimer's Association/Biomarkers, 2013), but the economical implications and technology requirements involved have complicated early diagnosis of diseases such as Alzheimer's (Álvarez et al., 2010; Alzheimer's Association/Biomarkers, 2013; Pareto et al., 2008; Wernickand & Aarsvold, 2004). Therefore, the application of noninvasive intelligent diagnosis techniques could play an important role in the classification and quick identification of neurological diseases. These new methods have the advantage of not requiring laboratory equipment and specialized personnel. Typical examples of such techniques include emotional response analysis (ERA) and automatic spontaneous speech analysis (ASSA) (Almeida & López-de-Ipiña, 2012; Bravo, Villarreal, Hervás, & Urzaiz, 2012; Faundez-Zanuy et al., 2012; Grimaldi & Manto, 2010; Hu et al., 2010; López-de-Ipiña et al., 2012). Spoken language is one of the most crucial elements of an individual's intellect. It allows easy communication between people and plays a role in a person's social life and personal and cultural identity. Moreover, spoken language is the most natural, effective, intuitive, and spontaneous means of communication among people (Haugrud, Crossley, & Vrbancic, 2010; Henry, Crawford, & Pshillips, 2004; Horley, Reid, & Burnham, 2010; Meilán, Martínez-Sánchez, Carro, Sánchez, & Pérez, 2012; Pakhomov, Hemmy, & Lim, 2012). This chapter describes the latest noninvasive techniques for the diagnosis, screening, evaluation, and early detection of common neurological diseases.

Neurological disorders

Neurological disorders are diseases of the nervous system that contribute 3% to the global burden of disease (Thakur, Albanese, Giannakopoulos, et al., 2016). The prevalence of these diseases is gradually increasing due to the psychological demand posed by economic, financial, environmental, and physical factors, among others. The need for early detection and treatment of neurological disorders cannot be overemphasized. Neurological disorders are of different categories, some of which include seizure, neurological infections, developmental disorders, cerebrovascular accidents, degenerative disorders, and headaches.

Seizures occur due to abnormal or uncontrolled electrical activities in the brain. A seizure may be foci (partial) or generalized. In foci seizure, only one half of the brain is affected, and it is usually characterized by sensory and motor activities (Wilke, van Drongelen, Kohrman, & He, 2010). A generalized seizure affects both halves of the cerebral hemisphere and is characterized by loss of consciousness. It may be triggered by fever, blood sugar, stroke, and so on (Stafstrom & Carmant, 2015). Epileptic seizure is a generalized seizure that occurs due to imbalance between glutamate-mediated excitation and GABA-mediated inhibition. It could be acquired, idiopathic, or due to predisposed genetic factors (Sirven, 2015).

The main infections of the nervous system include encephalitis, meningitis, and brain abscesses, which may be caused by bacteria or viruses (Parikh, Tucci, & Galwankar, 2012). Studies have reported increased morbidity and mortality in populations exposed to neurological infections (Silberberg & Katabira, 2006). Meningitis occurs due to inflammation of the meninges, which may be caused by streptococcus, *Escherichia coli*, *Histeria monocytogenes*, and other noninfectious processes (Runde, Anjum, & Hafner, 2021). When bacteria are present in the blood, they may find their way into the subarachnoid space and cross the blood-brain barrier to inflame the meninges. Viruses can infect the cerebrospinal fluid (CSF) through transmission along the neuronal pathway. In contrast to meningitis, encephalitis is inflammation of the brain. It is characterized by loss of consciousness, headaches, behavioral disturbances, fever, seizure, and abnormal brain waves (Said & Kang, 2021). Viral infections that are likely to cause encephalitis include chicken pox, influenza, herpes virus, arbovirus, and some bacteria (Greenberg, 2003).

A brain abscess is an area of necrosis in the brain that results from infection or trauma. *Staphylococcus* and *Streptococcus* bacteria have been implicated in the etiology of the disease (Patel & Clifford, 2014). Some of the developmental disorders that affect the nervous system include adrenoleukodystrophy, Barth syndrome, Fragile X syndrome, Rett syndrome, and Smith-Lemli-Opitz syndrome (Institute of Medicine Committee US on Nervous System Disorders in Developing Countries, 2001). Down syndrome, which occurs due to the alteration in the number of chromosomes, is also a developmental disorder.

Cerebrovascular accidents (CVAs) occur when the brain is traumatized and cerebral perfusion and vasculature are compromised. The major risk factor is hypertension, which is more prevalent in older individuals, although younger individuals with substance abuse problems may also have high blood pressure (Iadecola & Davisson, 2008). Other risk factors include obesity, smoking, and atrial fibrillation. CVA could be due to occlusion of thrombus or embolus in some cerebral arteries and decreased perfusion to areas of the brain (Mello, Judge, Kelly, Bradley, & Harbison, 2018).

Neurodegenerative diseases are characterized by impaired motor skills and cognitive and behavioral changes. Examples of some neurodegenerative diseases include amyloidosis, in which amyloid deposits as plaques in the brain parenchyma; Prion disease, in which normal cellular prion protein PrPc is converted into the pathogenic form PrPsc; tauopathies, in which tau protein accumulates in neuron and glial cells; and Alzheimer's disease, in which a progressive neurologic disorder that causes the brain to shrink (atrophy) and brain cells to dieresulting into dementia (Geschwind, 2015). Chronic traumatic encephalopathy, progressive supranuclear palsy, corticobasal degeneration, Argyrophilic grain disease, Pick's disease, synucleinopathies, and Lewy body dementia, multiple system atrophy, amyotrophic lateral sclerosis, Parkinson's disease, and multiple sclerosis are further examples of neurodegenerative diseases (Thakur et al., 2016). Headache and facial pain may be benign or life threatening. They may occur as migraine or because of intracranial and extracranial vessels, rhinosinusitis, or brain tumor (Kanner, 2012).

Cognitive analysis—Psychological evaluation and physiological signals

Cognitive analysis refers to varieties of analytical approaches, relating to human-like intelligence, required for understanding major concepts including health-related issues such as neurological disorders. This chapter describes several smart approaches to cognitive analysis, such as artificial intelligence, machine learning, semantics, and deep learning. These methods are noninvasive tools for early detection of neurological diseases that become more effective over time as more data is obtained and analyzed (McClelland, 2001). Physiological signals generated from brain activity, heart activity, and eye movements, help to create sufficient data for cognitive analysis.

Brain activity that mediates cognitive function and signaling patterns originating from the brainstem and midbrain nuclei as well as other brain areas involved in cognitive function is used to analyze neurological disorders (Stufflebeam & Rosen, 2007). Advanced neuroscience techniques such as electroencephalography (EEG), positron emission tomography (PET), and functional magnetic resonance imaging (fMRI) are used to examine these brain activities concurrently with ongoing cognitive assignment. Combination of these techniques with adequate experimental and data analysis (such as multivariate analysis and multimodality data integration) procedures will provide a smarter, noninvasive technique for the detection of neurological disorders (Xue, Chen, Lu, & Dong, 2010). Speech analysis is a valuable approach to diagnosing neurological disorders such as Alzheimer's disease (Meilán et al., 2012). Speech impairment, characterized by difficulty speaking and comprehending others (aphasia), commonly follows memory loss in Alzheimer's disease and is typically discovered utilizing ASSA and ERA. This impairment often originates from lesions in some brain areas that are often see on the EEG pattern, making diagnosis easy, noninvasive, and cost-effective (López-de-Ipiña et al., 2012).

Heart activity is another physiological signal that may provide a noninvasive approach to the detection of neurological disorders resulting from the significant role played by the sympathetic and parasympathetic nervous systems. The autonomic nervous system (ANS), which controls cardiac activities directly, has been used as an index to measure the regulation of heart rate asymmetry (HRA) (Alessandro, Giovanna, Dampney Roger, & Pietro, 2016). During neurological disorders affecting the ANS outflow, significant alterations are observed in heart activity due to imbalance in the sympathetic and parasympathetic system, which is easily detected via ECG (Ghiasi, Greco, Barbieri, et al., 2002). Eye-tracking signals such as pupillary dilation have been well correlated with cognitive tasks (Sirois & Brisson, 2014). The ANS controls the contraction and relaxation of the pupillary muscles (dilator and sphincter muscles), reflecting the neuromodulatory influence of neurotransmitters exerted by the brain (Verguts, Vassena, & Silvetti, 2015). The sympathetic nervous system (SNS) receives a response from the frontal cortex and the subcortical regions that mediate cognitive tasks to stimulate the contraction of the dilator muscle causing pupillary dilation (Beatty & Lucero-Wagoner, 2000). These eye-tracking signals can be analyzed using electrooculogram to provide robust data for detecting neurological abnormalities.

Noninvasive screening methods for speech analysis

Neurological disorders are increasing worldwide, thus creating a public health crisis. Early diagnosis is key to detect and prevent these diseases from manifesting. Some of the available procedures to assess or evaluate these disorders involve speech analysis. As described by Lopez-Poveda (2014), ASSA involves testing fluency using Praat software (Rocco et al., 2018), which has the potential to extract voice and unvoiced parts of a sound. It measures the duration, time, frequency, and quality of signals. The brightness of the sound is measured using the spectral centroid (McKechnie et al., 2018). ERA evaluates three characteristics in speech: acoustic characteristics, voice quality, and duration (Schwartz & Pell, 2012). Acoustic characteristics measure the pitch, intensity, standard deviation, and root mean square amplitude of intensity. Voice characteristics measure noise-to-harmonic ratio and harmonic-to-voice ratio. The duration characteristics measure the degree of voice breaks and unvoiced frame. The prosodic and paralinguistic feature sets obtained from the temporal domain of a speech signal are measured using emotional temperature analysis (Schwartz & Pell, 2012).

Computer-aided diagnosis (CAD) systems

A computer-aided diagnosis (CAD) system is a computer-oriented system that promotes quick decision-making by the physician by evaluating the information provided and appropriately analyzing anomalies within a short time (Doi, 2017; Li & Nishikawa, 2015). CAD is a rapidly evolving technique that encompasses different components such as artificial intelligence, medical imaging, and computer vision with the major aim of detecting health problems. Among the existing health problems, CAD has recorded significant progress in diagnosing tumors at the earliest stage possible before they progress to cancer (Giger et al., 2000). In the medical sector, CAD is mostly utilized for appropriate diagnosis; however, in some health conditions where clinicians are confident in their diagnosis, they may use the computer output as a secondary opinion or ignore it entirely. When a clinician is unsure about a decision, the computer output is frequently used as a confirmation tool to improve diagnosis. However, the performance of the computer may not necessarily equate to that of the physician, but a higher performance level can increase the overall percentage of diagnosis. The synergy between the physician's competence and the computer's output has resulted in increased benefits of CAD and prompted its wide application in clinical conditions.

The past two decades have witnessed an increasing research effort to generate and incorporate the use of CAD systems in various diseases. Thus, several CAD systems are commercially developed worldwide, such as CAD for use in mammograms (cancers), chest radiographs (lungs disease), and computerized tomography (Chan, Hadjiiski, Zhou, & Sahiner, 2008; van Ginneken, Hogeweg, & Prokop, 2009). In addition, the purpose of the CAD system has been extended to include image analysis procedures and quantitative techniques (size measurement and segmentation) that will assist physicians in diagnosis, management, response monitoring, and identifying disease risk using biomarkers obtained from the images alone or in synergy with other clinical information (Petrick et al., 2013).

Better precision (computers deliver more accurate quantitative measurements), standardization (more reproducible findings, especially for immediate categories and complex scoring systems), and efficiency (reduced time taken for diagnosis) are some of the advantages of CAD. However, CAD is expensive and may not be used as a routine diagnostic technique. Currently, CAD is a very slow process with low number of cases to confirm diagnosis, it requires a large volume and high-quality of data, there are a lot of variables to consider, and the algorithm used in developing the CAD system needs routine validation to ensure adequate diagnosis (Cirdan, 2018).

Artificial intelligence and machine learning techniques

In neurophysiology and neuroscience, intelligence is defined as the ability to learn and apply facts and skills. It is an executive function masterminded by the central nervous system, especially the human brain. Today, there exists a phenomenon in which nonhumans and nonliving things are made to think, reason, move, talk, and even solve both simple and complex tasks. Artificial intelligence is the name given to this phenomenon. Artificial intelligence, which was first developed in the 1950s, has spread to all aspects of life, including health (Ching et al., 2018), with predictions that it may eventually replace humans. Artificial intelligence is a branch of computer science concerned with the creation and design of systems capable of doing activities that would ordinarily necessitate human intelligence, such as problem solving, reasoning, and recognition. When contrasted to humans, artificial intelligence lacks commonsense reasoning or the human proclivity to assume or foresee the rationale and purpose of real-life circumstances.

Computers do not necessarily require programming to execute certain tasks explicitly. With machine learning, computers can "learn" how to achieve the task. Machine learning involves algorithm training that enables tasks to be carried out by learning from data instead of executing specific programmed tasks (Chartrand et al., 2017). To clarify, machine learning in health care involves teaching a machine to detect illnesses in the same way that a doctor would. To create a machine learning software, data is divided into a training set, where a human determines whether a desired outcome exists. The validation sets are the other training sets, in which the system applies what it has learned to determine whether desired results are present (Crown, 2015). When there are a small number of patient attributes of interest, however, machine learning approaches are used.

Deep learning-based techniques

Deep learning is a type of machine learning that involves several layers that can be used to change data. Significant credit assignment path depth is a feature of deep learning. Inputs can be converted to outputs via the credit assignment path. Image processing is a nice example. Artificial neural networks are used in deep learning. They employ a multilayer approach to extract higher-level features from raw data. It is possible to analyze difficult data using deep learning (Jiang et al., 2017). Classic neural networks, recurrent neural networks, generative adversarial networks, self-organizing maps, Boltzmann machines, deep reinforcement learning, autoencoder, backpropagation, gradient descent, and convolutional neural networks (CNNs) are examples of deep learning approaches. The CNN is a prominent type of deep learning in medicine (Jiang et al., 2017). CNNs are used to monitor preprocessing, data compilation, and increased complexity. They are based on the visual cortex order of neuron organization in animals. It is vital to remember that a deep learning process can figure out which traits belong at which level on its own. Deep learning techniques can be utilized for both supervised and unsupervised tasks (Ching et al., 2018).

Artificial intelligence and CAD systems for early detection of neurological disorders

Artificial intelligence research has exploded in popularity during the last decade. Researchers claim that a CAD system trained on a large amount of patient data, physiological signals, and images, as well as the integration of advanced signal processing and artificial intelligence techniques, can help neurologists, neurosurgeons, radiologists, and other health providers make diagnoses (Acharya, Hagiwara, & Adeli, 2018; Acharya, Oh, Hagiwara, Tan, & Adeli, 2018; Acharya, Oh, Hagiwara, Tan, Adeli, & Subha, 2018; Bairy et al., 2017; Bhat et al., 2014a, 2014b; Sridhar et al., 2017). Support tools focused on machine learning have a lot of promise for improving the precision, reliability, and quality of clinical decision-making and evaluation in clinical practice. Since accurate diagnostics and early-stage diagnosis are top priorities in medical practice, a lot of work is being put into developing new diagnostic support methods that will improve accuracy while reducing time and resources (Berardelli et al., 2013).

Early detection of neurological diseases can improve quality of life for patients and their families. It can also assist scientists in developing appropriate remedies. However, progress toward a scheme for early detection of mental illnesses is an economic requirement because mental illnesses are a huge public health concern with a large socioeconomic cost to society (Papakostas, Savio, Gran, & Kaburlasos, 2015). Machine learning algorithms, in particular, can discover essential patterns and understand the underlying properties of huge, complicated, and multivariate data representations. This information can then be utilized to build models, which are composed of a set of mathematical and statistical rules that can perform recognition tasks. Artificial intelligence-based systems are nearing, if not exceeding, human skills in a range of data analytic activities because of recent major advancements in processing capability and the increasing availability of enormous datasets from which to learn (Cummins, Matcham, Klapper, & Schuller, 2020).

Clinical tests are also a complicated inferential procedure in which a doctor must assess a complete taxonomy of symptoms as well as a range of circumstantial elements to better diagnose the precise ailment that a patient may be suffering from. Evidence from the literature on machine learning and emotional computing, on the other hand, indicate that artificial intelligence technologies can aid in the identification of a variety of neurological disorders, including bipolar disorder and major depressive disorder (Cummins et al., 2015; Girard & Cohn, 2015; Khorram, Gideon,

McInnis, & Provost, 2016; Pampouchidou et al., 2017; Ringeval et al., 2017, 2018). In neurology, EEG data, which provide information on the electrical activity of the brain, are a useful diagnostic tool. To understand these signals and create predictions, a variety of machine learning algorithms have been applied. Subasi, Kevric, and Abdullah Canbaz, (2019) reported a system to detect epileptic seizures in EEG documents using two machine learning approaches, support vector machines (SVMs) and genetic algorithms (GAs), with an accuracy of 99.38% on the EEG dataset employed. Avcu, Zhang, and Chan (2019) introduced CNNs that can identify seizures with 93.3% accuracy using only two channels.

Machine learning and connectome mapping are both well-suited to epilepsy diagnosis. Epilepsy is a disorder marked by aberrant changes in brain organization that cause hyperexcitability and seizures in certain brain regions. Machine learning has demonstrated significant improvements over the prior state of the art in detecting epileptiform structures and movement, as well as predicting surgical outcomes, using anatomical MRI images combined with diffusion tensor imaging (DTI) tractography. In a 2015 publication, Munsell et al. used SVMs, which is consistent with past studies (Subasi et al., 2019). They did, however, limit the clinical features used in predictive tasks by using a feature selection and regularization approach. Munsell et al. used a feature selection and regularization approach to overcome a key hurdle in machine learning research in neurology: a small number of test subjects compared to the large number of features identified by connectome mapping. With these methods, the SVM was able to discriminate epileptic patients from healthy controls with an accuracy of 80% and predict surgery success with a 70% accuracy, results that are comparable to professional clinical decision-making (Munsell, Wee, et al., 2015).

In particular, the study of connectome mapping has been a focus of machine learning research. Using fMRI techniques combined with DTI and white matter fiber tractography, researchers and clinicians will be able to see a wiring diagram of the brain's medium and wide connections. This image of brain connectivity can be used to quantify the relationships between functionally and physically linked objects, as well as to predict changes in the brain's resting state. Although genetic profiles may suggest a tendency to microscopic changes in brain structure, connectome mapping depicts the macroscopic manifestations of these changes by modeling the brain as a network of interconnected regions (Mohammad-Rezazadeh, Frohlich, Loo, & Jeste, 2016). Connectome mapping has been limited to a macroscopic level of brain wiring, with 1010 neurons interconnected by

1014 connections, due to current imaging and throughput capabilities. When combined with machine learning, these techniques have the potential to uncover new biomarkers for the diagnosis of epilepsy, attention deficit hyperactivity disorder (ADHD), autism spectrum disorders, and other neurological illnesses. For example, Prashanth, Dutta-Roy, Mandal, and Ghosh (2016) developed a machine learning technique that can predict Parkinson's disease with 96.40% accuracy in the early stages of the disease. The technique considers nonmotor characteristics and olfactory loss, as well as cerebrospinal fluid measures and data from dopaminergic imaging indicators. After a disease has been diagnosed, another machine learning system can now be used to forecast illness progression. Rastegar, Ho, Halliday, and Dzamko (2019) developed a strategy that employed blood cytokines from a single time point (baseline) to predict the result over the next 2 years.

Summary

The prevalence of neurological diseases contributes about 3% to the global burden of disease, which is gradually increasing due to the psychological demand posed by economic, financial, environmental, and physical factors, among others. Thus, early detection and treatment of neurological disorders cannot be overemphasized. Neurodegenerative diseases are characterized by impaired motor, cognitive, and behavioral changes. Some of neurodegenerative diseases are characterized by biochemical changes such as amyloid deposits as plaques in the brain parenchyma, normal cellular prion protein PrPc converted into pathogenic form PrPsc, tau protein accumulation in neurons and glial cells, and headache and facial pain, which may be benign or life-threatening conditions. CAD is a computer-oriented system that can facilitate fast decision-making by the physician by evaluating the physiological signals and biomarkers provided to the computer and appropriate analysis of anomalies within a short time. These systems combine techniques involving the use of diverse components like artificial intelligence, medical imaging, and computer vision with the major aim of detecting health problems. Over the last few decades, significant progress has been made in enormous research output and progress in the adoption of CAD systems in mammograms, chest radiographs, computerized tomography, ECG, and echocardiography through image analysis and quantitative interpretation techniques that will aid physicians' decision-making process.

Conclusion and future perspective

This chapter highlighted several areas of research carried out on the utilization of artificial intelligence in various neurological diseases. Several cases of neurodegenerative diseases are often diagnosed, managed, and treated using diverse physiological signals resulting in generation of data to be analyzed via artificial intelligence/machine learning techniques. Neurological images obtained from advanced neurological disease signals are processed and interpreted for making clinical decisions for speech, locomotive, and cognitive therapy. Neurological monitoring and recording devices rapidly detect and analyze difficult physiological signals and patterns to enhance early diagnosis. They also provide an in-depth analysis of complex pathways and biomarkers of several neurological diseases via noninvasive approaches. Further research is needed to improve diagnostic support to enhance the precision, reliability, accuracy, and quality of clinical decision-making and evaluation in clinical practice. In addition, the expensive nature and large data size requirements of the methods discussed need to be addressed and proper algorithm types need to be developed to enhance the utilization of this technology for routine diagnostic procedures.

References

Acharya, U. R., Hagiwara, Y., & Adeli, H. (2018). Automated seizure prediction. *Epilepsy & Behavior, 88*, 251–261.

Acharya, U. R., Hagiwara, Y., Deshpande, S. N., Suren, S., Wei Koh, J. E., Lih Oh, S., et al. (2019). Characterization of focal EEG signals: A review. *Future Generation Computer Systems, 91*, 290–299.

Acharya, U. R., Oh, S. L., Hagiwara, Y., Tan, J. H., & Adeli, H. (2018). Deep convolutional neural network for the automated detection and diagnosis of seizure using EEG signals. *Computers in Biology and Medicine, 100*, 270–278.

Acharya, U. R., Oh, S. L., Hagiwara, Y., Tan, J. H., Adeli, H., & Subha, D. P. (2018). Automated EEG-based screening of depression using deep convolutional neural network. *Computer Methods and Programs in Biomedicine, 161*, 103–113.

Adetunji, J. B., Adetunji, C. O., & Olaniyan, O. T. (2021). African walnuts: A natural depository of nutritional and bioactive compounds essential for food and nutritional security in Africa. In O. O. Babalola (Ed.), *Food security and safety*. Cham: Springer.

Adetunji, C. O., Ajayi, O. O., Akram, M., Olaniyan, O. T., Chishti, M. A., Inobeme, A., et al. (2021). Medicinal plants used in the treatment of influenza A virus infections. In K. Dua, S. Nammi, D. Chang, D. K. Chellappan, G. Gupta, & T. Collet (Eds.), *Medicinal plants for lung diseases*. Singapore: Springer.

Adetunji, C. O., Akram, M., Olaniyan, O. T., Ajayi, O. O., Inobeme, A., Olaniyan, S., et al. (2021a). Medicinal plants used in the treatment of pulmonary hypertension. In K. Dua, S. Nammi, D. Chang, D. K. Chellappan, G. Gupta, & T. Collet (Eds.), *Medicinal plants for lung diseases*. Singapore: Springer. https://doi.org/10.1007/978-981-33-6850-7_14.

Adetunji, C. O., Akram, M., Olaniyan, O. T., Ajayi, O. O., Inobeme, A., Olaniyan, S., et al. (2021b). Targeting SARS-CoV-2 novel Corona (COVID-19) virus infection using medicinal plants. In K. Dua, S. Nammi, D. Chang, D. K. Chellappan, G. Gupta, & T. Collet (Eds.), *Medicinal plants for lung diseases*. Singapore: Springer.

Adetunji, C. O., Inobeme, A., Olaniyan, O. T., Ajayi, O. O., Olaniyan, S., & Adetunji, J. B. (2021). Application of nanodrugs derived from active metabolites of medicinal plants for the treatment of inflammatory and lung diseases: Recent advances. In K. Dua, S. Nammi, D. Chang, D. K. Chellappan, G. Gupta, & T. Collet (Eds.), *Medicinal plants for lung diseases*. Singapore: Springer.

Adetunji, C. O., & Varma, A. (2020). Biotechnological application of trichoderma: A powerful fungal isolate with diverse potentials for the attainment of food safety, management of pest and diseases, healthy planet, and sustainable agriculture. In C. Manoharachary, H. B. Singh, & A. Varma (Eds.), *Soil biology: Vol. 61. Trichoderma: Agricultural applications and beyond*. Cham: Springer.

Akram, M., Jabeen, F., Daniyal, M., Zainab, R., Haq, U. U., Adetunji, C. O., et al. (2020). Genetic engineering of novel products of health significance: Recombinant DNA technology. In C. Egbuna, & T. G. Dable (Eds.), *Functional foods and nutraceuticals*. Cham: Springer. https://doi.org/10.1007/978-3-030-42319-3_26.

Alessandro, S., Giovanna, C.-B., Dampney Roger, A. L., & Pietro, C. (2016). Brain–heart interactions: Physiology and clinical implications. *Philosophical Transactions of the Royal Society A*, *374*, 20150181.

Almeida, A., & López-de-Ipiña, D. (2012). Assessing ambiguity of context data in intelligent environments: Towards a more reliable context managing system. *Sensors*, *12*, 4934–4951.

Álvarez, I., Górriz, J. M., Ramírez, J., Salas-Gonzalez, D., López, M., Segovia, F., et al. (2010). Projecting independent components of SPECT images for computer aided diagnosis of Alzheimer's disease. *Pattern Recognition Letters*, *31*, 1342–1347.

Alzheimer's Association/Biomarkers. (2013). Available online: http://www.alz.org/research/funding/global_biomarker_consortium.asp. (Accessed 15 May 2013).

Avcu, M. T., Zhang, Z., & Chan, D. W. S. (2019). Seizure detection using least eeg channels by deep convolutional neural network. In *ICASSP 2019–2019 IEEE International Conference on Acoustics, Speech and Signal Processing (ICASSP)* (pp. 1120–1124).

Bairy, G. M., Lih, O. S., Hagiwara, Y., Subha, D. P., Faust, O., Niranjan, U. C. A., et al. (2017). Automated diagnosis of depression electroencephalograph signals using linear prediction coding and higher order spectra features. *Journal of Medical Imaging and Health Informatics*, 7(8), 1857–1862.

Beatty, J., & Lucero-Wagoner, B. (2000). The pupillar system. In J. T. Cacioppo, L. G. Tassinary, & G. G. Berntson (Eds.), *Handbook of psychophysiology* (pp. 142–162). Cambridge: Cambridge University Press.

Berardelli, A., Wenning, G. K., Antonini, A., Berg, D., Bloem, B. R., et al. (2013). EFNS/MDS-ES recommendations for the diagnosis of Parkinson's disease. *European Journal of Neurology, 2013*.

Bhat, S., Acharya, U. R., Adeli, H., Bairy, G. M., & Adeli, A. (2014a). Autism: Cause factors, early diagnosis and therapies. *Reviews in the Neurosciences*, *25*(6), 841–850.

Bhat, S., Acharya, U. R., Adeli, H., Bairy, G. M., & Adeli, A. (2014b). Automated diagnosis of autism: In search of a mathematical marker. *Reviews in the Neurosciences*, *25*(6), 851–861.

Bravo, J., Villarreal, V., Hervás, R., & Urzaiz, G. (2012). Using a communication model to collect measurement data through mobile devices. *Sensors*, *12*, 9253–9272.

Chan, H. P., Hadjiiski, L., Zhou, C., & Sahiner, B. (2008). Computer-aided diagnosis of lung cancer and pulmonary embolism in computed tomography—A review. *Academic Radiology*, *15*, 535–555.

Chartrand, G., Cheng, P. M., Vorontsov, E., Drozdzal, M., Turcotte, S., Pal, C. J., et al. (2017). Deep learning: A primer for radiologists. *Radiographics*, *37*(7), 2113–2131.

Ching, T., Himmelstein, D. S., Beaulieu-Jones, B. K., Kalinin, A. A., Do, B. T., Way, G. P., et al. (2018). *Opportunities and obstacles for deep learning in biology and medicine*. bioRxiv.

Cirdan. (2018). *Computer aided diagnosis pathology: Pros & cons*. Available at https://www.cirdan.com/dr-liron-pantanowitz-talks-about-computer-aided-diagnosis-in-pathology-pros-and-cons.

Crown, W. H. (2015). Potential application of machine learning in health outcomes research and some statistical cautions. *Value in Health*, *18*(2), 137–140.

Cummins, N., Matcham, F., Klapper, J., & Schuller, B. (2020). *Artificial intelligence to aid the detection of mood disorders*. Elsevier. https://doi.org/10.1016/B978-0-12-817133-2.00010-0.

Cummins, N., Scherer, S., Krajewski, J., Schnieder, S., Epps, J., & Quatieri, T. F. (2015). A review of depression and suicide risk assessment using speech analysis. *Speech Communication*, *71*, 10–49.

Das, S., Sharangi, A. B., Egbuna, C., Jeevanandam, J., Ezzat, S. M., Adetunji, C. O., et al. (2020). Health benefits of isoflavones found exclusively of plants of the Fabaceae Family. In C. Egbuna, & T. G. Dable (Eds.), *Functional foods and nutraceuticals*. Cham: Springer. https://doi.org/10.1007/978-3-030-42319-3_22.

Doi, K. (2017). Computer-aided diagnosis in medical imaging: Historical review, current status and future potential. *Computerized Medical Imaging and Graphics*, *31*(4), 198–211.

Egbuna, C., Gupta, E., Ezzat, S. M., Jeevanandam, J., Mishra, N., Akram, M., et al. (2020). Aloe species as valuable sources of functional bioactives. In C. Egbuna, & T. G. Dable (Eds.), *Functional foods and nutraceuticals*. Cham: Springer.

Fang, C., Li, C., Cabrerizo, M., Barreto, A., Andrian, J., Rishe, N., et al. (2018). Gaussian discriminant analysis-based dual high-dimensional decision spaces for the diagnosis of mild cognitive impairment in Alzheimer's disease. *International Journal of Neural Systems*, *28*(8), 1850017.

Faundez-Zanuy, M., Hussain, A., Mekyska, J., Sesa-Nogueras, E., Monte-Moreno, E., Esposito, A., et al. (2012). Biometric applications related to human beings: There is life beyond security. *Cognitive Computation*, *5*, 136–151.

Geschwind, M. D. (2015). Prion diseases. *Continuum (Minneapolis, Minn.)*, *21*(6 Neuroinfectious disease), 1612–1638.

Ghiasi, S., Greco, A., Barbieri, R., et al. (2002). Assessing autonomic function from electrodermal activity and heart rate variability during cold-pressor test and emotional challenge. *Scientific Reports*, *10*, 5406.

Giger, M. L., et al. (2000). Computer-aided diagnosis in mammography. In *Handbook of medical imaging* (2nd ed., pp. 915–1004). Europe: SPIE Digital Library.

Girard, J. M., & Cohn, J. F. (2015). Automated audiovisual depression analysis. *Current Opinion in Psychology*, *4*, 75–79.

Greenberg, S. B. (2003). Viral meningitis and encephalitis. In *Office Practice of Neurology* (pp. 466–469).

Grimaldi, G., & Manto, M. (2010). Neurological tremor: Sensors, signal processing and emerging applications. *Sensors*, *10*, 1399–1422.

Haugrud, N., Crossley, M., & Vrbancic, M. (2010). Clustering and switching strategies during verbal fluency performance differentiate Alzheimer's disease and healthy aging. *Journal of Speech, Language, and Hearing Research*, *53*, 1132–1146.

Henry, J. D., Crawford, J. R., & Pshillips, L. H. (2004). Verbal fluency performance in dementia of the Alzheimer's type: A meta-analysis. *Neuropsychologia*, *42*, 1212–1222.

Horley, K., Reid, A., & Burnham, D. (2010). Emotional prosody perception and production in dementia of the Alzheimer's type. *Journal of Speech, Language, and Hearing Research*, *53*, 1132–1146.

Hu, W. T., McMillan, C., Libon, D., Leight, S., Forman, M., Lee, V. M. Y., et al. (2010). Multimodal predictors for Alzheimer disease in nonfluent primary progressive aphasia. *Neurology, 75*, 595–602.

Iadecola, C., & Davisson, R. L. (2008). Hypertension and cerebrovascular dysfunction. *Cell Metabolism*, 7(6), 476–484.

Institute of Medicine Committee (US) on Nervous System Disorders in Developing Countries. (2001). *Neurological, psychiatric, and developmental disorders: Meeting the challenge in the developing world.* Washington, DC: National Academies Press. 5, Developmental Disabilities.

Jahmunah, V., Lih Oh, S., Rajinikanth, V., Ciaccio, E. J., Hao Cheong, K., Arunkumar, N., et al. (2019). Automated detection of schizophrenia using nonlinear signal processing methods. *Artificial Intelligence in Medicine, 100*, 101698.

Jiang, F., Jiang, Y., Zhi, H., Dong, Y., Li, H., Ma, S., et al. (2017). Artificial intelligence in healthcare: Past, present and future. *Stroke and Vascular Neurology*, *2*(4), 230–243.

Kanner, R. (2012). Facial pain: After the dentist and ENT have finished. *Neurology Clinical Practice*, *2*(3), 172–178.

Khedher, L., Illán, I. A., Górriz, J. M., Ramírez, J., Brahim, A., & Meyer-Baese, A. (2017). Independent component analysis-support vector machine-based computer-aided diagnosis system for Alzheimer's with visual support. *International Journal of Neural Systems*, *27*(3), 1650050.

Khorram, S., Gideon, J., McInnis, M., & Provost, E. M. (2016). Recognition of depression in bipolar disorder: Leveraging cohort and person-specific knowledge. In *Proceedings of INTERSPEECH 2016, 17th annual conference of the international speech communication association* (pp. 1215–1219).

Li, Q., & Nishikawa, R. M. (Eds.). (2015). *Computer-aided detection and diagnosis in medical imaging.* New York: Taylor & Francis, CRC Press.

López-de-Ipiña, K., Alonso, J. B., Solé-Casals, J., Barroso, N., Faundez, M., Ecay-Torres, M., et al. (2012). New approaches for Alzheimer's disease diagnosis based on automatic spontaneous speech analysis and emotional temperature. *Ambient Assisted Living: Home Care, 7657*, 407–414.

Lopez-Poveda, E. A. (2014). Why do I hear but not understand? Stochastic undersampling as a model of degraded neural encoding of speech. *Frontiers in Neuroscience, 8*, 348.

López-Sanz, D., Garcés, P., Álvarez, B., Delgado-Losada, M. L., López-Higes, R., & Maestú, F. (2017). Network disruption in the preclinical stages of Alzheimer's disease: From subjective cognitive decline to mild cognitive impairment. *International Journal of Neural Systems*, *27*(8), 1750041.

McClelland, J. L. (2001). Cognitive neuroscience. In *International encyclopedia of the Social & Behavioral Sciences* (pp. 2133–2140). https://doi.org/10.1016/b0-08-043076-7/03406-9.

McKechnie, J., Ahmed, B., Gutierrez-Osuna, R., Monroe, P., McCabe, P., & Ballard, K. J. (2018). Automated speech analysis tools for children's speech production: A systematic literature review. *International Journal of Speech-Language Pathology*, *20*(6), 583–598.

Meilán, J. J., Martínez-Sánchez, F., Carro, J., Sánchez, J. A., & Pérez, E. (2012). Acoustic markers associated with impairment in language processing in Alzheimer's disease. *The Spanish Journal of Psychology, 15*, 487–494.

Mello, S., Judge, C., Kelly, R., Bradley, D., & Harbison, J. (2018). A systematic review of the causes and management of nonthrombotic embolic stroke of tissue origin. *Stroke Research and Treatment, 2018*, 8092862.

Mohammad-Rezazadeh, I., Frohlich, J., Loo, S. K., & Jeste, S. S. (2016). Brain connectivity in autism spectrum disorder. *Current Opinion in Neurology*, *29*(2), 137–147.

Munsell, B. C., Wee, C. Y., et al. (2015). Evaluation of machine learning algorithms for treatment outcome prediction in patients with epilepsy based on structural connectome data. *NeuroImage, 118*, 219–230.

Ortega-Zamorano, F., Jerez, J. M., Gómez, I., & Franco, L. (2017). Layer multiplexing FPGA implementation for deep back-propagation learning. *Integrated Computer-Aided Engineering, 24*(2), 171–185.

Pakhomov, S. V. S., Hemmy, L. S., & Lim, K. O. (2012). Automated semantic indices related to cognitive function and rate of cognitive decline. *Neuropsychologia, 50*, 2165–2175.

Pampouchidou, A., Simos, P., Marias, K., Meriaudeau, F., Yang, F., Pediaditis, M., et al. (2017). Automatic assessment of depression based on visual cues: A systematic review. *Transactions of Affective Computing*.

Papakostas, G. A., Savio, A., Gran, M., & Kaburlasos, V. G. (2015). A lattice computing approach to Alzheimer's disease computer assisted diagnosis based on MRI data. *Neurocomputing, 150*, 37–42.

Pareto, D., Aguiar, P., Pavia, J., Gispert, J., Cot, A., Falcon, C., et al. (2008). Assessment of SPM in per-fusion brain SPECT studies. A numerical simulation study using boot-strap resampling methods. *IEEE Transactions on Biomedical Engineering, 55*, 1849–1853.

Parikh, V., Tucci, V., & Galwankar, S. (2012). Infections of the nervous system. *International Journal of Critical Illness and Injury Science, 2*(2), 82–97.

Patel, K., & Clifford, D. B. (2014). Bacterial brain abscess. *The Neurohospitalist, 4*(4), 196–204.

Petrick, N., Sahiner, B., Armato, S. G., III, Bert, A., Correale, L., Delsanto, S., et al. (2013). Evaluation of computer-aided detection and diagnosis systems. *Medical Physics, 40*(8), 087001.

Prashanth, R., Dutta-Roy, S., Mandal, P. K., & Ghosh, S. (2016). High-accuracy detection of early parkinson's disease through multimodal features and machine learning. *International Journal of Media and Information Literacy, 90*, 13–21.

Rastegar, A. D., Ho, N., Halliday, G. M., & Dzamko, N. (2019). Parkinson's progression prediction using machine learning and serum cytokines. *Npj Parkinson Disease, 5*(1), 14.

Ringeval, F., et al. (2017). AVEC 2017: Real-life depression, and affect recognition workshop and challenge. In *Proceedings of the 7th annual workshop on audio/visual emotion challenge* (pp. 3–9).

Ringeval, F., et al. (2018). AVEC 2018 workshop and challenge: Bipolar disorder and cross-cultural affect recognition. In *Proceedings of the 2018 on audio/visual emotion challenge and workshop* (pp. 3–13).

Rocco, D., Pastore, M., Gennaro, A., Salvatore, S., Cozzolino, M., & Scorza, M. (2018). Beyond verbal behavior: An empirical analysis of speech rates in psychotherapy sessions. *Frontiers in Psychology, 9*, 978.

Roli, O. I., Adetunji, C. O., Mishra, P., Adetunji, J. B., & Mishra, R. R. (2020). Caffeine: Nutraceutical and health benefit of caffeine-containing commodities and products. In P. Mishra, R. R. Mishra, & C. O. Adetunji (Eds.), *Innovations in food technology*. Singapore: Springer.

Runde, T. J., Anjum, F., & Hafner, J. W. (2021). Bacterial meningitis. [Updated 2020 Dec 4]. In *StatPearls [Internet]*. Treasure Island, FL: StatPearls Publishing. Available from: https://www.ncbi.nlm.nih.gov/books/NBK470351/.

Said, S., & Kang, M. (2021). Viral encephalitis. [Updated 2020 Aug 10]. In *StatPearls [Internet]*. Treasure Island, FL: StatPearls Publishing. Available from: https://www.ncbi.nlm.nih.gov/books/NBK470162/.

Schwartz, R., & Pell, M. D. (2012). Emotional speech processing at the intersection of prosody and semantics. *PLoS One*, 7(10), e47279.

Silberberg, D., & Katabira, E. (2006). Neurological disorders. In D. T. Jamison, R. G. Feachem, M. W. Makgoba, et al. (Eds.), *Disease and mortality in sub-Saharan Africa* (2nd ed.). Washington, DC: The International Bank for Reconstruction and Development/The World Bank. (chapter 23). Available from: https://www.ncbi.nlm.nih.gov/books/NBK2295/.

Sirois, S., & Brisson, J. (2014). Pupillometry. *Wiley Interdisciplinary Reviews: Cognitive Science, 5*(6), 679–692. https://doi.org/10.1002/wcs.1323.

Sirven, J. I. (2015). Epilepsy: A spectrum disorder. *Cold Spring Harbor Perspectives in Medicine, 5*(9), a022848.

Sridhar, C., Bhat, S., Acharya, U. R., Adeli, H., & Bairy, G. M. (2017). Diagnosis of attention deficit hyperactivity disorder using imaging and signal processing techniques. *Computers in Biology and Medicine, 88*, 93–99.

Stafstrom, C. E., & Carmant, L. (2015). Seizures and epilepsy: An overview for neuroscientists. *Cold Spring Harbor Perspectives in Medicine, 5*(6), a022426.

Stufflebeam, S. M., & Rosen, B. R. (2007). Mapping cognitive function. *Neuroimaging Clinics of North America, 17*(4), 469–ix.

Subasi, A., Kevric, J., & Abdullah Canbaz, M. (2019). Epileptic seizure detection using hybrid machine learning methods. *Neural Computing and Applications, 31*(1), 317–325.

Thakur, K. T., Albanese, E., Giannakopoulos, P., et al. (2016 Mar 14). Neurological disorders. In V. Patel, D. Chisholm, T. Dua, et al. (Eds.), *Vol. 4. Mental, neurological, and substance use disorders: Disease control priorities* (3rd ed.). Washington, DC: The International Bank for Reconstruction and Development/The World Bank. (chapter 5). Available from: https://www.ncbi.nlm.nih.gov/books/NBK361950/.

Valenzuela, O., Jiang, X., Carrillo, A., & Rojas, I. (2018). Multi-objective genetic algorithms to find most relevant volumes of the brain related to Alzheimer's disease and mild cognitive impairment. *International Journal of Neural Systems, 28*(9), 1850022.

van Ginneken, B., Hogeweg, L., & Prokop, M. (2009). Computer-aided diagnosis in chest radiography: Beyond nodules. *European Journal of Radiology, 72*, 226–230.

Verguts, T., Vassena, E., & Silvetti, M. (2015). Adaptive effort investment in cognitive and physical tasks: A neurocomputational model. *Frontiers in Behavioral Neuroscience, 9*, 57.

Wernickand, M. N., & Aarsvold, J. N. (2004). *Emission tomography: The fundamentals of PET and SPECT.* New York, NY: Elsevier.

Wilke, C., van Drongelen, W., Kohrman, M., & He, B. (2010). Neocortical seizure foci localization by means of a directed transfer function method. *Epilepsia, 51*(4), 564–572.

World Health Organization. (2017). *World Health Organization.* Available from: https://www.who.int/mediacentre/news/releases/2007/pr04/e/. (Accessed 12 December 2018).

Xue, G., Chen, C., Lu, Z. L., & Dong, Q. (2010). Brain imaging techniques and their applications in decision-making research. *Xin li xue bao Acta Psychologica Sinica, 42*(1), 120–137.

Yıldırım, Ö., Baloglu, U. B., & Acharya, U. R. (2018). A deep convolutional neural network model for automated identification of abnormal EEG signals. *Neural Computing and Applications*, 1–12.

CHAPTER 20

An insight into applications of deep learning in neuroimaging

M.A. Jabbar[a] and Syed Saba Raoof[b]
[a]Department of CSE (AI & ML), Vardhaman College of Engineering, Hyderabad, India
[b]Department of CSE, VIT University, Vellore, India

Introduction

Medical imaging is the use of various technologies that help to visualize human organs to assist physicians to identify disease and even diagnose disease at early stages. Computed tomography (CT) scans, X-ray, magnetic resonance imaging (MRI), ultrasound, endoscopy, neuroimaging, and mammogram are examples of medical imaging technologies. This chapter reviews neuroimaging and case studies of brain disorders and outlines data-based feature learning approaches. For example, (Biswal, Zerrin Yetkin, Haughton, & Hyde, 1995) proposed a seed-based correlation method, (Bell & Sejnowski, 1995; McKeown et al., 1998) proposed an individual component study, (Raichle et al., 2001; van den Heuvel & Hulshoff Pol, 2010) developed a resting-state transformation by following uniform patterns, (Brookes et al., 2011; Potluru & Calhoun, 2008) implemented a method to identify features from input images, and (Sui, Adali, Yu, Chen, & Calhoun, 2012; Van Essen et al., 2012) proposed analysis for canonical correlation. The key challenge of the neuroimaging domain is the requirement to preprocess and examine a continuous data stream. Some of the exclusive techniques to process and analyze neuroimage data include advanced normalization tools (ANT) (Avants, Tustison, & Song, 2009), BrainSuite (Shattuck & Leahy, 2002), FreeSurfer (Fischl et al., 2002), the FMRIB Software Library (FSL) (Jenkinson, Beckmann, Behrens, Woolrich, & Smith, 2012), and statistical parametric mapping (SPM) (Friston, Ashburner, Kiebel, Nichols, & Penny, 2007). These techniques require comprehensive statistical optimizations and thus they are time-consuming and expensive processes. An alternative to these time-consuming techniques are supervised deep learning (DL) techniques, which involve fully connected layers capable of learning features from an input image that doesn't require preprocessing.

Artificial Intelligence for Neurological Disorders
https://doi.org/10.1016/B978-0-323-90277-9.00014-6

Deep learning concepts

This section discusses various DL models used for neuroimaging.

Recurrent neural network (RNN)

This is a feed-forward network that contains periodic memory loops that fetch data from previous layers. Every node in an RNN functions as processing units and all the layers in the network share the same weight to minimize the parameter complexity. An RNN functions according to random item election, and then based on the active and inactive mode of adjacent nodes, it determines the network weight. It includes a separate memory unit to store the information of the node. It is mainly useful and applied where the result relies on preceding computations. The architecture of RNN can be found in (CSTWiki, 2022).

Convolutional neural network (CNN)

This is one among the most broadly implemented algorithms. It employs several perceptrons destined to minimize the preprocessing units; consequently, inspired through the functioning of the human brain (LeCun, Bottou, Bengio, & Haffner, 1998). A CNN consists of single or multiple convolutional layers, along with fully connected layers, and pooling layers. Feature maps are designed to organize the convolutional layers and these maps are used to store the image regions, where all the neurons of the feature map relate to neurons of the previous layer and with a receptive field to cover the entire image. The convolutional layers work to expand the abstract features, pooling layer functions to minimize the dimensions, whereas fully connected layers function as hidden layers. Thus, CNNs are more effective than RNNs. The architecture of CNN can be referred from (Khvostikov et al., 2018).

Self-organizing map (SOM)

The SOM, also known as the Kohonen map, was developed by Teuvo Kohonen in 1984. It is a supervised model and a variant of an artificial neural network (ANN) that is trained by employing unsupervised learning to generate a smaller dimensional distinct representation of an input space called a map. Thus, SOM works well for the dimension reduction process. As the name implies, it is well known for its application in developing topographic maps of input value (Vincent, Larochelle, Lajoie, Bengio, & Manzagol, 2010). Mapping starts in this algorithm by weight vector initialization, then the random selection is done to select the sample vector, and then searching is performed to identify the best weight to represent the sample vector. The neighbors of the identified/selected weights are also considered, thus, SOM generates the map to increase and frames in various shapes. Typically, it forms a two-dimensional space in square, rectangle, hexagonal, or L shapes. The architecture of RNN can be found in (Larochelle, Erhan, Courville, Bergstra, & Bengio, 2007).

Boltzmann machine (BM)

Many of the DL algorithms work in a certain direction, whereas a BM doesn't follow any certain direction. Every single node is connected mutually in a circular manner and the BM creates different parameters for different networks rather than utilizing similar parameters. It is a stochastic neural network algorithm possessing the ability to learn and represent from internal depictions and solve complex problems. The architecture of RNN can be found in (https://laptrinhx.com/restricted-boltzmann-machine-how-to-create-a-recommendation-system-for-movie-review-988466013/).

Restricted Boltzmann machine (RBM)

The RBM was developed by Geoffrey E. Hinton (https://www.compthree.com/blog/autoencoder/). It is a specific variant of the Boltzmann machine; the main difference is that visible and hidden layers connections in an RBM are restricted. Thus, implementation of RBM is simpler compared to a BM. An RBM consists of two layers (i.e., visible and hidden layers) where all the nodes of each layer are connected to the nodes of another layer. BM nodes of the same graph cannot be connected; thus this makes the training more accurate. RBM is a stochastic neural network that implies that every neuron in the network possesses arbitrary nature once activated. The hidden layer is responsible for activation during forwarding move, whereas the visible layer is responsible for generating the input value on backward move. Here the generated value will differ from the original input value since there is no link among visible neurons. The architecture of RNN can be found in (Hinton, Osindero, & Teh, 2006).

Autoencoder (AE)

An AE is a specific type of feedforward neural network that consists of two parts (i.e., encoder and decoder) and functions for encoding and decoding input data. Encoding is done based upon input value and generates hidden depictions of input data. Decoding is done on the outcome, that is, original data is constructed from these generated hidden depictions (https://towardsdatascience.com/restricted-boltzmann-machines-simplified-eab1e5878976). An AE consists of three layers: input, output, and hidden layers. AEs are considered as the best algorithms for dimensionality reduction and to identify the intrinsic features of the pattern. AEs are automated feature extractors; performed by employing loss function, which supports the representation of specified features such as sparse representation thus forming sparse autoencoders (http://www.ai-junkie.com/ann/som/som1.html). There are four types of AEs: sparse, stacked, denoising, and contractive AEs. The architecture of RNN can be found in (https://towardsdatascience.com/kohonen-self-organizing-maps-a29040d688da).

Neuroimaging

Neuroimaging is a technology used to visualize activity or parts of the brain and nervous system. In other words, it is the study of the function of the brain and various actions affecting the brain. There are two types of neuroimaging: structural and functional/physiological imaging. Brain visualization, intracerebral disease, and massive brain damage can be analyzed through structural imaging, whereas functional/physiological imaging gives information about metabolic diseases and accurate microanatomy information. The fusion of multimodal imaging methods plays a vital role in medicine and research. MRI is the most widely used structural imaging technique, as it provides the best three-dimensional resolution images. Unlike structural imaging, functional imaging doesn't provide complete spatial localization. PET, electroencephalography (EEG), and fMRI are functional imaging techniques. EGG provides neural activity measures, PET includes radioisotope supervision, and fMRI is a new, innovative technology that is widely accepted by many researchers and physicians for mapping functional brain activities.

fMRI includes two major techniques: blood oxygen level-dependent (BOLD) imaging and structural MRI (sMRI). The fMRI raw data need to be preprocessed before analytical comparison; one of the most important preprocessing steps is to consider that all the fMRI slices are obtained at slightly different times. Therefore time-slicing operations are carried out on raw data and further various preprocessing operations are performed to eradicate artifacts from data, such as rifts and motion artifacts.

Deep learning case studies in neurological disorders

In this section, we discuss various neurological disorders.

Alzheimer's disease (AD)

This is a progressive neurological brain disorder that is frequently caused by dementia and mostly affects people over the age of 65 years. The study of AD was initiated in 2005 when (Mueller et al., 2005) proposed the Alzheimer's Disease Neuroimaging Initiative (ADNI). AD is characterized by the progressive impairment of neurons and their connections, which results in decline and loss of cognitive functions. Frisoni et al. (Frisoni et al., 1996) implemented a linear discriminant analysis (LDA) classifier for the classification of AD. An MRI dataset was utilized for experimental analysis and achieved 87% accuracy. Freeborough et al. (Freeborough & Fox, 1998) utilized texture features from MRI images and applied a linear discriminant function and achieved 91% accuracy for detecting AD. Hampel et al. (Hampel et al., 2002) used thresholding and linear measurement of various structures. Li et al. (Li et al., 2007) implemented an SVM classifier and used surface-based measures of hippocampus features for classification purposes. Their results showed accuracy of 95%. Lerch et al. (Lerch et al., 2008) implemented various methods like LDA, quadratic Discriminant Analysis (QDA), and logistic regression, compared results of three methods, and obtained results between 90% to 100%. Magnin et al. (Magnin et al., 2009) proposed a method for detecting AD by applying regions of interest (ROIs), Gaussian mixture (GM) distribution, and an SVM classifier. Miller et al. (Miller et al., 2009) implemented the LDA method along with hippocampus morphometric measures. They used MRI scans were for experimental results and obtained 77% accuracy. Graña et al. (Graña et al., 2011) recorded the highest accuracy to date, 100%, by implementing an SVM classifier for the classification of MRI images into AD and non-AD. Van Leemput et al. (Sabuncu & Van Leemput, 2012) used different features for classification like cortical thickness, mesh vertices, and volumetric measures. For experimental analysis, the researcher used an MRI dataset and achieved 93% of area under the curve (AUC). Wu et al. (Wu et al., 2013) proposed a multivariate algorithm for identifying AD and achieved 100% of accuracy. Lee et al. (Lee, Kim, Kim, Kim, & Seong, 2014) extracted various features from input images for better classification results like cortical thickness, texture features, and hippocampus shape. They used an LDA classifier to achieve accuracy of 88%. Beheshti et al. (Beheshti & Demirel, 2015) applied an SVM classifier on an MRI dataset for the classification of AD. Choi et al. (Choi, 2017) implemented a deep CNN known as PDNet, which comprises four three-dimensional convolutional layers. Esmaeilzadeh et al. (Esmaeilzadeh, Yang, & Adeli, 2018) employed a three-dimensional CNN algorithm and utilized MRI images and statistical features like patient age and gender for experimental analysis. Sivaranjini et al. (Sivaranjini & Sujatha, 2019) developed an AlexNet system for the classification and identification of AD using an MRI dataset.

Parkinson's disease (PD)

This is the most common brain disorders after AD. It is characterized by gradual neuron deterioration and impairment, which is generally caused when the brain's chemical messenger production blocks or stops. In 1817, James Parkinson first identified (Parkinson, 1817) PD as a chronic brain disorder. Spadotto et al. (Spadotto, Guido, Papa, & Falcão, 2010) introduced an optimum-path forest (OPF) classifier for the detection of PD. Pan et al. (Pan, Iplikci, Warwick, & Aziz, 2012) employed an SVM along with radial basis function (RBF) to detect tremor in PD patients. Braatz et al. (Braatz & Coleman, 2015) implemented a statistical model based on biological theory to analyze the variations among PD patients. Gharehchopogh et al. (Gharehchopogh & Mohammadi, 2013) employed an ANN and multilayer perceptrons to detect PD effects. Pasluosta et al. (Pasluosta, Gassner, Winkler, Klucken, & Eskofier, 2015) employed Internet of Things (IoT) mechanisms for the identification of symptoms and to diagnose the disease. Oung et al. (Oung et al., 2015) surveyed various approaches employed for detecting motor disorders. Pereira et al. (Pereira et al., 2015) applied a CNN to examine the data concern of manuscripts to recognize PD. Peker et al. (Peker, Sen, & Delen, 2015) employed a complex neural network and utilized sound-based features to detect and diagnose PD. Das et al. (Das, 2010) compared various classification techniques available for detecting PD. Hariharan et al. (Hariharan, Polat, & Sindhu, 2014) proposed a new method for weighting features by employing GM to classify the input PD data and achieved 100% accuracy. Ekker et al. (Ekker, Janssen, Nonnekes, Bloem, & de Vries, 2016) analyzed visual restoration by utilizing neurorestoration telemedicine among PD patients.

Attention-deficit/hyperactive disorder (ADHD)

This is a disorder characterized by a running structure of hyperactivity, inattention, and impulsivity that disrupts brain function and development. The major disorders of ADHD are hyperactivity/impulsivity and inattention. Few patients suffer from one of the disorders, whereas others suffer from both disorders. Kuang et al. (Kuang & He, 2014) proposed a method to distinguish healthy brains from ADHD brains. Hao et al. (Hao, He, & Yin, 2015) applied a deep Bayesian network (DBN) for dimensionality reduction purposes and feature extraction tasks. Te proposed model detected keen regions like prefrontal, visual cortex, and cingulated. Zhu et al. (Zhu et al., 2008) employed the Fisher discriminative analysis (FDA) method for the classification of ADHD, and experimental analysis fMRI images were utilized to obtain 85% of accuracy. Dai et al. (Dai, Wang, Hua, & He, 2012) implemented various image processing methods for extraction of features such as cortical thickness and gray matter were extracted. Results show an accuracy of 61%. Dey et al. (Dey, Rao, & Shah, 2014) developed an innovative model to automatically diagnose ADHD. Ghiassian et al. (Ghiassian, Greiner, Jin, & Brown, 2013) proposed histograms of oriented gradients (HOG) to analyze the diagnosis of ADHD. Guo et al. (Guo, An, Kuang, Zhao, & He, 2014) detected voxels' connection by employing the social colony network analytical technique and achieved 63% accuracy.

Autism spectrum disorder (ASD)

This is a major neural disorder characterized by impairment of social communication. According to a 2015 study, ASD affects approximately 62.2 million people globally. Many researchers have studied this disorder and developed DL applications for its diagnosis. Akhil et al. (Dekhil et al., 2018) developed a computerized system to diagnose ASD that consists of 34 SAE to activate spatial regions. Heinsfeld et al. (Heinsfeld, Franco, Craddock, Buchweitz, & Meneguzzi, 2018) employed the DAE method for reducing multivariable heterogeneous data and enhancing the analysis of ASD. Parisot et al. (Parisot et al., 2017) proposed an artificial graph convolutional network (GCN) for neurological disorder analysis among tomography and non-tomography information. The proposed GCN method was utilized to determine unlabeled data. Choi et al. (Choi, 2017) applied the variational autoencoder (VAE) method to demonstrate functional connectivity in two-dimensional features, which were utilized to detect ASD complicated brain areas. Li et al. (Li et al., 2018) implemented an innovative deep neural network (DNN) along with transfer learning for improving the classification process of ASD. Thiagarajan et al. (Anirudh & Thiagarajan, 2017) developed a new version of GCN (i.e., bootstrapped GCN) for sensitivity reduction that they trained on the random graph. Yao et al. (Yao et al., 2019) developed a multi-scalar GCN to prevent spatial drawbacks. This multi-scalar GCN was used to detect ROI and to construct patterns of multi-scalar functional connectivity. It was also used for analyzing multi-scalar brain features. Bi et al. (Bi et al., 2018) implemented multiple neural network models to enhance the classification of ASD. Dvornek et al. (Dvornek, Ventola, Pelphrey, & Duncan, 2017) employed the long short-term memory (LSTM) technique for ASD classification where the rs-fMRI time series dataset was utilized for experimental analysis.

Schizophrenia analysis

Schizophrenia is a serious mental health condition that affects a person's ability to think, feel, and behave clearly. It is a lifelong condition, but treatment can help manage symptoms. Schizophrenia causes devastating effects on the brain and the daily activities of patients. MRI is one of the neuroimaging techniques used to explore brain abnormalities in schizophrenia.

Various DL techniques have been employed to obtain an accurate diagnosis of schizophrenia. In (Plis et al., 2014), the authors implemented a model using DBN on a multi-site dataset consisting of data from 198 schizophrenia patients and 191 healthy people. The accuracy obtained with 10-fold cross validations by the proposed model was 91%.

Authors in (Ulloa, Plis, Erhardt, & Calhoun, 2015) developed a model using DBN for schizophrenia. The proposed model improved the classification of healthy controls and schizophrenia patients from their structural MRI. The dataset contained data from 198 schizophrenia patients and 191 controls. A grid search of hyper-parameters within nested 10-fold cross validation was used for experimental evaluation. The proposed model obtained an accuracy of 75%. Table 1 lists the various DL techniques used for schizophrenia analysis.

Dementia diagnosis

Dementia is increasingly prevalent as the global population continues to age. A major challenge in dementia is achieving a timely and accurate diagnosis. Neuroimaging with computer-aided algorithms has become popular to

Table 1 Deep learning techniques used for schizophrenia analysis.

Ref.	Classifier used	Accuracy recorded
Plis et al. (2014)	DBN	91%
Ulloa et al. (2015)	DBN	75%
Niu, Lin, Qiu, Kuang, and Calhoun (2019)	Softmax	90.79%
Yan et al. (2017)	Softmax	84.75%
Qureshi, Oh, and Lee (2019)	Softmax	98.09%
Pinaya et al. (2016)	Softmax	73.6%
Oh et al. (2019)	Softmax	84.43%
Salvador et al. (2019)	Two-step Ridge classifier	87%
Yan et al. (2019)	Softmax	83.2%
Oh, Oh, Lee, Chae, and Yun (2020)	Softmax	70%
Latha and Kavitha (2019)	Softmax	90%
Reiter (2020)	Softmax	82.35%
Plis et al. (2018)	Softmax	NA
Zeng et al. (2018)	Softmax	85%
Wang, Bezerianos, Cichocki, and Li (2020)	Softmax	82.42%
Yang et al. (2019)	Softmax	82.83%
Li et al. (2020)	Sigmod	93.8%
Zhou, Cui, and Jiang (2018)	Softmax	76.82%
Srinivasagopalan, Barry, Gurupur, and Thankachan (2019)	Sigmod	84.4%
Zhu, Fu, Yang, Liang, and Tan (2020)	Softmax	61%
Kim, Calhoun, Shim, and Lee (2016)	Softmax	85.3%

diagnose dementia. DL algorithms have been widely used for dementia diagnosis. Table 2 lists the various DL techniques applied for dementia diagnosis.

Open-source tool kits for deep learning

DL techniques are being successfully applied to a variety of fields, especially in health care. Researchers and companies have developed various datasets so that DL models can be built. Table 3 lists various open-source tool kits used for implementing DL algorithms. These tools are useful for the implementation of DL models for health care.

Table 2 Deep learning techniques used for dementia diagnosis.

Ref.	Classifier used	Accuracy recorded
Akhila, Markose, and Aneesh (2017)	CNN	70.97%
Dolph, Alam, Shboul, Samad, and Iftekharuddin (2017)	SFTA + Feedforward Neural Network	97.5%
Faturrahman, Wasito, Hanifah, and Mufidah (2017)	SAE + DNN	56.6%
Suk et al. (2017)	DBN	91.6%
Islam and Zhang (2017)	DSERN	90.08%
Liu, Shang, Zheng, and Wen (2016)	CNN	73.75%
Ithapu et al. (2015)	ANN	96.7%
Choi (2018)	Two-Level Randomization	–

Table 3 Open-source toolkits used for deep learning.

S. No.	Toolkit name	Ref.
1	Caffe	Jia et al. (2014)
2	Deep learning 4j	Eclipse Deep Learning 4j Development Team (2017)
3	Keras	Chollet et al. (2015)
4	MXNet	Chen et al. (2015)
5	TensorFlow	Abadi et al. (2016)
6	Pytorch	Paszke et al. (2017)
7	Theano	Alrfou et al. (2016)
8	Torch	Ronan, Kavukcuoglu, and Farabet (2011)
9	CNT	Seide and Agarwal (2016)

Challenges and future directions

Despite the success of applying DL in the field of neuroimaging, some challenges remain.

Training DL classifiers with small datasets will lead to the problem of overfitting, which is a major challenge. Many strategies have been developed to overcome the overfitting problem, such as dropout, early stopping, and regularization (Dash, Pani, Rodrigues, & Majhi, 2022; Sujata Dash, Acharya, Mittal, & Abraham, 2020).

Data size is another issue in DL. These models require large datasets to achieve accurate classification and obtaining these datasets can be challenging (Dash, Pani, Abraham, & Liang, 2021; Dash, Thulasiram, & Thulasiram, 2017; Pani, Dash, Bukhari, & Flammini, 2021).

Missing values in data can be problematic, but deep network architectures are developed to overcome this issue (Dash, Abraham, & Rehman, 2018; Dash, Biswas, Banerjee, & Rahman, 2019; Dash, Thulasiram, & Thulasiram, 2019).

Multimodal data is another challenge in neuroimaging. DL techniques allow intermediate fusion of learned representations.

Additional challenges in implementing DL methods include developing proper algorithms and hardware (Dash, Abraham, Luhach, Mizera-Pietraszko, & Rodrigues, 2019; Dash, Chakraborty, Giri, Pani, & Frnda, 2021; Dash, Dash, Chakraborty, Giri, & Pani, 2021; Rahman et al., 2019) and building and safeguarding large, labeled, and open datasets.

Conclusion

In this chapter, we reviewed recent studies on DL techniques applied for the analysis of neuroimaging. We focused on analyses of five neurological disorders: Parkinson's disease, Alzheimer's disease, autism spectrum disorder, attention deficit hyperactivity disorder, and schizophrenia. DL models are successful in the analysis of neuroimages and have performed well for the diagnosis of neurological disorders. These methods require huge data for training, thus the lack of open-source datasets is a challenge for the automatic diagnosis of neurological disorders. Further research on DL techniques for medical diagnosis with human-machine interaction will become a symbiosis in the medical field.

References

Abadi, M., Agarwal, A., Barham, P., Brevdo, E., Chen, Z., Citro, C., et al. (2016). *Tensorflow: Large-scale machine learning on heterogeneous distributed systems.*

Akhila, J. A., Markose, C., & Aneesh, R. P. (2017). Feature extraction and classification of dementia with neural network. In *Proc. Int. Conf. Intell. Comput., Instrum. Control Technol* (pp. 1446–1450).

Alrfou, R., Alain, G., Almahairi, A., Angermueller, C., Bahdanau, D., Ballas, N., et al. (2016). *Theano: A python framework for fast computation of mathematical expressions.* arXiv: Symbolic Computation.

Anirudh, R., & Thiagarajan, J. J. (2017). *Bootstrapping graph convolutional neural networks for autism spectrum disorder classification.*

Avants, B. B., Tustison, N., & Song, G. (2009). Advanced normalization tools (ants). *Insight Journal, 2*, 1–35.

Beheshti, I., & Demirel, H. (2015). Probability distribution function-based classification of structural MRI for the detection of Alzheimer's disease. *Computers in Biology and Medicine, 64*, 208–216.
Bell, A. J., & Sejnowski, T. J. (1995). An information-maximization approach to blind separation and blind deconvolution. *Neural Computation*, 7, 1129–1159. https://doi.org/10.1162/neco.1995.7.6.1129.
Bi, X., Liu, Y., Jiang, Q., Shu, Q., Sun, Q., & Dai, J. (2018). The diagnosis of autism spectrum disorder based on the random neural network cluster. *Frontiers in Human Neuroscience, 12*(257).
Biswal, B., Zerrin Yetkin, F., Haughton, V. M., & Hyde, J. S. (1995). Functional connectivity in the motor cortex of resting human brain using echo-planar MRI. *Magnetic Resonance in Medicine, 34*, 537–541. https://doi.org/10.1002/mrm.1910340409.
Braatz, E. M., & Coleman, R. A. (2015). A mathematical model of insulin resistance in Parkinson's disease. *Computational Biology and Chemistry, 56*, 84–97.
Brookes, M., Woolrich, M., Luckhoo, H., Price, D., Hale, J., Stephenson, M., et al. (2011). Investigating the electrophysiological basis of resting state networks using magnetoencephalography. *Proceedings. National Academy of Sciences. United States of America, 108*, 16783–16788. https://doi.org/10.1073/pnas.1112685108.
Chen, T., Li, M., Li, Y., Lin, M., Wang, N., Wang, M., et al. (2015). *Mxnet: A flexible and efficient machine learning library for heterogeneous distributed systems*. CoRR. abs/1512.01274.
Choi, H. (2017). *Functional connectivity patterns of autism spectrum disorder identified by deep feature learning*.
Choi, H. (2018). Deep learning in nuclear medicine and molecular imaging: Current perspectives and future directions. *Nuclear Medicine and Molecular Imaging, 52*(2), 109–118.
Chollet, F., et al. (2015). *Keras*. https://keras.io.
CSTWiki. (2022). http://cstwiki.wtb.tue.nl/index.php?title=File:RNN-unrolled.png.
Dai, D., Wang, J., Hua, J., & He, H. (2012). Classification of ADHD children through multimodal magnetic resonance imaging. *Frontiers in Systems Neuroscience, 6*, 63.
Das, R. (2010). A comparison of multiple classification methods for diagnosis of Parkinson disease. *Expert Systems with Applications, 37*(2), 1568–1572.
Dash, S., Abraham, A., Luhach, A. K., Mizera-Pietraszko, J., & Rodrigues, J. J. P. C. (2019). Hybrid chaotic firefly decision-making model for Parkinson's disease diagnosis. *International Journal of Distributed Sensor Networks, 15*(12). SCI, Scopus indexed, IF-1.787.
Dash, S., Abraham, A., & Rehman, A. (2018). Kernel based chaotic firefly algorithm for diagnosing Parkinson's disease. In *18th international conference on hybrid intelligent systems (HIS 2018), advances in intelligent systems and computing, Porto, Portugal* (pp. 176–188).
Dash, S., Biswas, S., Banerjee, D., & Rahman, A. (2019). Edge and fog computing in healthcare—A review. *Scalable Computing: Practice and Experience, 20*(2), 191–205. http://www.scpe.org. 10.12694/scpe.v20i2.1504. ISSN 1895-1767.
Dash, S., Chakraborty, C., Giri, S. K., Pani, S. K., & Frnda, J. (2021). BIFM: big-data driven intelligent forecasting model for COVID-19. *IEEE Access, 9*, 97505–97517. https://doi.org/10.1109/ACCESS.2021.3094658.
Dash, S., Dash, C. S., Chakraborty, C., Giri, S. K., & Pani, S. K. (2021). Intelligent computing on time-series data analysis and prediction of Covid-19 pandemics. *Pattern Recognition Letters, 151*, 69–75. https://doi.org/10.1016/j.patrec.2021.07.027. Elsevier.
Dash, S., Pani, S., Abraham, A., & Liang, Y. (2021). Advanced soft computing techniques in data science. In *Vol. 89. IoT and cloud computing, series: Studies in big data* (p. 443). Springer. ISSN: 2197-6503, ISBN 978-3-030-75656- 7,ISBN 978-3-030-75657-4 (eBook) https://doi.org/10.1007/978-3-030-75657-4.
Dash, S., Pani, S. K., Rodrigues, J. J. P. C., & Majhi, B. (2022). *Deep learning, machine learning and IoT in biomedical and health informatics: Techniques and applications* (p. 382). CRC Press, Taylor & Francis Group. ISBN 9780367544256.

Dash, S., Thulasiram, R., & Thulasiram, P. (2017). An enhanced chaos-based firefly model for parkinson's disease diagnosis and classification. In *IEEE ICIT conference, Bhubaneswar, Odisha* (pp. 159–164). https://doi.org/10.1109/ICIT.2017.43. Published in IEEE Xplore.

Dash, S., Thulasiram, R., & Thulasiram, P. (2019). A modified firefly based meta-search algorithm for feature selection: A predictive model for medical data. *International Journal of Swarm Intelligence, 10*(2), 2. IGI Global, USA.

Dekhil, O., Hajjdiab, H., Shalaby, A., Ali, M. T., Ayinde, B., Switala, A., et al. (2018). Using resting state functional MRI to build a personalized autism diagnosis system. *PLoS One, 13*, e0206351.

Dey, S., Rao, A. R., & Shah, M. (2014). Attributed graph distance measure for automatic detection of attention deficit hyperactive disordered subjects. *Frontiers in Neural Circuits, 8*, 94.

Dolph, C. V., Alam, M., Shboul, Z., Samad, M. D., & Iftekharuddin, K. M. (2017). Deep learning of texture and structural features for multiclass Alzheimer's disease classification. In *Proc. Int. Joint Conf. Neural Netw* (pp. 2259–2266).

Dvornek, N. C., Ventola, P., Pelphrey, K. A., & Duncan, J. S. (2017). Identifying autism from resting-state fMRI using long short-term memory networks. In *International workshop on machine learning in medical imaging* (pp. 362–370). Springer: Quebec City QC.

Eclipse Deep Learning 4j Development Team. (2017). *Deeplearning4j: Open-source distributed deep learning for the JVM.* Apache Software Foundation License 2.0 http://deeplearning4j.org.

Ekker, M. S., Janssen, S., Nonnekes, J., Bloem, B. R., & de Vries, N. M. (2016). Neurorehabilitation for Parkinson's disease: Future perspectives for behavioural adaptation. *Parkinsonism & Related Disorders, 22*(Supplement 1), S73–S77.

Esmaeilzadeh, S., Yang, Y., & Adeli, E. (2018). *End-to-end Parkinson disease diagnosis using brain MR-images by 3D-CNN.*

Faturrahman, M., Wasito, I., Hanifah, N., & Mufidah, R. (2017). Structural MRI classification for Alzheimer's disease detection using deep belief network. In *Proc. Int. Conf. Inf. Commun. Technol. Syst* (pp. 37–42).

Fischl, B., Salat, D. H., Busa, E., Albert, M., Dieterich, M., Haselgrove, C., et al. (2002). Whole brain segmentation: Automated labeling of neuroanatomical structures in the human brain. *Neuron, 33*(3), 341–355.

Freeborough, P. A., & Fox, N. C. (1998). MR image texture analysis applied to the diagnosis and tracking of Alzheimer's disease. *IEEE Transactions on Medical Imaging, 17*, 475–479.

Frisoni, G. B., Beltramello, A., Weiss, C., Geroldi, C., Bianchetti, A., & Trabucchi, M. (1996). Linear measures of atrophy in mild Alzheimer disease. *AJNR. American Journal of Neuroradiology, 17*, 1421913–1421923.

Friston, K., Ashburner, J., Kiebel, S., Nichols, T., & Penny, W. (2007). *Statistical parametric mapping: The analysis of functional brain images.* Academic Press.

Gharehchopogh, F. S., & Mohammadi, P. (2013). A case study of Parkinsons disease diagnosis using artificial neural networks. *International Journal of Computers and Applications, 73*(19), 1–6.

Ghiassian, S., Greiner, R., Jin, P., & Brown, M. (2013). Learning to classify psychiatric disorders based on fMR images: Autism vs healthy and ADHD vs healthy. In *Proc. 3rd NIPS Workshop Mach. Learn. Interpretation NeuroImag* (pp. 1–7).

Graña, M., Termenon, M., Savio, A., Gonzalez-Pinto, A., Echeveste, J., Pérez, J. M. M., et al. (2011). Computer aided diagnosis system for Alzheimer disease using brain diffusion tensor imaging features selected by Pearson's correlation. *Neuroscience Letters, 502*, 225–229.

Guo, X., An, X., Kuang, D., Zhao, Y., & He, L. (2014). ADHD-200 classification based on social network method. In *Proc. Int. Conf. Intell. Comput* (pp. 233–240).

Hampel, H., Teipel, S. J., Bayer, W., Alexander, G. E., Schwarz, R., Schapiro, M. B., et al. (2002). Age transformation of combined hippocampus and amygdala 1503 volume improves diagnostic accuracy in Alzheimer's disease. *Journal of the Neurological Sciences*, *194*, 150415–150419.

Hao, A. J., He, B. L., & Yin, C. H. (2015). Discrimination of ADHD children based on deep Bayesian network. In *2015 international conference on biomedical image and signal processing* (pp. 1–6).

Hariharan, M., Polat, K., & Sindhu, R. (2014). A new hybrid intelligent system for accurate detection of Parkinson's disease. *Computer Methods and Programs in Biomedicine*, *113*(3), 904–913.

Heinsfeld, A. S., Franco, A. R., Craddock, R. C., Buchweitz, A., & Meneguzzi, F. (2018). Identification of autism spectrum disorder using deep learning and the ABIDE dataset. *NeuroImage: Clinical*, *17*, 16–23.

Hinton, G. E., Osindero, S., & Teh, Y. W. (2006). A fast learning algorithm for deep belief nets. *Neural Computation*, *18*, 1527–1554.

Islam, J., & Zhang, Y. (2017). A novel deep learning based multi-class classification method for Alzheimer's disease detection using brain MRI data. In *Proceedings of the international conference on brain informatics* (pp. 213–222).

Ithapu, V. K., Singh, V., Okonkwo, O. C., Chappell, R. J., Dowling, N. M., & Johnson, S. C. (2015). Imaging-based enrichment criteria using deep learning algorithms for efficient clinical trials in mild cognitive impairment. *Alzheimer's & Dementia*, *11*(12), 1489–1499.

Jenkinson, M., Beckmann, C. F., Behrens, T. E., Woolrich, M. W., & Smith, S. M. (2012). FSL. *NeuroImage*, *62*(2), 782–790.

Jia, Y., Shelhamer, E., Donahue, J., Karayev, S., Long, J., Girshick, R., Guadarrama, S., & Darrell, T. (Eds.). (2014). Caffe: Convolutional architecture for fast feature embedding. In *ACM international conference on multimedia*.

Khvostikov, A., Aderghal, K., Benois-Pineaub, J., Krylova, A., Cathelined, G., & for the Alzheimer's Disease Neuroimaging Initiative I. (2018). *3D CNN-based classification using sMRI and MD-DTI images for Alzheimer disease studies*. arXiv:1801.05968v1 [cs.CV] 18 Jan 2018.

Kim, J., Calhoun, V. D., Shim, E., & Lee, J.-H. (2016). Deep neural network with weight sparsity control and pre-training extracts hierarchical features and enhances classification performance: Evidence from whole-brain resting state functional connectivity patterns of schizophrenia. *NeuroImage*, *124*, 127–146. https://doi.org/10.1016/j.neuroimage.2015.05.018.

Kuang, D., & He, L. (2014). Classification on ADHD with deep learning. In *Proceedings of the international conference on cloud computing and big data* (pp. 27–32).

Larochelle, H., Erhan, D., Courville, A., Bergstra, J., & Bengio, Y. (2007). An empirical evaluation of deep architectures on problems with many factors of variation. In *Proceedings of the 24th international conference on machine learning* (pp. 473–480).

Latha, M., & Kavitha, G. (2019). Detection of schizophrenia in brain MR images based on segmented ventricle region and deep belief networks. *Neural Computing and Applications*, *31*(9), 5195–5206. https://doi.org/10.1007/s00521-018-3360-1.

LeCun, Y., Bottou, L., Bengio, Y., & Haffner, P. (1998). Gradient-based learning applied to document recognition. *Proceedings of the IEEE*, *86*, 2278–2324.

Lee, G.-Y., Kim, J., Kim, J. H., Kim, K., & Seong, J.-K. (2014). Online learning for classification of Alzheimer disease based on cortical thickness and hippocampal shape analysis. *Healthcare Informatics Research*, *20*, 61–68.

Lerch, J. P., Pruessner, J., Zijdenbos, A. P., Collins, D. L., Teipel, S. J., Hampel, H., et al. (2008). Automated cortical thickness measurements from MRI can accurately separate Alzheimer's patients from normal elderly controls. *Neurobiology of Aging*, *29*, 23–30.

Li, X., Dvornek, N. C., Papademetris, X., Zhuang, J., Staib, L. H., Ventola, P., et al. (2018). 2-channel convolutional 3D deep neural network (2CC3D) for fMRI analysis: ASD classification and feature learning. In *2018 IEEE 15th international symposium on biomedical imaging (ISBI 2018)* (pp. 1252–1255). IEEE: Washington DC.

Li, S., Shi, F., Pu, F., Li, X., Jiang, T., Xie, S., et al. (2007). Hippocampal shape analysis of Alzheimer disease based on machine learning methods. *AJNR. American Journal of Neuroradiology, 28*, 1339–1345.

Li, G., Wang, C., Han, D.-P., Zhang, Y.-P., Peng, P., Calhoun, V. D., et al. (2020). Deep principal correlated auto-encoders with application to imaging and genomics data integration. *IEEE Access, 8*, 20093–20107. https://doi.org/10.1109/ACCESS.2020.2968634.

Liu, J., Shang, S., Zheng, K., & Wen, J.-R. (2016). Multi-view ensemble learning for dementia diagnosis from neuroimaging: An artificial neural network approach. *Neurocomputing, 195*, 112–116.

Magnin, B., Mesrob, L., Kinkingnéhun, S., Pélégrini-Issac, M., Colliot, O., Sarazin, M., et al. (2009). Support vector machine-based classification of Alzheimer's disease from whole-brain anatomical MRI. *Neuroradiology, 51*, 73–83.

McKeown, M. J., Makeig, S., Brown, G. G., Jung, T. P., Kindermann, S. S., Bell, A. J., et al. (1998). Analysis of fMRI data by blind separation into independent spatial components. *Human Brain Mapping, 6*, 160–188.

Miller, M. I., Priebe, C. E., Qiu, A., Fischl, B., Kolasny, A., Brown, T., et al. (2009). Collaborative computational anatomy: An MRI morphometry study of the human brain via diffeomorphic metric mapping. *Human Brain Mapping, 30*, 2132–2141.

Mueller, S. G., Weiner, M. W., Thal, L. J., Petersen, R. C., Jack, C., Jagust, W., et al. (2005). The Alzheimer's disease neuroimaging initiative. *NeuroImage: Clinical, 15*, 869–877. https://doi.org/10.1016/j.nic.2005.09.008.

Niu, Y.-W., Lin, Q.-H., Qiu, Y., Kuang, L.-D., & Calhoun, V. D. (2019). Sample augmentation for classification of schizophrenia patients and healthy controls using ICA of FMRI data and convolutional neural networks. In *2019 tenth international conference on intelligent control and information processing (ICICIP)* (pp. 297–302). IEEE. https://doi.org/10.1109/ICICIP47338.2019.9012169.

Oh, K., Kim, W., Shen, G., Piao, Y., Kang, N.-I., Oh, I.-S., et al. (2019). Classification of schizophrenia and normal controls using convolutional neural network and outcome visualization. *Schizophrenia Research, 212*, 186–195. https://doi.org/10.1016/j.schres.2019.07.034.

Oh, J., Oh, B.-L., Lee, K.-U., Chae, J.-H., & Yun, K. (2020). Identifying schizophrenia using structural MRI with a deep learning algorithm. *Frontiers in Psychiatry, 11*, 16. https://doi.org/10.3389/fpsyt.2020.00016.

Oung, Q. W., Muthusamy, H., Lee, H. L., Basah, S. N., Yaacob, S., Sarillee, M., et al. (2015). Technologies for assessment of motor disorders in Parkinson's disease: A review. *Sensors, 15*(9), 21710–21745.

Pan, S., Iplikci, S., Warwick, K., & Aziz, T. Z. (2012). Parkinson's disease tremor classification, a comparison between support vector machines and neural networks. *Expert Systems with Applications, 39*(12), 10764–10771.

Pani, S., Dash, S., dos Santos, W., Bukhari, S. A. C., & Flammini, F. (2021). *Assessing covid-19 and other pandemics and epidemics using computational modelling and data analysis* (p. 431). Springer. ISBN-13: 978-3030797522.

Parisot, S., Ktena, S. I., Ferrante, E., Lee, M., Moreno, R. G., Glocker, B., et al. (2017). Spectral graph convolutions for population-based disease prediction. In *International conference on medical image computing and computer-assisted intervention* (pp. 177–185). Springer: Quebec City QC.

Parkinson, J. (1817). An essay on the shaking palsy. *The Journal of Neuropsychiatry and Clinical Neurosciences, 20*(4), 223–236.

Pasluosta, C. F., Gassner, H., Winkler, J., Klucken, J., & Eskofier, B. M. (2015). An emerging era in the management of Parkinson's disease: Wearable technologies and the internet of things. *IEEE Journal of Biomedical and Health Informatics*, 1873–1881.

Paszke, A., Gross, S., Chintala, S., Chanan, G., Yang, E., DeVito, Z., et al. (2017). *Automatic differentiation in pytorch.*

Peker, M., Sen, B., & Delen, D. (2015). Computer-aided diagnosis of Parkinson's disease using complex-valued neural networks and MRMR feature selection algorithm. *Journal of Healthcare Engineering*, *6*(3), 281–302.

Pereira, C. R., Pereira, D. R., da Silva, F. A., Hook, C., Weber, S. A. T., Pereira, L. A. M., et al. (2015). A step towards the automated diagnosis of Parkinson's disease: Analyzing handwriting movements. In *IEEE 28th international symposium on computer-based medical systems* (pp. 171–176).

Pinaya, W. H., Gadelha, A., Doyle, O. M., Noto, C., Zugman, A., Cordeiro, Q., et al. (2016). Using deep belief network modelling to characterize differences in brain morphometry in schizophrenia. *Scientific Reports*, *6*(1), 1–9. https://doi.org/10.1038/srep38897.

Plis, S. M., Amin, M. F., Chekroud, A., Hjelm, D., Damaraju, E., Lee, H. J., et al. (2018). Reading the (functional) writing on the (structural) wall: Multimodal fusion of brain structure and function via a deep neural network based translation approach reveals novel impairments in schizophrenia. *NeuroImage*, *181*, 734–747. https://doi.org/10.1016/j.neuroimage.2018.07.047.

Plis, S. M., Hjelm, D. R., Salakhutdinov, R., Allen, E. A., Bockholt, H. J., Long, J. D., et al. (2014). Deep learning for neuroimaging: A validation study. *Frontiers in Neuroscience*, *8*, 1–11.

Potluru, V. K., & Calhoun, V. D. (2008). Group learning using contrast NMF: Application to functional and structural MRI of schizophrenia. In *Circuits and systems, 2008. ISCAS 2008. IEEE international symposium on (Seattle, WA)* (pp. 1336–1339).

Qureshi, M. N. I., Oh, J., & Lee, B. (2019). 3D-CNN based discrimination of schizophrenia using resting-state FMRI. *Artificial Intelligence in Medicine*, *98*, 10–17. https://doi.org/10.1016/j.artmed.2019.06.003.

Rahman, A., Dash, S., Luhach, A. K., Chilamkurti, N., Baek, S., & Nam, Y. (2019). A neuro-fuzzy approach for user behaviour classification and prediction. *Journal of Cloud Computing: Advances, Systems and Applications*, *8*, 17. https://doi.org/10.1186/s13677-019-0144-9, Springer.

Raichle, M. E., MacLeod, A. M., Snyder, A. Z., Powers, W. J., Gusnard, D. A., & Shulman, G. L. (2001). A default mode of brain function. *Proceedings. National Academy of Sciences. United States of America*, *98*, 676–682. https://doi.org/10.1073/pnas.98.2.676.

Reiter, J. (2020). *Developing an interpretable schizophrenia deep learning classifier on FMRI and SMRI using a patient-centered deepSHAP.*

Ronan, C., Kavukcuoglu, K., & Farabet, C. (2011). Torch7: A Matlab-like environment for machine learning. In *BigLearn, NIPS workshop*.

Sabuncu, M. R., & Van Leemput, K. (2012). The relevance voxel machine (RVoxM): A self-tuning Bayesian model for informative image-based prediction. *IEEE Transactions on Medical Imaging*, *31*, 2290–2306.

Salvador, R., Canales-Rodríguez, E., Guerrero-Pedraza, A., Sarró, S., Tordesillas-Gutiérrez, D., Maristany, T., et al. (2019). Multimodal integration of brain images for MRI-based diagnosis in schizophrenia. *Frontiers in Neuroscience*, *13*, 1203. https://doi.org/10.3389/fnins.2019.01203.

Seide, F., & Agarwal, A. (2016). Cntk: Microsoft's open-source deep-learning toolkit. In *Proceedings of the 22Nd ACM SIGKDD international conference on knowledge discovery and data mining* (p. 2135). New York, NY, USA: ACM.

Shattuck, D. W., & Leahy, R. M. (2002). Brainsuite: An automated cortical surface identification tool. *Medical Image Analysis*, *6*(2), 129–142.

Sivaranjini, S., & Sujatha, C. (2019). Deep learning based diagnosis of Parkinson's disease using convolutional neural network. *Multimedia Tools and Applications*, *79*, 15467–15479.

Spadotto, A. A., Guido, R. C., Papa, J. P., & Falcão, A. X. (2010). Parkinson's disease identification through optimum-path forest. In *IEEE international conference of the engineering in medicine and biology society* (pp. 6087–6090).

Srinivasagopalan, S., Barry, J., Gurupur, V., & Thankachan, S. (2019). A deep learning approach for diagnosing schizophrenic patients. *Journal of Experimental & Theoretical Artificial Intelligence*, *31*(6), 803–816. https://doi.org/10.1080/0952813X.2018.1563636.

Sui, J., Adali, T., Yu, Q., Chen, J., & Calhoun, V. D. (2012). A review of multivariate methods for multimodal fusion of brain imaging data. *Journal of Neuroscience Methods*, *204*, 68–81. https://doi.org/10.1016/j.jneumeth.2011.10.031.

Sujata Dash, B. R., Acharya, M., Mittal, A., & Abraham, A. K. (2020). Deep learning techniques for biomedical and health informatics. In *Studies in big data series* Springer. SSN:2197-6511, ISBN:978-3-030-33966-1.

Suk, H.-I., Lee, S.-W., Shen, D., & Alzheimer's Disease Neuroimaging Initiative. (2017). Deep ensemble learning of sparse regression models for brain disease diagnosis. *Medical Image Analysis*, *37*, 101–113.

Ulloa, A., Plis, S., Erhardt, E., & Calhoun, V. (2015). Synthetic structural magnetic resonance image generator improves deep learning prediction of schizophrenia. In *2015 IEEE 25th international workshop on machine learning for signal processing (MLSP)* (pp. 1–6).

van den Heuvel, M., & Hulshoff Pol, H. (2010). Exploring the brain network: A review on resting-state fMRI functional connectivity. *European Neuropsychopharmacology*.

Van Essen, D. C., Ugurbil, K., Auerbach, E., Barch, D., Behrens, T., Bucholz, R., et al. (2012). The human connectome project: A data acquisition perspective. *NeuroImage*, *62*(4), 2222–2231.

Vincent, P., Larochelle, H., Lajoie, I., Bengio, Y., & Manzagol, P. A. (2010). Stackeddenoising autoencoders: Learning useful representations in a deep network with a local denoising criterion. *Journal of Machine Learning Research*, *11*, 3371–3408.

Wang, T., Bezerianos, A., Cichocki, A., & Li, J. (2020). Multi kernel capsule network for schizophrenia identification. *IEEE Transactions on Cybernetics*. https://doi.org/10.1109/TCYB.2020.3035282.

Wu, X., Li, J., Ayutyanont, N., Protas, H., Jagust, W., Fleisher, A., et al. (2013). The receiver operational characteristic for binary classification with multiple indices and its application to the neuroimaging study of Alzheimer's disease. *IEEE/ACM Transactions on Computational Biology and Bioinformatics*, *10*, 173–180.

Yan, W., Calhoun, V., Song, M., Cui, Y., Yan, H., Liu, S., et al. (2019). Discriminating schizophrenia using recurrent neural network applied on time courses of multi-site FMRI data. *eBioMedicine*, *47*, 543–552. https://doi.org/10.1016/j.ebiom.2019.08.023.

Yan, W., Plis, S., Calhoun, V. D., Liu, S., Jiang, R., Jiang, T.-Z., et al. (2017). Discriminating schizophrenia from normal controls using resting state functional network connectivity: A deep neural network and layer-wise relevance propagation method. In *2017. IEEE 27th international workshop on machine learning for signal processing (MLSP)* (pp. 1–6). IEEE. https://doi.org/10.1109/MLSP.2017.8168179.

Yang, B., Chen, Y., Shao, Q.-M., Yu, R., Li, W.-B., Guo, G.-Q., et al. (2019). Schizophrenia classification using FMRI data based on a multiple feature image capsule network ensemble. *IEEE Access*, 7, 109956–109968. https://doi.org/10.1109/ACCESS.2019.2933550.

Yao, D., Liu, M., Wang, M., Lian, C., Wei, J., Sun, L., et al. (2019). Triplet graph convolutional network for multi-scale analysis of functional connectivity using functional MRI. In *International workshop on graph learning in medical imaging* (pp. 70–78). Shenzhen: Springer.

Zeng, L.-L., Wang, H., Hu, P., Yang, B., Pu, W., Shen, H., et al. (2018). Multi-site diagnostic classification of schizophrenia using discriminant deep learning with functional connectivity MRI. *eBioMedicine*, *30*, 74–85. https://doi.org/10.1016/j.ebiom.2018.03.017.

Zhou, A., Cui, Y., & Jiang, T. (2018). Multisite schizophrenia classification based on brain onetime atlas by deep learning. In *2018. 5th IEEE international conference on cloud computing and intelligence systems (CCIS)* (pp. 451–455). IEEE. https://doi.org/10.1109/CCIS.2018.8691336.

Zhu, Y., Fu, S., Yang, S., Liang, P., & Tan, Y. (2020). Weighted deep forest for schizophrenia data classification. *IEEE Access*, *8*, 62698–62705. https://doi.org/10.1109/ACCESS.2020.2983317.

Zhu, C.-Z., et al. (2008). Fisher discriminative analysis of resting-state brain function for attention-deficit/hyperactivity disorder. *NeuroImage*, *40*(1), 110–120.

CHAPTER 21

Incremental variance learning-based ensemble classification model for neurological disorders

Ricky Mohanty[a] and Subhendu Kumar Pani[b]
[a]School of Information System, ASBM University, Bhubaneswar, India
[b]Krupajal Computer Academy, Bhubaneswar, India

Introduction

Electroencephalogram (EEG) classification is a multidomain task that involves efficient design of data acquisition, pre-processing, feature extraction, feature selection, classification, and post-processing blocks. For a highly efficient EEG classification system, these blocks must work in tandem with maximum fan-out and fan-in capabilities. This indicates that signals from one block must be seamlessly interfaceable with processes designed in consecutive blocks. Fig. 1 shows a typical EEG classification system wherein discrete wavelet transform is used for feature extraction along with sub-band relative energy calculations for feature enhancements. These features are given to a feature normalization unit, which is governed using Eq. (21.1), and assists in standardizing feature ranges.

$$N_f = \frac{I_f - \min(f)}{\max(f) - \min(f)} \tag{1}$$

where, N_f, I_f, and f represent normalized value, intrinsic value, and feature value of the wavelet coefficients.

These features are given to a feature selection model wherein optimum features are selected. Models like principal component analysis (PCA), independent component analysis (ICA), intrinsic mode functions (IMFs), and others are used for this purpose. These models aim at increasing variance between features of different neurological samples of different classes to improve efficiency of classification. The extracted features are given to a classification engine wherein methods like artificial neural network models (ANNs), convolutional neural networks (CNNs), recurrent neural

Artificial Intelligence for Neurological Disorders
https://doi.org/10.1016/B978-0-323-90277-9.00015-8

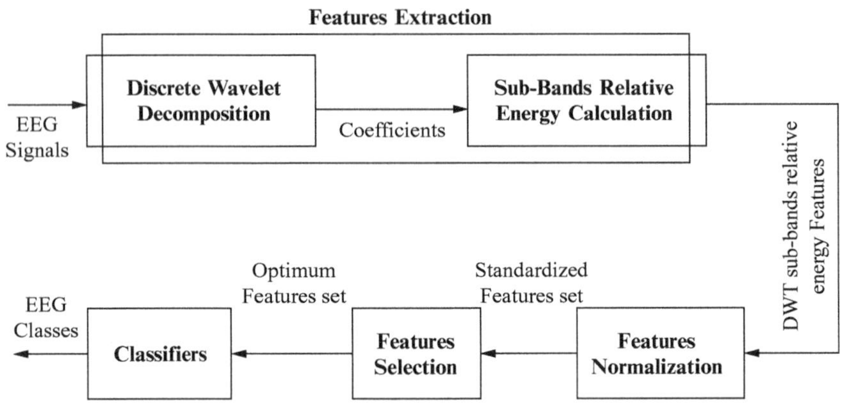

Fig. 1 A typical EEG classification model.

networks (RNNs), random forest (RF), naïve Bayes (NB) gated recurrent units (GRUs), and long short-term memory (LSTM) models are used. These methods vary in terms of performance metrics like accuracy of classification, precision, recall, F-measure, end-to-end delay, and other parameters. Moreover, these methods require a large amount of data for efficient classification, which results in class imbalance, cold start, and other issues. We review these algorithms and their issues in the section that follows. To remove these issues, this chapter proposes a novel incremental variance learning-based ensemble classification model for neurological disorders, which we describe in Section 3. The model utilizes a combination of ensemble classification and incremental learning to efficiently classify EEG signals into different neurological classes, including acute spinal cord injury, amyotrophic lateral sclerosis, Alzheimer's disease, and epilepsy. We present the evaluation of the performance of the proposed method in terms of accuracy and end-to-end delay in Section 4. Finally, this chapter concludes with some interesting observations about the proposed model and recommendations for improving its efficiency.

Literature review

Classification deals with the problem of constructing methods and algorithms to enable machine implementation of various recognition tasks that a human normally performs. The motivation is to find ways in which the recognition tasks can be performed faster, more accurately, and maybe more economically than a human can accomplish. The purpose of PC also includes tasks that humans are not good at, like reading bar codes, separating sounds from one another, and so on.

The goal of pattern recognition research is to devise ways and means of automating certain decision-making processes that lead to classification and recognition. For example, a sorting factory can use these algorithms to sort red apples from green apples traveling on a conveyor belt using a video camera, acting as the "eye" of the machine, and a computer, which will process the input information from the camera to determine which apple is red or green. Pattern recognition has been an intensely studied field for the past few decades, as is amply borne out by numerous.

The object that is inspected for the "recognition" process is called a *pattern*. We normally refer to a pattern as a *description* of an object that we want to recognize. In this chapter, we are interested in spatial patterns like humans, apples, fingerprints, electrocardiograms, or chromosomes. In most cases, a pattern recognition problem is a problem of discriminating between different populations. Hence, the recognition process turns into classification. To determine which class an object belongs to, we must first find which *features* are going to determine this classification.

In constructing a pattern classification system, that is, a system able to obtain an unknown incoming pattern and classify it into one (or more) of several given classes, we clearly want to employ all the available related information that was previously accumulated. We assume that some sample patterns with known classification are available. These patterns with their typical attributes form a training set that provides relevant information about how to associate between input information and decision-making. By using the training set the pattern recognition system may learn various types of information like statistical parameters, relevant features, and so on. The goal of pattern recognition is classification of a pattern. From the original information that we obtain about the pattern, we first identify the features and then use a *feature extractor* to measure them. These measurements are then sent to a *classifier*, which performs the actual classification (determines into which of the existing classes to classify the pattern). When the number of classes is known and when the training patterns are such that there is geometrical separation between the classes, we can often use a set of decision functions to classify an unknown pattern. For example, a case when we have two classes Class1 and Class2 that exist in R over n and a hyperplane $d(x) = 0$ that separates the two classes. Then we can use the decision function $d(x)$ as a linear classifier and classify the new pattern by.

$d(x) > 0 \geq x$ belongs to Class1
$d(x) < 0 \geq x$ belongs to Class2

The hyperplane $d(x) = 0$ is called a decision boundary. When a set of hyperplanes can separate between m given classes in R over n, these classes are linearly separable. Sometimes, we cannot classify elements only by using linear decision functions; in this case, we can either use generalized functions (which are nonlinear) or use a nonlinear classifier to transform the problem to a space of a much higher dimension where classification is done using linear boundaries. Many times, the training pattern of various classes overlap, for example, when they are originated by some statistical distributions (el Halabi, Daou, Achkar, Hayek, & Börcsök, 2019; Siddiqui, Morales-Menendez, Huang, & Hussain, 2020). In this case a statistical approach is needed, especially when the various distribution functions of the classes are known. A statistical classifier must also evaluate the risk associated with every classification, which measures the probability of misclassification. The Bayes classifier based on the Bayes formula from probability theory minimizes the total expected risk. To use the Bayes classifier, one must know beforehand the pattern distribution function for each class. If these distributions are unknown, they must at least be approximated using the training patterns. There are times when classification is performed with some degree of uncertainty. Either the classification outcome itself may be in doubt or the classified pattern x may belong in some degree to more than one class. For example, a medium apple cannot be considered to fully belong to class "big," although at the same time it cannot be fully accepted in the class "small" (provided only these two classes exist). So, we naturally introduce fuzzy classification where a pattern is a member of every class with some degree of membership between 0 and 1.

Unlike the previous approaches, the syntactic pattern recognition utilizes the structure of the patterns. Instead of carrying an analysis based strictly on quantitative characteristics of the pattern, we emphasize the interrelationships between the primitives, the components that compare the pattern. Usual patterns that are subject to syntactic pattern recognition research are characters, fingerprints, chromosomes, and so on. The neural network approach assumes as other approaches before that a set of training patterns and their correct classification is given. The architecture of the net, which includes input layer, output layer, and hidden layers, may be very complex. It is characterized by a set of weights and activation function that determine how any information (input data) is being transmitted to the output layer. The neural network is trained by training patterns and adjusts the weights until the correct classifications are obtained. It is then used to classify arbitrary unknown patterns. There are several neural net classifiers but the most used and most simple one is the perceptron. A wide variety of deep learning and machine learning networks are proposed for EEG classification. For instance, the work in (el Halabi et al., 2019; Liu, Jiang, & Feng, 2021; Siddiqui et al., 2020; Thara, Premasudha, & Nayak, 2021) proposes a stacked bidirectional LSTM, a feature dimension reduction model with locality preserving projections, black box models, and ANN models for detection of various neurological issues. These models assist in improving accuracy of EEG classification via adding multiple other parameters like accelerometer, brain scans, and more. Standalone systems like the ones described in (Shoeibi, Ghassemi, Khodatars, et al., 2021) propose use of linear classification models with various feature extraction units. Similarly, models in (Billeci et al., 2019; Fu, Tian, Shi, et al., 2020; Galanopoulou, Ferastraoaru, Correa, et al., 2020; Pernice et al., 2020) propose use of sparse common space pattern, Fisher's linear discriminant analysis (LDA), support vector machine (SVM), thresholding model, and epileptiform discharge study models for detection of neurological disorders in COVID and other patient types. These models are extended using (Chowdhury, Hossain, Fattah, & Shahnaz, 2019; Halabi, El, Zeid Daou, Achkar, & Boercsoek, 2020; Zhang et al., 2020) wherein decision trees (DTs), discriminant analysis (DA), SVM, wavelet neural network, and 1D CNN are proposed. These models aim at large-scale feature extraction for improving efficiency of EEG classification. Moreover, models proposed in (Forooghifar, Aminifar, & Cammoun, 2019; Meisel et al., 2020; Müller, Caporro, & Gast, 2020) utilize wristband-based sensors, self-aware monitoring, and nonlinear models for improving efficiency of EEG classification.

Based on these models, it can be observed that deep learning models have good performance but require large delays while classifying EEG signals. To reduce the probability of this drawback, we propose a novel EEG classification model next.

References	Machine learning techniques	Associated methods
Valliani, Ranti, and Oermann (2019) McBee et al. (2018) Bartels, Shi, and Lu (2010) Chin-Teng, Yu-Kai, and Shi-An (2012) Gao, Yang, Lin, Wang, and Zheng (2010) Halder et al. (2007) Hsu, Lin, Hsu, Chen, and Chen (2012) Lawhern, Hairston, McDowell, Westerfield, and Robbins (2012) Lawhern, Hairston, and Robbins (2013)	Support vector machine	ICA, BSS, autoregressive model
O'Regan, Faul, and Marnane (2013) Phothisonothai et al. (2012)	Artificial neural network	ICA, spectral analysis
Chin-Teng et al. (2012) Gao et al. (2010)		
Sovierzoski et al. (2009)		
	Fuzzy inference system	Differential devolution Adaptive noise cancellation
Nicolaou and Nasuto (2007)	Clustering	Kurtosis
Gao et al. (2010)	K-NN	Polynomial fitting, Hjort descriptor
	Bayesian Model Genetic programming	Spectral power Power spectral analysis, kurtosis

Proposed incremental variance learning-based ensemble classification model for neurological disorders

As observed from the literature survey, classification of EEG signals into neurological conditions has been done in multiple ways. These include deep learning models, transfer learning models, linear classification models, and so on. Each of these models require large amounts of data for training and validating the underlying classifier. To overcome this issue, we propose a novel incremental variance learning-based ensemble classification model for neurological disorders. Fig. 1 shows the architecture of the proposed model, including data flow from the collected EEG dataset and final classification and incremental learning. The model uses a tagged EEG dataset with different neurological disease classes including acute spinal cord injury, amyotrophic lateral sclerosis, Alzheimer's disease, and epilepsy. The EEG signal is given to a feature extraction block, wherein wavelet, cosine, and Fourier transforms are applied. These transforms evaluate spatial, temporal, and spectral features of the input EEG waveform.

Discrete wavelet transform

As the traditional method of Fourier transform is used very often in analysis of sound of the bird, it proved to be less efficient due to its trade off between temporal resolution and frequency resolution. Wavelet transform can be an alternative solution to this problem, which is a comparatively recent advancement in digital signal processing by (Graps, 1995) although it has been found to have been invented separately in diverse fields of mathematics, quantum analysis and in electrical engineering (Patil, Chandel, & Gupta, 2014). Wavelets have been applied in many areas, such as compression in time series data, filtering of noise from data, and feature detection (Burrus, Gopinath, & Guo, 1997). The representation of the signal in the Fourier transform is mapped into a fundamental wave of sine and cosine. The wavelet transform also utilizes a logic, that is, elements are scale-invariant defined, as clearly understood that the basis seems to be the same at all scales, and the basis are space localized. The outcome is that in the wavelet representation, the signal at different resolutions can be seen in different window sizes, as it can be visualized just as a building and its windows at the same instant of time. At a large scale, the group of buildings can be seen, and this can be viewed to obtain the global features. To look for the window of the building, closer focus is needed and to get local features. A closer look can be taken to view hooks on the window. Different scales can be used to view all groups of buildings, a single building or window, and even a hook on a window. The major difference between Fourier and wavelet analysis is that the flexible window size is sufficient for wide-spectrum quasi-stationary signals such as EEG data (for low frequencies, large windows are used for low frequencies and small windows are used for higher frequencies). Mother wavelets Ψ_m referred as the basis feature and various choices to be obtained experimentally for particular application. Examples of some mother wavelets include the simplest Haar wavelet, which is a discontinuous step function. One of the disadvantages of discontinuity in few domains like audio signal is not suitable, whereas its advantage lies in random transitions like failure of machine. The apt wavelets are the Daubechies wavelets (db*N*) showing that the smoothness of the wavelets increases as *N* increases (dB) for the EEG data. To construct other elements of the wavelet basis, the mother wavelet is scaled and translated by factors *i* and *j* using:

$$\Psi_{i,j} = \frac{1}{\sqrt{|i|}} \Psi\left(\frac{t-j}{i}\right) \tag{2}$$

The measure of stretching or compression of the mother wavelet is based on parameter $i=0$ (depending on whether *i* is greater than or less than 1). Thus, high-frequency components that are introduced to the wavelet family, as *i* is small as result of wavelets, can capture high frequencies of the signals. In a similar fashion, to obtain low frequency signals, low-frequency components are introduced to the family of wavelets. The amount of shifting of the wavelet along the horizontal axis is determined by parameter *j*. If $j > 1$, the wavelet shifts to the right, and if $j < 1$, the wavelet shifts to the left. As a result, the onset of that wavelet specifies parameter *j*. Subsequently, the scaled wavelet is defined as daughter wavelets, whereas main wavelets are called as the wavelet function (mother wavelet) and function of scaling (also called the father wavelet).

Wavelet Packet Decomposition: Wavelet-based features are also used in environmental sound event detection in (Kandel, Schwartz, & Jessell, 2000). First, wavelet packet decomposition trees of each signal are derived. Then, features such as spectral centroid, sparsity, node energy, and spectral spread are extracted from the child nodes of the wavelet tree. The application of wavelets to the digital signal gives rise to separating the data into an approximation (high frequency) part and a detail (low frequency) part of signal into a matrix. Due to this, wavelets can be used as low-pass and high-pass filters. Analyses of this filtered segment can be performed by the wavelet again with a scale with shorter value (typically half of the scale), giving rise to daughter wavelet. Usually, the approximation parts of the signal contain actual information, which is why this part must be analyzed again instead of both the detail part and approximation part (coefficients). However, this cannot take place separately, so another technique can be used, like wavelet packet decomposition (Fagerlund & Laine, 2014). This method produces a tree of wavelet decompositions, where there are M levels at the tree, which is again starting with head at 2M that produces a rich spectral analysis. The levels of decomposition are measured with the requirement of the application of wavelet packet decomposition. The method is applied to the bird sound for analysis again even after filtering for segmentation. The segmentation and feature extraction of the signal after filtering the sound can be achieved by this approach to determine the approximation part at each section. The experiment or any convention through which can be estimated is by using information theory. The information contained in the approximation part of the signal as well as the detail parts contains noise that can be removed. In the area of information theory, the amount of uncertainty or disorder in a system is defined as Shannon entropy (Mohanty & Pani, 2020). The amount of information retained in the provided signal is due to this Shannon entropy (Mohanty, Solanki, Mallick, & Pani, 2020). The concept of entropy for using a wavelet to evaluate accuracy at an optimum level for the use of selective wavelet is high (Dash, 2014; Dash & Behera, 2016; Sen & Dash, 2014, 2015). The system like we use this computation at each node to choose whether to retain a node and stopped creating the tree up to the point where all of the nodes contained noise were removed by this computation, meaning that the signal was fully described (Figs. 2 and 3).

The temporal features are evaluated using discrete wavelet transform (DWT) block, which uses the following Eq. (3):

$$DWT_i = \sum_{i=1}^{N} \frac{(EEG_i + EEG_{i+1})}{2} \tag{3}$$

Similarly, the spatial components are evaluated via discrete cosine transform (DCT) using Eq. (4) as follows:

$$DCT_i = \frac{1}{N} * \sum_{i=1}^{N} EEG_i * \cos\left(2 * pi * \frac{i}{N}\right) \tag{4}$$

While frequency components are evaluated using discrete frequency transform (DFT) using Eq. (5) as follows:

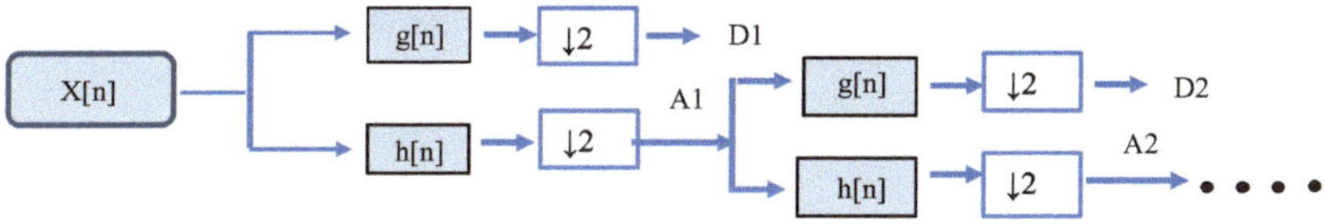

Fig. 2 Discrete wavelet transform sub-band decomposition.

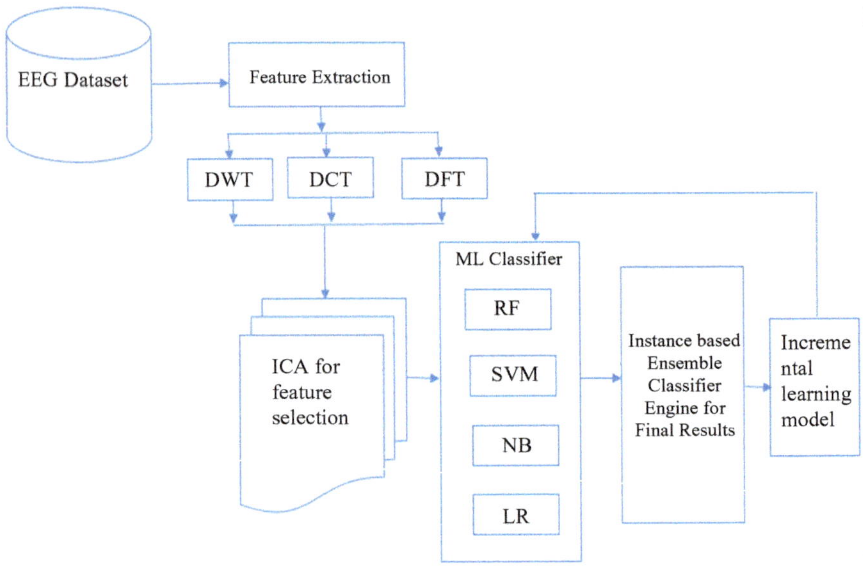

Fig. 3 Proposed EEG classification model.

$$DFT_l = \sum_{i=1}^{N} EEG_i * \exp\left(-2 * pi * j * l * \frac{i}{N}\right) \tag{5}$$

where EEG and N represent input EEG signal and its length, respectively.

These signals are given to an independent component analysis (ICA) block, wherein natural gradient descent algorithm is used. This algorithm aims at maximizing weight of the extracted features, and the feature weight at each iteration is evaluated using Eq. (6) as follows:

$$W_{j+1} = W_j - \varnothing * \frac{\partial_j(W_j)}{\partial(W_j)} * W_j^T * W_j \tag{6}$$

where $W, \varnothing$, and ∂ represent weight of the EEG features, learning rate, and differential analysis operation, respectively. The weight is evaluated using Eq. (7) as follows:

$$W_j = -P(F_j) * \exp\left(\frac{1}{P(F_j)}\right) \tag{7}$$

where P and F represent probability and feature value from the extracted feature vector, respectively. These features are combined with their neurological classes to form training and validation sets. The obtained sets might be limited in number and are given to an ensemble classification model for obtaining coarse accuracy. This accuracy is evaluated using the following process:

- Training and validation sets are given to the following classifier sequence:
  - RF containing number of forests = 1000, and depicting number of correct classes as $RF_{correct}$
  - LR with limited-memory Broyden Fletcher Goldfarb Shannon solver depicting number of correct classes as $LR_{correct}$
  - NB classifier with high posterior probability, and number of correctly classified instance as $NB_{correct}$
  - SVM with a value of error tolerance of 0.1%, and number of correctly classified instances as $SVM_{correct}$
- All the correctly classified instances are combined into a single array using Eq. (21.8),

$$Final_{correct} = Unique\left(\bigcup \begin{pmatrix} NB_{correct}, SVM_{correct}, \\ RF_{correct}, LR_{correct} \end{pmatrix}\right) \tag{8}$$

- Validation accuracy is evaluated using value of $Final_{correct}$ and comparing it with ground truth dataset.
- Each classification instance is stored along with the used classifier set for that particular instance
- For example, if instance 1 was correctly classified by NB and RF, then the entry is stored as $I_1 \in (NB, RF)$
- Features are extracted for new EEG signals, and their correlation values are estimated with existing features
- Correlation values are estimated using Eq. (9) as follows:

$$Corr_j = \frac{\sum_{i=1}^{N_{f\,Val}} F_{Val_i} - F_{new_i}}{\sqrt{\sum_{i=1}^{N_{f\,Val}} \left(F_{Val_i} - F_{new_i}\right)^2}} \tag{9}$$

where "j" is classified index, which indicates use of NB, RF, SVM, and LR classifiers, respectively. F_{Val_i} and F_{new_i} are i^{th} validation selected feature set and feature set for the new EEG signal, respectively, and $N_{f_{Val}}$ represents total number of features selected by the ICA model. Instance with maximum value of correlation is used, and its classifiers are evaluated. Input EEG signal is classified using these classifiers, and the classification result is used for incremental learning.

The incremental learning module stores classified EEG signal into the training set if,

$$Corr_{max} > 0.9999$$

And it stores EEG signal into validation set if,

$$Corr_{max} > 0.999$$

Moreover, the models are retrained once more than 10% of data is added either to training or validation sets. This increases overall accuracy of classification as well as speed of operation. The accuracy and delay values are compared on different datasets including PhysioNet, IEEE DataPort, Child Mind Institute, and IMAGENET of The Brain datasets. Comparison of the same format is followed here (Dash, Abraham, & Rehman, 2018; Dash, Patra, & Tripathy, 2012a, 2012b; Dash, Thulasiram, & Thulasiram, 2017; Patra, Dash, & Tripathy, 2013). These results are compared with various state-of-the-art methods, and the evaluation is tabulated in the next section of this text.

Result and comparison

This section provides a detailed description of how the proposed technique is simulated. The experiments were implemented in MATLAB 2015b software. The test platform used is a 6th generation Intel Core i3, 2.2-GHz CPU, and 4-GB RAM processor with a Windows 7 operating system.

Extraction of features: In the proposed system, we used the attribute extrication techniques DCT, DWT, and DFT. In DWT, the approximate coefficient of the attribute vector is 22 in dimension with the use Daubechies 4 and decomposition level of 10 which rendered a moderate result.

To estimate performance of the proposed model, its accuracy and delay values were compared with (Zhang et al., 2020) and (Thara et al., 2021) models for PhysioNet, IEEE DataPort, Child Mind Institute, and IMAGENET of The Brain datasets. These datasets contain tagged EEG signals for amyotrophic lateral sclerosis, Bell's palsy, cerebral aneurysm, Alzheimer's disease, ataxia, brain tumors, epilepsy, and seizures. A total of 10 k EEG samples were used for initial training, and the dataset was incrementally grown for accuracy improvement. Table 1 shows the accuracy performance of the proposed model.

These results show that the proposed model's accuracy increases exponentially with respect to number of testing samples, which is mainly due to the incremental learning process. The proposed model is observed to have more than 94% accuracy for 300 k sample evaluations, which makes it useful for clinical applications. Moreover, the model outperforms (Zhang et al., 2020) by 14%, and (Thara et al., 2021) by 5% when compared on the same datasets, which further assists in validating its real-time performance. Fig. 4 shows this performance, wherein accuracy of the model is visualized w.r.t.

Table 1 Accuracy performance of different models.

Testing set size	Accuracy (%) (Zhang et al., 2020) (R1)	Accuracy (%) (Thara et al., 2021) (R2)	Accuracy (%) (Proposed)
10,000	75.30	81.20	82.37
20,000	75.40	81.50	82.58
30,000	75.43	81.90	82.81
40,000	75.50	82.60	83.21
50,000	75.59	82.90	83.42
60,000	76.20	82.95	83.76
70,000	76.35	83.10	83.92
80,000	76.38	83.15	83.96
90,000	76.56	83.76	84.38
100,000	76.73	84.06	84.63
110,000	76.91	84.35	84.88
120,000	77.08	84.65	85.13
130,000	77.26	84.95	85.37
140,000	77.44	85.25	85.62
210,000	78.66	87.34	92.23
220,000	78.84	87.64	92.49
230,000	79.01	87.94	92.75
240,000	79.19	88.24	93.02
250,000	79.36	88.54	93.28
260,000	79.54	88.84	93.54
270,000	79.71	89.14	93.81
280,000	79.89	89.43	94.07
290,000	80.07	89.73	94.33
300,000	80.24	90.03	94.60

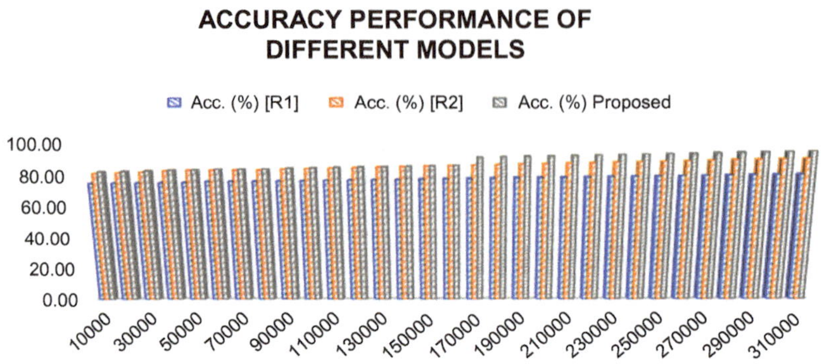

Fig. 4 Accuracy of different models.

number of test samples. It is observed that the proposed model's accuracy performance is initially closer to performance of (Thara et al., 2021), but as number of training samples are increased this performance showcases an exponential rise, which makes the proposed model applicable for large-scale clinical applications.

We conducted a similar evaluation for delay performance. Table 2 shows those results.

Results show that the proposed model's delay increases linearly w.r.t. number of testing samples, which is mainly due to the incremental learning process. The proposed model is observed to have a more than 500-ms delay for 300 k sample evaluations, which makes it useful for high-speed clinical applications. Moreover, the model outperforms (Thara et al., 2021) by 14% and (Zhang et al., 2020) by 5% when compared on the same datasets, which further assists in validating its real-time performance. The work in (Thara et al., 2021) is slower than (Zhang et al., 2020), mainly due to computational complexity. Fig. 5 shows this performance, wherein delay of the model is visualized w.r.t. number of test samples. It is observed that the proposed model's delay performance is always better than (Zhang et al., 2020) and (Thara et al., 2021), but as number of training samples are increased this performance showcases an exponential improvement, which makes the proposed model applicable for large-scale clinical applications.

Based on this performance, the proposed model is useful for high-speed and high-scalability EEG classification.

Table 2 Delay performance of proposed model.

Testing set size	Delay (ms) (Zhang et al., 2020) (R1)	Delay (ms) (Thara et al., 2021) (R2)	Delay (ms) (Proposed)
10,000	37.65	54.13	15.30
20,000	75.40	108.67	30.68
30,000	113.15	163.80	46.16
40,000	151.00	220.27	61.88
50,000	188.98	276.33	77.55
60,000	228.60	331.80	93.40
70,000	267.23	387.80	109.17
80,000	305.53	443.47	124.83
90,000	344.51	502.54	141.18
100,000	383.67	560.37	157.34
110,000	423.00	618.60	173.60
120,000	462.51	677.23	189.96
130,000	502.19	736.25	206.41
210,000	825.96	1222.80	341.46
220,000	867.22	1285.41	358.77
230,000	908.65	1348.42	376.18
240,000	950.26	1411.83	393.68
250,000	992.05	1475.63	411.28
260,000	1034.01	1539.84	428.98
270,000	1076.15	1604.44	446.77
280,000	1118.46	1669.44	464.65
290,000	1160.95	1734.84	482.63
300,000	1203.61	1800.64	500.71

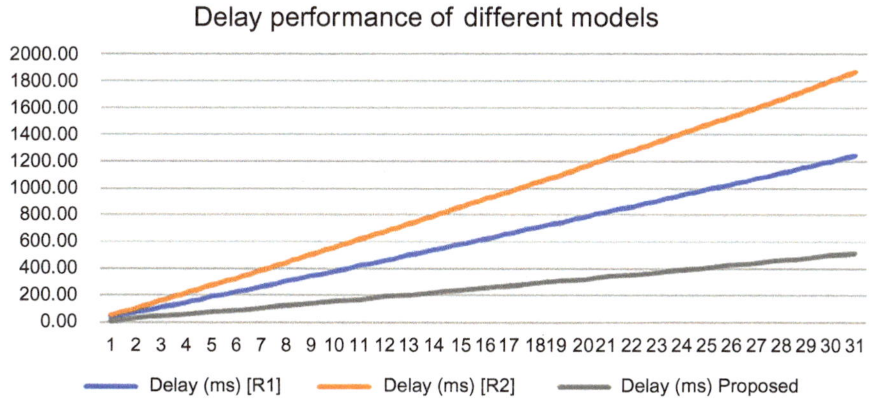

Fig. 5 Delay of different models.

Conclusion and future scope

Due to use of incremental learning and ensemble classification, the proposed model has better performance in terms of accuracy and delay evaluation. From the accuracy performance, it was observed that the proposed model has exponentially increasing values w.r.t. number of testing samples. Again, this is mainly due to the incremental learning and ensemble process. The proposed model is observed to have more than 94% accuracy for 300 k samples, which makes it useful for clinical applications. The model also outperforms (Thara et al., 2021) by 5%, and (Zhang et al., 2020) by 14%, when compared on the same datasets. This further assists in validating its real-time performance for clinical applications. Furthermore, from the delay performance it can be observed that the proposed model's delay increases linearly w.r.t. number of testing samples, which is mainly due to the incremental learning process. The proposed model is observed to have a more than 500-ms delay for 300 k sample evaluations, which makes it useful for high-speed clinical applications. In the future, transfer learning models like Q-learning, RNN, auto-encoder (AE), and generative adversarial network (GAN) can be used to further improve performance of EEG classification. The performance can be evaluated on larger datasets as well.

References

Bartels, G., Shi, L. C., & Lu, B. L. (2010). Automatic artifact removal from EEG—A mixed approach based on double blind source separation and support vector machine. In *Conference proceedings: Annual international conference of the IEEE engineering in medicine and biology society. IEEE engineering in medicine and biology society. Conference* (pp. 5383–5386).

Billeci, L., Tonacci, A., Varanini, M., Detti, P., de Lara, G. Z. M., & Vatti, G. (2019). Epileptic seizures prediction based on the combination of EEG and ECG for the application in a wearable device. In *2019 IEEE 23rd international symposium on consumer technologies (ISCT)* (pp. 28–33). https://doi.org/10.1109/ISCE.2019.8900998.

Burrus, C. S., Gopinath, R. A., & Guo, H. (1997). Generalizations of the basic multiresolution wavelet system. In *Introduction to wavelets and wavelet transforms: A primer* (pp. 98–145). Prentice-Hall, Inc.

Chin-Teng, L., Yu-Kai, W., & Shi-An, C. (2012). A hierarchal classifier for identifying independent components. In *Paper presented at the neural networks (IJCNN), The 2012 international joint conference on The 2012 international joint conference on neural networks (IJCNN)* (pp. 1–9).

Chowdhury, T. T., Hossain, A., Fattah, S. A., & Shahnaz, C. (2019). Seizure and non-seizure EEG signals detection using 1-D convolutional neural network architecture of deep learning algorithm. In *2019 1st international conference on advances in science, engineering and robotics technology (ICASERT)* (pp. 1–4). https://doi.org/10.1109/ICASERT.2019.8934564.

Dash, S. (2014). Hill-climber based Fuzzy-Rough feature extraction with an application to cancer classification. *Journal of Network and Innovative Computing*, *2*(2014), 178–186. ISSN 2160-2174.

Dash, S., Abraham, A., & Rehman, A. (2018). Kernel based chaotic firefly algorithm for diagnosing Parkinson's disease. In *18th international conference on hybrid intelligent systems (HIS 2018), advances in intelligent systems and computing, Porto, Portugal* (pp. 176–188).

Dash, S., & Behera, R. (2016). Sampling based hybrid algorithms for imbalanced data classification. *International Journal of Hybrid-Intelligent-Systems*, *13*(2016), 77–86. https://doi.org/10.3233/HIS-160226. IOS Press.

Dash, S., Patra, B., & Tripathy, B. K. (2012a). Study of classification accuracy of microarray data for cancer classification using multivariate and hybrid feature selection method. *IOSR Journal of Engineering*, *2*(8), 112–119. ISSN:2250–3021.

Dash, S., Patra, B., & Tripathy, B. K. (2012b). A hybrid data mining technique for improving the classification accuracy of microarray data set. *International Journal of Information Engineering and Electronic Business (IJIEEB)*, *2012*(2), 43–50.

Dash, S., Thulasiram, R., & Thulasiram, P. (2017). An enhanced Chaos-based firefly model for Parkinson's disease diagnosis and classification. In *IEEE ICIT conference, Bhubaneswar, Odisha* (pp. 159–164). Published in IEEE Xplore. https://doi.org/10.1109/ICIT.2017.43.

el Halabi, N., Daou, R. A. Z., Achkar, R., Hayek, A., & Börcsök, J. (2019). Monitoring system for prediction and detection of epilepsy seizure. In *2019 fourth international conference on advances in computational tools for engineering applications (ACTEA)* (pp. 1–7). https://doi.org/10.1109/ACTEA.2019.8851094.

Fagerlund, S., & Laine, U. K. (2014). Classification of audio events using permutation transformation. *Applied Acoustics*, *83*, 57–63.

Forooghifar, F., Aminifar, A., & Cammoun, L. (2019). A self-aware epilepsy monitoring system for real-time epileptic seizure detection. *Mobile Networks and Applications*, 1–5. https://doi.org/10.1007/s11036-019-01322-7.

Fu, R., Tian, Y., Shi, P., et al. (2020). Automatic detection of epileptic seizures in EEG using sparse CSP and fisher linear discrimination analysis algorithm. *Journal of Medical Systems*, *44*(2), 43. https://doi.org/10.1007/s10916-019-1504-1.

Galanopoulou, A. S., Ferastraoaru, V., Correa, D. J., et al. (2020). EEG findings in acutely ill patients investigated for SARS-CoV-2/COVID-19: A small case series preliminary report (published correction appears in Epilepsia Open. 2021 Jun; 6(2):455). *Epilepsia Open*, *5*(2), 314–324. https://doi.org/10.1002/epi4.12399.

Gao, J. F., Yang, Y., Lin, P., Wang, P., & Zheng, C. X. (2010). Automatic removal of-wake-up-call/eye-movement and blink artifacts from EEG signals. *Brain Topography*, *23*(1), 105–114. https://doi.org/10.1007/s10548-009-0131-4.

Graps, A. (1995). An introduction to wavelets. *IEEE Computational Science and Engineering*, *2*(2), 50–61.

Halabi, N., El, R. A., Zeid Daou, R., Achkar, A. H., & Boercsoek, J. (2020). Comparative study for classification methods to predict and detect epilepsy seizure. In *2020 IEEE 5th middle east and africa conference on biomedical engineering (MECBME)* (pp. 1–6). https://doi.org/10.1109/MECBME47393.2020.9265151.

Halder, S., Bensch, M., Mellinger, J., Bogdan, M., Kubler, A., Birbaumer, N., et al. (2007). Online artifact removal for brain-computer interfaces using support vector machines and blind source separation. *Computational Intelligence and Neuroscience*, 82069. 10 pages https://doi.org/10.1155/2007/82069.

Hsu, W.-Y., Lin, C.-H., Hsu, H.-J., Chen, P.-H., & Chen, I. R. (2012). Wavelet based envelope features with automatic EOG artifact removal: Application to single-trial EEG data. *Expert Systems with Applications*, *39*(3), 2743–2749.

Kandel, E. R., Schwartz, J. H., & Jessell, T. M. (2000). *Principles of neural science*. New York, NY, USA: McGraw-Hill.

Lawhern, V., Hairston, W. D., McDowell, K., Westerfield, M., & Robbins, K. (2012). Detection and classification of subject-generated artifacts in EEG signals using autoregressive models. *Journal of Neuroscience Methods*, *208*(2), 181–189.

Lawhern, V., Hairston, W. D., & Robbins, K. (2013). DETECT: A MATLAB toolbox for event detection and identification in time series, with applications to artifact detection in EEG signals. *PLoS One*, *8*(4). https://doi.org/10.1371/journal.pone.0062944, e62944.

Liu, Y., Jiang, B., & Feng, J. (2021). Classification of EEG signals for epileptic seizures using feature dimension reduction algorithm based on LPP. *Multimedia Tools and Applications*, *80*, 30261–30282. https://doi.org/10.1007/s11042-020-09135-7.

McBee, M. P., Awan, O. A., Colucci, A. T., Ghobadi, C. W., Kadom, N., Kansagra, A. P., et al. (2018). Deep learning in radiology. *Academic Radiology*, *25*, 1472–1480.

Meisel, C., El Atrache, R., Jackson, M., Schubach, S., Ufongene, C., & Loddenkemper, T. (2020). Machine learning from wristband sensor data for wearable, noninvasive seizure forecasting. *Epilepsia*, *61*, 2653–2666. https://doi.org/10.1111/epi.16719.

Mohanty, R., & Pani, S. K. (2020). Livestock health monitoring using a smart IoT enabled neural network recognition system. In *Bio-inspired neurocomputing* (pp. 131–149).

Mohanty, R., Solanki, S. S., Mallick, P. K., & Pani, S. K. (2020). A classification model based on an adaptive neuro-fuzzy inference system for disease prediction. In *903. Cognmitive big data intelligence with a meta-heuristic approach* (pp. 131–149).

Müller, M., Caporro, M., & Gast, H. (2020). Linear and nonlinear interrelations show fundamentally distinct network structure in preictal intracranial EEG of epilepsy patients. *Human Brain Mapping*, *41*, 467–483. https://doi.org/10.1002/hbm.24816.

Nicolaou, N., & Nasuto, S. J. (2007). Automatic artefact removal from event-related potentials via clustering. *Journal of VLSI Signal Processing Systems for Signal, Image, and Video Technology*, *48*, 173–183. In press.

O'Regan, S., Faul, S., & Marnane, W. (2013). Automatic detection of EEG artefacts arising from head movements using EEG and gyroscope signals. *Medical Engineering & Physics*, *35*(7), 867–874.

Patil, S., Chandel, G. S., & Gupta, R. (2014). Performance analysis of steganography based on 5-wavelet families by 4 levels-DWT. *International Journal of Computer Science and Network Security (IJCSNS)*, *14*(12), 56–61.

Patra, B. N., Dash, S., & Tripathy, B. K. (2013). Neural techniques for improving the classification accuracy of microarray data set using rough set feature selection method. *International Journal of Computer Trends and Technology*, *4*(3), 424–429. ISSN: 2231–2803.

Pernice, R., et al. (2020). Synergistic and redundant brain-heart information in patients with focal epilepsy. In *2020 11th conference of the European study group on cardiovascular oscillations (ESGCO)* (pp. 1–2). https://doi.org/10.1109/ESGCO49734.2020.9158196.

Phothisonothai, M., et al. (2012). Artifactual component classification from MEG data using support vector machine. In *Biomedical engineering international conference (BMEiCON), 2012*. https://doi.org/10.1109/BMEiCon.2012.6465462.

Sen, S., & Dash, S. (2014). Application of meta learning algorithms for the prediction of diabetes disease. *International Journal of Advance Research in Computer Science and Management Studies*, *2*(12), 394–401. ISSN: 2321-7782.

Sen, S., & Dash, S. (2015). Empirical evaluation of classifiers' performance using data mining algorithm. *International Journal of Computer Trends and Technology (IJCTT)*, *21*(Number 3), 2015.

Shoeibi, A., Ghassemi, N., Khodatars, M., et al. (2021). Epileptic seizure detection using deep learning techniques: A review. *International Journal of Environmental Research and Public Health*, *18*(11), 5780–5813.

Siddiqui, M. K., Morales-Menendez, R., Huang, X., & Hussain, N. (2020). A review of epileptic seizure detection using machine learning classifiers. *Brain Inform*, 7(1), 1–5. https://doi.org/10.1186/s40708-020-00105-1.

Sovierzoski, M. A., et al. (2009). Binary neural classifier of raw EEG data to separate spike and sharp wave of the eye blink artifact. In *2009 Fifth international conference on natural computation, in press*.

Thara, D. K., Premasudha, B. G., & Nayak, R. S. (2021). Electroencephalogram for epileptic seizure detection using stacked bidirectional LSTM_GAP neural network. *Evolutionary Intelligence*, *14*, 823–833. https://doi.org/10.1007/s12065-020-00459-9.

Valliani, A., Ranti, D., & Oermann, E. K. (2019). Deep learning and neurology: A systematic review. *Neurology and Therapy*, *8*, 351–365.

Zhang, Z., Ren, Y., Sabor, N., Pan, J., Luo, X., Li, Y., et al. (2020). DWT-net: Seizure detection system with structured EEG montage and multiple feature extractor in convolution neural network. *Journal of Sensors*, 1–13. https://doi.org/10.1155/2020/3083910. Article ID 3083910.

CHAPTER 22

A systematic review of adaptive machine learning techniques for early detection of Parkinson's disease

Sujata Dash
Department of Computer Application, Maharaja Sriram Chandra Bhanja Deo University, Baripada, Odisha, India

Introduction

Parkinson's disease (PD) is an important degenerative disease of the central nervous system, affecting the quality of life of millions of seniors worldwide (Mohamad, 2021). PD is a chronic and progressive neurodegenerative disease (Burke & O'Malley, 2013) that generally affects people older than 50 years (Ranjan & Swetapadma, 2018). Its symptoms begin slowly and intensify over time. Parkinson's patients may show symptoms such as tremors when they are at rest. These tremors can affect the hands or legs, resulting in limb rigidity and gait and balance problems. PD symptoms are categorized into two types: motor and nonmotor (Gunduz, 2019). Motor symptoms include movement disorders, shaking, walking issues (Viz Hub, 2021), stiffness, and postural instability (Quan, Ren, & Luo, 2021), whereas nonmotor symptoms include cognitive dysfunction, mood disorder (Rana, Ahmed, Chaudry, & Vasan, 2015), depression, and anxiety (Sveinbjornsdottir, 2016). PD is the second most common neurodegenerative disease worldwide after Alzheimer's disease. In 2019, its incident rate ranged from 40.37 to 53.89 per 100,000 population per year in the United States alone (NHS, 2021). The number of people affected by PD has exceeded 10 million worldwide. Diagnosis of PD at an early stage is integral to mitigating its complications. However, no conclusive medical test is available for early diagnosis. It should be noted that the timely detection of PD facilitates rapid treatment and alleviates symptoms significantly, as reported in Burke and O'Malley (2013). Therefore, detection of PD at an earlier stage is certainly a key element to slowing its progression and giving patients the

Artificial Intelligence for Neurological Disorders
https://doi.org/10.1016/B978-0-323-90277-9.00018-3

opportunity to access disease-modifying therapy. To date, there is no way to diagnose PD (Burke & O'Malley, 2013). However, there are various symptom and diagnostic tests that can be used in combination. Several biomarkers have been investigated by scientists for early identification of PD to. Various methods based on different kinds of measurements including speech data (Quan et al., 2021; Ranjan & Swetapadma, 2018), gait patterns (Rana et al., 2015), force tracking data (Sveinbjornsdottir, 2016), smell identification data (NHS, 2021), and spontaneous cardiovascular oscillations (Singh, Pillay, & Choonara, 2007) have been proposed to expedite the detection of PD. Speech disorders in PD patients can be recorded via smartphone and assessed by a sawtooth waveform inspired pitch estimator (SWIPE) scheme (Illner, Sovka, & Rusz, 2020) for discriminating PD patients from healthy patients. However, at a lower signal-to-noise ratio level, an improved algorithm is needed to obtain efficient robustness to noise. Solana-Lavalle, Galán-Hernández, and Rosas-Romero (2020) developed an early detection algorithm for PD based on reduced vocal features. It has been shown that the wrappers subset selection is suitable because of the low dimensionality of the selected feature and improved PD detection capability. El Maachi, Bilodeau, and Bouachir (2020) introduced a PD detection system using a 1D convolutional neural network (CNN) based on gait signals. However, the performance of PD detection based on both speech and gait analyses is generally limited by the sensitivity to background noise in speech recording, causing a large number of false alarms and missed detections. In addition, gait tracking and inspection require specialized devices and sufficient space for walking (Gupta, Bansal, & Joshi, 2020). The authors of Wagner, Fixler, and Resheff (2017) suggested a method based on wavelet to analyze data collected from smartwatches worn by 19 patients affected by PD. This method showed good ability in detecting symptoms of tremor, bradykinesia, and dyskinesia. Arroyo-Gallego et al. (2017) introduced an approach to detect motor impairment in PD based on mobile touchscreen typing. Essentially, the proposed algorithm uncovers signs of PD motor by analyzing touchscreen typing features that include descriptive statistics such as covariance, skewness, and kurtosis and time information. Dinov et al. (2016) incorporated multi-source data including imaging, genetics, and clinical and demographic information in developing models for PD prediction. Other approaches employed handwriting measurement for PD (Drotár et al., 2015; Drotár et al., 2016; Gupta et al., 2020). Gupta et al. (2020) proposed a PD diagnosis approach based on handwriting measurements gathered from patients with PD. It has been shown that improved PD diagnosis is obtained when age and sex information are taking into consideration in the decision process (El Maachi et al., 2020; Gupta et al., 2020).

Accurate and early detection of PD is vital due to its ability to provide crucial information to slow the progression of the disease. Over the years, various data-driven methods have been developed to improve the detection of PD. In contrast to model-based detection techniques, where prior availability of an analytical model is required, data-driven techniques require only the availability of historical data.

Feature engineering for identifying clinical biomarkers

Feature engineering is the difficult part of deploying such systems, as it is expensive to identify the relevant features in the data. When automatic feature extraction methods and techniques (e.g., CNN, RNN) were proposed, most researchers used deep learning (DL) and neural networks to detect PD due to automatic feature extraction, ability to learn more complex patterns, and high accuracy.

This chapter comprehensively summarizes literature in which machine learning (ML) models were applied for the diagnosis of PD. It discusses data sources, data types, machine learning models, and associated outcomes. The application of ML to clinical and nonclinical data of different modalities has often led to high diagnostic accuracies in human participants. As such, the adaptation of ML algorithms and novel biomarkers in clinical settings can assist in more accurate and informed decision making.

Many functional learning algorithms have been used by researchers for diagnosing PD patients using dysphonic measurements as the symptoms. A Gaussian radial basis kernel (Shahbakhi, Far, & Tahami, 2014) used as the learning function for support vector machine (SVM) for predicting PD dataset after applying search model to reduce the size of search space. Hence, four different dysphonic features, namely, Recurrence Period Density Entropy, Harmonics-to-Noise Ratio, Pitch Period Entropy, and Detrended Fluctuation Analysis, were selected to identify PD patients. For early detection of PD, a hybrid instance-based learning model has been proposed (Cai et al., 2018) by combining a chaos-based bacterial foraging optimization (CBFO) with a fuzzy-based k-nearest neighbor (FKNN). The model showed better performance than other optimization methods when simulated with vocal measurements of PD patients. A hybrid model combining an enhanced chaos-based firefly algorithm (ECFA) with radial basis function (RBF) kernel-based SVM (Dash, Thulasiram, &

Thulasiraman, 2017) developed an efficient model by predicting discriminant speech patterns from a PD dataset and helped to develop telediagnosis and telemonitoring models. An efficient classification model is designed in Dash and Behera (2016) by identifying relevant features from a microarray dataset employing a hybrid model by combining a meta search method based on information theory with a chaotic firefly algorithm (CFA). The experimental outcomes of the research prove the quality of the hybrid model. It is observed so far from the literature that a naïve Bayes algorithm can predict the binding residues of DNA/RNA, whereas Isler, Narin, Ozer, and Perc (2019) reported a novel kernel density estimation (KDE)-based naïve Bayes algorithm that can also predict the residues for binding proteins in protein sequences. An adaptive block-wise naïve Bayes kernel machine model that operates in multistages to improve the estimation of genomic biomarkers in a diseased dataset is discussed in Narin, Isler, Ozer, and Perc (2018).

Population-based metaheuristics for biomarker selection

Smart algorithms and optimization techniques have been used for a long time to diagnose diseases. Recent examples of research (Helbing et al., 2015) include adaptive neural networks for diagnosing diabetes, multi-stage classification of congestive heart failure based on short-term heart rate variability, (Isler et al., 2019) and early prediction of paroxysmal atrial fibrillation based on short-term heart rate variability (Narin et al., 2018). Recently, metaheuristic search algorithms have been used efficiently for solving optimization problems of different domains. These metaheuristic algorithms mimic the characteristics of living and nonliving things and are not dependent on the characteristics of the given optimization problem. This helps to broaden (Dos Santos Coelho, de Andrade Bernert, & Mariani, 2011) the field applications. There are two types of metaheuristic search algorithms used for feature selection process: single solution-based metaheuristics (SBM) and population-based metaheuristics (PBM). The difference between these two methods is that the former uses a single solution for manipulation during the search, whereas the latter uses a population of solutions for searching operation. Hill climbing, Tabu search, and simulated annealing are some of the representative examples of SBM (Goldberg, 1989) which suffer local optimum. Unlike SBM methods, PBM methods adopt an iterative improvement of the population of solutions. Some examples of PBM algorithms are genetic algorithm (GA) (Goldberg, 1989), particle swarm optimization (PSO) (Kennedy & Eberhart, 1995), differential evolution (DE) (Storn & Price, 1997), and bat algorithm (BA)

(Sharma, Luhach, & Jyoti, 2016). All these algorithms start with an initial population of random solutions at the beginning of the computation and then use iterations for subsequent evaluations.

Amongst these algorithms, the most important one in the family of swarm intelligence is FA, which has been used in several areas of approximation problems including engineering practice (Dos Santos Coelho et al., 2011; Dos Santos Coelho & Mariani, 2012; Yang, 2011). The FA is conceived on the idealized behavior of the flashing characteristic of fireflies. The literature study shows that (Yang, 2009) FA has surpassed the GA and PSO. The two important characteristics of a metaheuristic search algorithm that play a crucial role in achieving global optima are exploitation and exploration. Many researchers have proposed several methods (Brajević & Ignjatović, 2018; Xiao, Liu, Guo, & Zhou, 2017) to equalize them to improve the performance of metaheuristic algorithms. Recently, a paradigm of mathematics known as "chaos theory" was combined with the domain of stochastic optimization algorithms to increase efficiency. The three dynamic characteristics of chaos—(Dash, Abraham, & Atta-Ur-Rahman, 2018; Fister, Perc, Kamal, & Fister, 2015) quasi-stochastic property, sensitivity toward initial conditions, and ergodicity—are the key factors in stochastic optimization algorithms.

The random parameters of the probabilistic algorithms are replaced by chaotic maps (Fister et al., 2015; Mingjuan, Su, Su, & Xu, 2014). Different types of chaotic maps are used for tuning the attractiveness parameter β in the FA to enhance the convergence rate and accuracy, and a chaos-enhanced FA (Mingjuan et al., 2014) is introduced to automate the tuning of parameters. Yang (2011) developed a chaos-based FA applying a logistic map for attractiveness and absorption coefficient in place of Gaussian or Lévy flight distributed random variables and applied it to a global optimization problem. A novel metaheuristic algorithm, the chaotic crow search algorithm (CCSA), was developed by Sayed, Hassanien, and Azar (2017) to optimize feature selection problems. The CCSA employed 10 different types of chaotic maps to enhance (Askarzadeh, 2016) the classification performance and identify a reduced feature set. Dos Santos Coelho et al. (2011) (Yang, 2011) developed a modified FA by combining a chaotic map to solve reliability and redundancy-based optimization problems. In this case, the logistic map improves the quality of the absorption coefficient as well as the randomized parameter and moreover shows outstanding performance over other optimization techniques like dynamic programming, integer programming, and mixed-integer programming. A Tinkerbell chaotic map (Dos Santos Coelho & Mariani, 2012) was combined with the chaotic firefly and was tested against a multi-loop proportional-integral-derivative (PID) controller with promising results. The hybrid model was also compared with GA, PSO, standard FA, and modified FA. The randomizing and attractiveness parameters were enhanced with tent chaotic map (Dash, Abraham, Luhach, Mizera-Pietraszko, & Rodrigues, 2020) for solving the economic load dispatch application, and the reported results showed good convergence characteristics on test cases when compared with other soft computing techniques reported in the literature. In Gandomi, Yang, Talatahari, and Alavi (2013), the attractiveness and absorption coefficient parameters of chaos-based FA were enhanced with chaotic maps of 12 different types and applied to a global optimization problem. The result was promising and showed that some of the chaotic maps have surpassed the results of the standard FA. Similarly, a metaheuristic CFA (Kazem, Sharifi, Hussain, Saberi, & Hussain, 2013) was proposed for finding optimal solutions for support vector regression (SVR) parameters that was used to forecast the pricing of stock market considering all three parameters of FA, that is, randomized parameter, attractiveness, and absorption coefficient, which were augmented with a logistic chaotic map. The efficiency of the algorithm proved to be better than that of chaotic genetic algorithm-based SVR (SVR-GA), firefly-based SVR (SVR-FA), ANNs, and adaptive neuro-fuzzy inference systems (ANFIS). In addition, the components of FA responsible for the movement of FA, that is, attractiveness and absorption coefficient, were augmented with a sinusoidal chaotic map (Abdel-Raouf, Abdel-Baset, & El-henawy, 2014) for a parallel calculating numerical integration in an engineering problem. The simulation results showed a high convergence rate, high accuracy, and robustness of the proposed CFA. In Yang (2014) various probability distributions like uniform, Gaussian, and Lévy flights as well as logistic and Kent chaotic maps were used to develop a randomized FA, wherein the randomized parameters were enhanced with the probability distributions and chaotic maps and produced a promising solution. CFA (Kaveh, Mahdipour Moghanni, & Javadi, 2019) was adopted for optimization of skeletal structure design.

Application of machine learning methods for diagnosing PD

ML techniques are being increasingly applied in the healthcare sector. As its name implies, ML allows for a computer program to learn and extract meaningful representation from data in a semi-automatic manner. Recently, ML has emerged as a promising field of research in PD diagnosis, both in academia and industry (Pereira et al., 2018). Owing to its data-driven approaches, ML has brought a paradigm shift in the way relevant information in PD biomarkers are extracted and analyzed. Furthermore, ML techniques provide pertinent information that offers guidance related to PD classification and diagnosis to speed up decision making. Various ML techniques have been applied in the literature to address the PD detection problem. For the diagnosis of PD, ML models have been applied to a multitude of data modalities, including handwritten patterns (Drotár et al., 2015; Pereira et al., 2018) movement (Pham & Yan, 2018), neuroimaging (Choi, Ha, Im, Paek, & Lee, 2017; Segovia, Górriz, Ramírez, Martínez-Murcia, & Castillo-Barnes, 2019), voice (Ma, Ouyang, Chen, & Zhao, 2014; Sakar et al., 2013), cerebrospinal fluid (CSF) (Maass et al., 2020), cardiac scintigraphy (Nuvoli et al., 2020), serum (Váradi, Nehéz, Hornyák, Viskolcz, & Bones, 2019), and optical coherence tomography (OCT) (Nunes et al., 2019). ML also allows for combining different modalities, such as magnetic resonance imaging (MRI) and single-photon emission computed tomography (SPECT) data (Wang et al., 2017) in the diagnosis of PD. By using ML approaches, we may therefore identify relevant features that are not traditionally used in the clinical diagnosis of PD and rely on these alternative measures to detect PD in preclinical stages or atypical forms. For instance, in Pham and Yan (2018), dysphonia measurements have been used to detect patients with PD from healthy people. SVM is applied to only four dysphonic features for PD classification due to its ability to extract nonlinearity by using nonlinear kernels. In Quan et al. (2021), three common ML algorithms, namely Random Forest (RF) or SVM and neural network, have been applied to detect PD based on acoustic analysis of speech. It has been shown the promising results of RF an SVM in early PD detection. Das (2010) compared the performance of decision trees, regression, DMneural, and neural networks in detecting PD. The best accuracy of 92.9% was obtained using a neural network algorithm. Recently, DL-based techniques have attracted attention in PD diagnosis due to their capacity to handle Big Data and achieve high accuracy with free-assumption on data distribution (Ashour, El-Attar, Dey, El-Kader, & Abd El-Naby, 2020; Ranjan & Swetapadma, 2018). Ashour et al. (2020) applied a long short-term memory (LSTM) algorithm to detect freezing of gait (FOG), which an abnormal gait pattern that can accompany PD. It has been shown that LSTM outperforms SVM in detecting FOG.

In a traditional clinical setup, the physician will ask the patient to perform some mental and physical tasks (e.g., moving and walking around) (Cavallo, Moschetti, Esposito, Maremmani, & Rovini, 2019) or take the MRI and/or positron emission tomography-computed tomography (PET/CT) scan of the brain. However, it is challenging to differentiate PD from other neurological disorders, and it depends on the radiologist's experience to distinguish and identify it precisely. Therefore, a computer-aided diagnosis (CAD) system helps the radiologist interpret MRI scans. In 2003, Rana et al. (2015) developed a CAD system to monitor body acceleration to detect the FOG in PD patients. Several studies were conducted to implement ML approaches to detect PD and differentiate it from other common neurological diseases. Table 1 shows the performance of various ML models.

The adaptive searching characteristic of metaheuristic search algorithms increases the possibility of finding an optimal solution from the feature space. Several metaheuristic algorithms have been employed effectively in many complex optimization problems. Some of them are PSO (Behera & Dash, 2016; Storn & Price, 1997), Crow search algorithm (CSA) (Sayed, Hassanien, Azar, 2019), gray wolf optimizer (GWO) (Storn & Price, 1997), teaching-learning-based optimization (TLBO) (Črepinšek, Liu, & Mernik, 2012), harmony search (HS) (Woo, Hoon, & Loganathan, 2001), BA (Zawbaa, Emary, Parv, & Sharawi, 2016), moth-flame optimization (MFO) (Osaba, Yang, Diaz, Lopez-Garcia, & Carballedo, 2016), and animal migration optimization (AMO) (Li, Zhang, & Yin, 2014).

Methodology and result analysis

Various ML and DL techniques have been used in diagnostic and early detection models of neurological disorders. Much effort and practice is needed to build a good DL network. Usually, heuristic methods are effective in tuning the hyperparameters. The studies in Dash et al. (2017) and Goldberg (1989) show effective results when a genetic algorithm is employed in the proposed model. A genetic algorithm is a metaheuristic based on the process of natural selection. This type of algorithm belongs to a large class of evolutionary algorithms. It is used for achieving high-quality, optimized solutions for search problems, using operations such as mutation, selection, and crossover. To build a neural network, it is necessary to configure the following parameters: number of neurons, number of layers, choice of the activation function, and network optimizer. The implementation of the algorithm consists of the following steps.

(1) Initialization of N random networks to create a population.

(2) Evaluation of each network. This step takes a long time, as it is necessary to train each neural network, then determine how well it performs when classifying the test set.

Table 1 Different machine learning techniques and their outcomes.

Study	Objective of the study	Dataset	Experimental findings
(Grover et al., 2018)	The author proposed a method for the prediction of Parkinson's disease severity using deep neural networks and a patients' voice dataset from the UCI repository. The performance of the proposed method is better than that of other methods that used ANFIS and SVR	UCI's Parkinsons Telemonitoring Data Set	PD severity prediction based on Motor UPDRS Score. The predicted results are 94.4422% and 62.7335% for training and testing the dataset, respectively
(Lacy, Smith, & Lones, 2018)	The objective of this research was to implement echo state networks (ESNs) in place of traditional ANN and RNN techniques to diagnose PD. ESNs have been implemented in traditional signal forecasting applications and give better results than other existing systems	Test subjects were recorded performing a subtly modified version of the finger tapping protocol laid out in the UPDRS. Patient were instructed to tap the index finger on the thumb 10 times as quickly and as big as possible. The task was performed twice, once with the dominant hand before being repeated with the nondominant hand. The task duration was set at 30 s rather than the 10-tap limit enforced by the UPDRS, thereby allowing more time for bradykinesia symptoms to be exhibited	The proposed method determined the optimal configuration of ESN parameters for the classification problems, and it established whether directly modeling the separation waveform. Model training with ESN was very quick and made the classifier more flexible to adapt with new data source. AUC = 84.00%

Continued

Table 1 Different machine learning techniques and their outcomes—cont'd

Study	Objective of the study	Dataset	Experimental findings
Al Sayaydeha and Falah (2019)	The objective of this study was to diagnose PD by using an enhanced fuzzy min-max neural network (EFMM) as a classification technique. The most relevant features were identified after feature selection using the OneR Attribute evaluator and the reduced dataset was sent to the EFMM neural network to construct a knowledge. Then, the testing dataset was used to examine the performance of the EFMM network at various stages of learning. The proposed system obtained accuracy of 0.94	University of California at Irvine (UCI) machine learning repository	OneR attribute evaluator was used for feature selection and an EFMM neural network was used for the classification process. Performance evaluation was performed with 10-fold cross validation. ACC = 94.2%
(Maass et al., 2020)	The objective of this study was to validate a predefined SVM for the diagnosis of PD. During SVM model validation, 82 PD patients, 68 age-matched controls, and 7 normal pressure hydrocephalus (NPH) patients were used	82 PD patients, 68 age-matched controls and 7 additional Normal Pressure Hydrocephalus (NPH) patients	The proposed method was able classify PD and AMC patients. The study found a significant correlation between iron levels and disease duration, even if controlled for age. AUROC = 88.00%

(Karapinar, 2020)	The aim of this study was to develop a model using feature selection and classification process of ML for early diagnosis of PD. The SVM, ANN, and CART methods of ML were used. The CART decision tree produces either a regression tree or classification tree depending on the output variable. The proposed method achieved 0.938 accuracy	The dataset includes 48 healthy phonetics and 147 Parkinson's Disease phonetics	ACC = 93.84%
(Dastjerd, Sert, Ozyer, & Alhajj, 2019)	The objective of this study was to develop a PD prediction method using a fuzzy classification algorithm. The experiment involved 10-fold cross validation. The proposed model achieved 100% accuracy	Parkinson Speech Dataset from UCI repository	The fuzzy rough nearest neighborhood algorithm showed worse performance than the other classifiers. The FURIA gave the best performance. ACC = 100%
(Celik & Omurca, 2019)	The aim of this study was to develop a model for the prediction of PD. Different methods of ML algorithms were used and compared to different methods. SVM method gave the best result with accuracy of 0.75	1208 speech dataset from UCI repository	The extended feature spaces improved the classification accuracies for all classification algorithms. ACC = 75.49%

Continued

Table 1 Different machine learning techniques and their outcomes—cont'd

Study	Objective of the study	Dataset	Experimental findings
(Haq et al., 2019)	The aim of this study was to develop a PD prediction model based on ML algorithms. L1-SVM was used for features selection because it can select highly related and suitable features for the classification of PD patients and healthy controls	194 voice recording dataset from the repository of the University of Oxford	ACC = 99.00%
(Kuresan, Samiappan, & Masunda, 2019	The authors developed a classification system for the diagnosis of PD using Mel-frequency cepstral coefficients (MFCCs), DWT, and the fusion of MFCC and DWT for features extraction. The hidden Markov model SVM was applied for the classification of PD patients and healthy controls	20 voice samples from 40 people	The fusion of MFCC and WPT using the HMM classifier gave the best performance. ACC = 95.00%
(Choi, Kim, Yoon, Lee, & Lee, 2019)	The aim of this study was to develop a model that can classify multiple diseases. The authors proposed a model for predicting both Alzheimer's disease and Parkinson's	PET images (243 CE, 62 PD, 393 NC) from ADNI	AUC = 94.00%

	disease using a CNN with ReLU activation function. The model was trained with 636 PET images, including 243 images from Alzheimer's disease patients and 393 images from healthy individuals. Normal people. It included 62 PD images to differentiate PD patients with dementia		
(Mozhdehfarahbakhsh et al., 2021)	The aim of this study was to predict the different stages of PD using a CNN	MRI data from PPMI repository	ACC = 94.00%
Dash, Abraham, and Atta-ur-Rahman (2019)	A novel chaos-based stochastic model was proposed by combining the characteristics of chaotic firefly algorithm with a kernel-based naïve Bayes (KNB) algorithm for early diagnosis of PD The dynamics of chaos optimization algorithm enhanced the firefly algorithm by introducing six types of chaotic maps to increase the diversification and intensification capability of the algorithm	Voice measurement dataset from UCI repository	Stratified 10-fold cross validation was used for model validation. ACC = 89%, AUC = 91%, and model building time = 7.14 s

Continued

Table 1 Different machine learning techniques and their outcomes—cont'd

Study	Objective of the study	Dataset	Experimental findings
(Mohammed, He, & Lin, 2021)	The objective of this study was to diagnose PD using a CNN-based network model which is highly capable of discriminating PD patients based on Single Photon Emission Computed Tomography (SPECT) images from healthy controls. Image normalization process is carried out to enhance the regions of interests (ROIs) necessary for the network to learn distinguishing features from them	This study analyzed 2723 SPECT images, including 1364 from the healthy control group and 1359 from the PD group	Two classes were identified: class 0 for the healthy control group and class 1 for PD patients. A 10-fold cross validation was used to evaluate the performance of the network model. Results show accuracy of 99.34%, sensitivity of 99.04%, and specificity of 99.63%, outperforming all previously published results
(Mohammadi, Mehralian, Naseri, & Sajedi, 2021)	The objective of this study was to review datasets developed for PD diagnosis and the use of ML classification models, such as SVM, XGBoost, and MLP, to accurately predict PD according to vocal symptoms characteristics provided in the UCI database, which suffers a low number of samples compared to features and thus is imbalanced. The results of comparative studies demonstrated that the proposed classic classification models can outperform various DL methods	Vocal symptoms characteristics dataset from UCI Machine Learning database including 564 PD patients and 192 control/normal cases	Logistic regression to stack the outputs of SVM, XGBoost, MLP, and autoencoder-preceded SVM provides an accurate classification of PD patients. Results were validated by a fivefold cross-validation method on min-max normalized data. Logistic regression and voting algorithms achieved accuracy of 97.22%.

(Solana-Lavalle & Rosas-Romero, 2021)	The authors proposed a new method to classify 3D MRI scans to assist in PD diagnosis. Separate studies were conducted for men and women because men are more prone to PD than women are	Parkinson's Progression Markers Initiative Dataset	A group of 64 female controls and 117 PD female patients were compared using VBM. Another group of 95 male controls and 33 prodromal male PD patients were also compared using VBM. In men, accuracy = 99.01%, sensitivity = 99.35%, specificity = 100%, and precision = 100%. In women, accuracy = 96.97%, sensitivity = 100%, specificity = 96.15%, and precision = 97.22%
(Karaman, Çakın, Alhudhaif, & Polat, 2021)	The aim of this study was to develop deep CNNs for automated PD identification based on biomarkers derived from voice signals. The developed CNN methods consisted of two main stages: data pre-processing and finetuning-based transfer learning	mPower voice database	The performances of ResNet, DenseNet, and SequeezeNet, were compared. The accuracy of the ResNet and DenseNet were very closed, but SequeezeNet gave poor accuracy. The proposed model successfully identified PD with an accuracy of 89.75%, sensitivity of 91.50%, and precision of 88.40% for DenseNet-161 architecture, which was identified as the most suitable fine-tuning architecture

(3) Sorting of all networks in the population, according to the prediction accuracy of the test sample. A certain percentage of the best networks are retained to be part of the next generation and to create descendants. There will also be several networks with a low level of accuracy, potentially helping to find combinations between the worst and best neural networks.

(4) Reproduction. The algorithm selects two different members of the population and creates one or more descendants, where each descendant is a combination of a random set of parameters of its parents; for example, one descendant may have the same number of layers as one of its parents, and the rest of the parameters from its other parent.

(5) After it has been decided which networks should be stored, some parameters in each set of networks are randomly changed.

El-Hasnony, Barakat, and Mostafa (2020) introduced a novel model for extracting clinically useful information for PD assessment and utilizing learning algorithms to achieve reliable decision support systems. In this study, a proposed ANFIS + PSOGWO model for PD prediction using sets of vocal (speech) features in fog-cloud has introduced. The fog infrastructure provides real-time data processing and analysis and overcoming the cloud limitations. The PSOGWO algorithm merges the exploration and exploitation capabilities of gray wolf optimization (GWO) and particle swarm optimization (PSO), respectively. A chaotic tent map for initialization prevents the code from getting locked into a local optima issue. The proposed model was evaluated through different dataset with different metrics to approve its efficiency. Outcomes show the suggested model produces good results compared to other algorithms; it achieved accuracy of 87.5 for predicting PD.

In this work (Dash, Acharya, Mittal, Abraham, & Kelemen, 2020), researchers presented a new hybrid kernel-based probabilistic chaotic metaheuristic feature selection model. They employed six different chaotic mappings to develop chaos-based firefly algorithms combined with a nonparametric kernel density estimated naïve Bayes classifier for selecting the most discriminative features to develop a robust, reliable, and generalized diagnostic model for PD patients. The performance of the six chaotic mappings was compared based on their best fitness values and p values to determine the best chaotic combination with the CFA-KNB model that selects the most reliable features from the PD dataset. The performances of all the six chaotic CFA-KNB models with respect to discriminative feature set, classification performance, robustness, and generalization were measured through five well-known classifiers. The experiment proved that for very small clinical dataset, a good set of discriminative features matters more than considering a small set of features that are not able to characterize the whole problem.

Characteristics of chaotic maps

The chaos optimization theory is a nonlinear phenomenon (Dash et al., 2017; Dash et al., 2018; Fister et al., 2015; Gandomi et al., 2013; Xiao et al., 2017) that develops the trajectories to traverse all the states nonrepetitively in a certain range, hence providing an optimal search configuration for the problem. The chaos theory transforms the candidate variables from chaotic sequences to solution space by strictly adhering to the initial condition. Three of the characteristics of chaotic optimization theory, namely, randomness, ergodicity, and regularity of chaotic motion, help to secure a global optimal solution avoiding the limitation of local optima. Hence, the integration of chaotic maps into the metaheuristic algorithm enhances the efficiency of nature-inspired metaheuristics (Dash et al., 2017; Gandomi et al., 2013) by using deterministic chaotic maps instead of random variables. Sharpening the chaotic maps in CFA structures will lead to greater convergence rate, accuracy, and robustness. These introduce greater diversity to the search process preventing it from being localized (Table 2).

Proposed model

The smartphone has become an indispensable tool in communication and computing architecture. Many researchers are investigating the application of mobile computing in the healthcare sector. A variety of wearable IoT health sensors are currently available and these devices can be connected in what is called a Body Sensors Network (BSN). A BSN can connect with a smartphone or personal computer can share data collected via sensors. A smartphone has many sensors, and these sensors can be used to collect individual health information. Because a smartphone has enough processing capability, it can act as an edge node. It can perform some operation on the sensors' data and it can then transmit the health data to the cloud for remote monitoring and data processing. Edge computing has recently been applied in the IoT-based healthcare system, reducing latency and allowing for performance of real-time operations. It extends healthcare services to individuals wherever they may be (e.g., at work, playing a sport, in the hospital). The main advantages of using edge computing in the healthcare system are

Table 2 Performance of hybrid (selection models + classification models) models.

Study	Model	Accuracy (%)
Sarkar et al. (Sakar et al., 2019)	(MRMR) Feature selection, SVM-RBF classifier	86
Gunduz (Gunduz, 2019)	Different feature selections + CNN	86.9
Ibrahim et al. (El-Hasnony et al., 2020)	CFS feature selection, PSOGWO + ANFIS	87.5
Dash et al. (Dash et al., 2018)	Chaotic firefly feature selection, CFA-KNB	89
(Grover et al., 2018)	Deep neural	94
(Mozhdehfarahbakhsh et al., 2021)	CNN	94
(Karapinar, 2020)	RFE + SVM	93.84
(Celik & Omurca, 2019)	SVM with the expanded features	75.49
(Haq et al., 2019)	SVM-RBF and SVM-linear	99
(Grover et al., 2018)	DNN	99
(Maass et al., 2020)	SVM	88

reduced latency, real-time processing, and the patients' mobility. Fig. 1 shows the three-tier architecture of an edge computing-based healthcare system.

The proposed system is to be used to detect PD via edge computing technology for analyzing patient speech recorded using a smartphone.

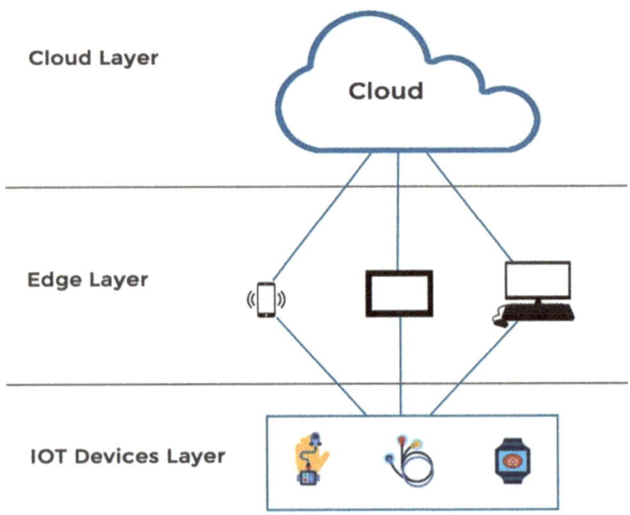

Fig. 1 Edge-based healthcare system architecture.

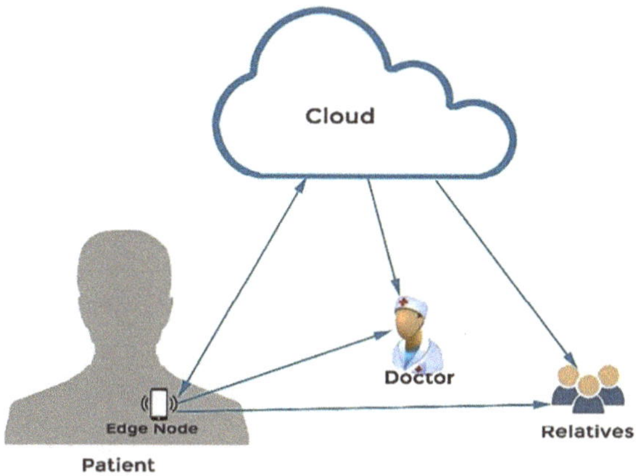

Fig. 2 Proposed system architecture.

A smartphone can act as the edge node because it has sufficient computing power. The smartphone can record the speech, which will be inputted to the classification model for the detection of PD.

The proposed system has four phases.

(1) Pre-process the data: Partition the audio speech into two audio files: voiced and un-voiced.

(2) Build the classification model: Build a classification model using the SVM method for the classification of PD patients and healthy controls.

(3) Develop the application: Build an Android application that can implement the classification model.

(4) Upload the data: Upload the recorded speech and the output of the classification model to the cloud server for further analysis.

Fig. 2 shows the proposed system architecture. The edge node can check the status of the patient randomly. The system will be able to detect PD in the early stage. The smartphone can send an alert message to the patient's doctor and relatives. The proposed system will be helpful for patients as well as for the healthcare system.

Conclusion

Feature engineering is paramount for increasing the efficiency level of ML models for early detection of PD. As clinical biomarker selection is an important aspect for enhancing the efficiency of the proposed model, selection techniques such as evolutionary and metaheuristics have been studied and employed by many researchers. The SVM classifier is mostly used to develop an ML-based prediction model for both Alzheimer's and Parkinson's. As DL algorithms are becoming more advanced, the CNN method is also widely used to develop prediction models for the early detection of neurological diseases. Nowadays, as the smartphones are also becoming more advanced and more powerful, they are increasingly being used in the healthcare system. For example, the data generated via the accelerometer sensors in smartphones can be used to diagnose Alzheimer's disease. The system proposed in this chapter uses the accelerometer sensor of a smartphone to generate a sequence of data while the patient is moving. This data will be used to identify the stage of Alzheimer's disease. The smartphone is used as the edge node of the proposed system, which can monitor the patient continuously as well as send alerts to the patient's doctor or relatives. The sensor's data is uploaded to cloud storage and thus can be used for further analysis. This proposed system will be helpful to PD patients and may be a valuable contribution to the healthcare system.

References

Abdel-Raouf, O., Abdel-Baset, M., & El-henawy, I. (2014). Chaotic firefly algorithm for solving definite integral. *International Journal of Information Technology and Computer Science*, *6*(6), 19–24. https://doi.org/10.5815/ijitcs.2014.06.03.

Al Sayaydeha, O. N., & Falah, M. (2019). Diagnosis of the Parkinson disease using enhanced fuzzy min-max neural network and oneR attribute evaluation method. In *International conference on advanced science and engineering* IEEE. https://doi.org/10.1109/ICOASE.2019.8723870.

Arroyo-Gallego, T., Ledesma-Carbayo, M. J., Sanchez-Ferro, A., Butterworth, I., Mendoza, C. S., Matarazzo, M., et al. (2017). Detection of motor impairment in Parkinson's disease via mobile touchscreen typing. *IEEE Transactions on Bio-Medical Engineering*, *64*(9), 1994–2002. https://doi.org/10.1109/TBME.2017.2664802.

Ashour, A. S., El-Attar, A., Dey, N., El-Kader, H. A., & Abd El-Naby, M. M. (2020). Long short term memory based patient-dependent model for fog detection in Parkinson's disease. *Pattern Recognition Letters*, *131*, 23–29. https://doi.org/10.1016/j.patrec.2019.11.036.

Askarzadeh, A. (2016). A novel metaheuristic method for solving constrained engineering optimization problems: Crow search algorithm. *Computers & Structures*, *169*, 1–12. https://doi.org/10.1016/j.compstruc.2016.03.001.

Behera, R. N., & Dash, S. (2016). A particle swarm optimization based hybrid recommendation system. *International Journal of Knowledge Discovery in Bioinformatics*, *6*(2), 1–10. https://doi.org/10.4018/IJKDB.2016070101.

Brajević, I., & Ignjatović, J. (2018). An upgraded firefly algorithm with feasibility-based rules for constrained engineering optimization problems. *Journal of Intelligent Manufacturing*, *30*(6), 2545–2574. https://doi.org/10.1007/s10845-018-1419-6.

Burke, R. E., & O'Malley, K. (2013). Axon degeneration in Parkinson's disease. *Experimental Neurology*, *246*, 72–83. https://doi.org/10.1016/j.expneurol.2012.01.011.

Cai, Z., Gu, J., Wen, C., Zhao, D., Huang, C., Huang, H., et al. (2018). An intelligent Parkinson's disease diagnostic system based on a chaotic bacterial foraging optimization enhanced fuzzy KNN approach. *Computational and Mathematical Methods in Medicine, 2018*, 2396952. https://doi.org/10.1155/2018/2396952.

Cavallo, F., Moschetti, A., Esposito, D., Maremmani, C., & Rovini, E. (2019). Upper limb motor pre-clinical assessment in Parkinson's disease using machine learning. *Parkinsonism & Related Disorders, 63*, 111–116. https://doi.org/10.1016/j.parkreldis.2019.02.028.

Celik, E., & Omurca, S. I. (2019). Improving Parkinson's disease diagnosis with machine learning methods. In *2019 Scientific meeting on electrical-electronics biomedical engineering and computer science (EBBT)* (pp. 1–4). https://doi.org/10.1109/EBBT.2019.8742057.

Choi, H., Ha, S., Im, H. J., Paek, S. H., & Lee, D. S. (2017). Refining diagnosis of Parkinson's disease with deep learning-based interpretation of dopamine transporter imaging. *NeuroImage: Clinical, 16*, 586–594. https://doi.org/10.1016/j.nicl.2017.09.010.

Choi, H., Kim, Y. K., Yoon, E. J., Lee, J., & Lee, D. S. (2019). Cognitive signature of brain FDG pet based on deep learning: Domain transfer from Alzheimer's disease to Parkinson's disease. *European Journal of Nuclear Medicine and Molecular Imaging, 47*(2), 403–412. https://doi.org/10.1007/s00259-019-04538-7.

Črepinšek, M., Liu, S.-H., & Mernik, L. (2012). A note on teaching-learning-based optimization algorithm. *Information Sciences, 212*, 79–93. https://doi.org/10.1016/j.ins.2012.05.009.

Das, R. (2010). A comparison of multiple classification methods for diagnosis of Parkinson disease. *Expert Systems with Applications, 37*(2), 1568–1572. https://doi.org/10.1016/j.eswa.2009.06.040.

Dash, S., Abraham, A., & Atta-Ur-Rahman. (2018). *Kernel based chaotic firefly algorithm for diagnosing Parkinson's disease*. HIS.

Dash, S., Abraham, A., & Atta-ur-Rahman. (2019). Kernel based chaotic firefly algorithm for diagnosing Parkinson's disease. In *International conference on hybrid intelligent* Springer. https://doi.org/10.1007/978-3-030-14347-3_18.

Dash, S., Abraham, A., Luhach, A. K., Mizera-Pietraszko, J., & Rodrigues, J. J. (2020). Hybrid chaotic firefly decision making model for Parkinson's disease diagnosis. *International Journal of Distributed Sensor Networks*. https://doi.org/10.1177/1550147719895210.

Dash, S., Acharya, B. R., Mittal, M., Abraham, A., & Kelemen, A. (2020). *Deep learning techniques for biomedical and health informatics*. Cham: Springer International Publishing, ISBN:978-3-030-33966-1.

Dash, S., & Behera, R. (2016). Sampling based hybrid algorithms for imbalanced data classification. *International Journal of Hybrid-Intelligent-Systems, 13*, 77–86. https://doi.org/10.3233/HIS-160226. IOS Press.

Dash, S., Thulasiram, R., & Thulasiraman, P. (2017). An enhanced chaos-based firefly model for Parkinson's disease diagnosis and classification. In *2017 international conference on information technology (ICIT)* (pp. 159–164). https://doi.org/10.1109/ICIT.2017.43.

Dastjerd, N. K., Sert, O. C., Ozyer, T., & Alhajj, R. (2019). Fuzzy classification methods based diagnosis of Parkinson's disease from speech test cases. *Current Aging Science, 12*(2), 100–120. https://doi.org/10.2174/1874609812666190625140311.

Dinov, I. D., Heavner, B., Tang, M., Glusman, G., Chard, K., Darcy, M., et al. (2016). Predictive big data analytics: A study of Parkinson's disease using large, complex, heterogeneous, incongruent, multi-source and incomplete observations. *PLoS One, 11*(8), e0157077. https://doi.org/10.1371/journal.pone.0157077.

Dos Santos Coelho, L., de Andrade Bernert, D. L., & Mariani, V. C. (2011). A chaotic firefly algorithm applied to reliability-redundancy optimization. In *2011 IEEE congress of evolutionary computation (CEC)* (pp. 517–521). https://doi.org/10.1109/CEC.2011.5949662.

Dos Santos Coelho, L., & Mariani, V. C. (2012). Firefly algorithm approach based on chaotic Tinkerbell map applied to multivariable PID controller tuning. *Computers & Mathematics with Applications, 64*(8), 2371–2382. https://doi.org/10.1016/j.camwa.2012.05.007.

Drotár, P., Mekyska, J., Rektorová, I., Masarová, L., Smékal, Z., & Faundez-Zanuy, M. (2015). Decision support framework for Parkinson's disease based on novel handwriting markers. *IEEE Transactions on Neural Systems and Rehabilitation Engineering: A Publication of the IEEE Engineering in Medicine and Biology Society, 23*(3), 508–516. https://doi.org/10.1109/TNSRE.2014.2359997.

Drotár, P., Mekyska, J., Rektorová, I., Masarová, L., Smékal, Z., & Faundez-Zanuy, M. (2016). Evaluation of handwriting kinematics and pressure for differential diagnosis of Parkinson's disease. *Artificial Intelligence in Medicine, 67*, 39–46. https://doi.org/10.1016/j.artmed.2016.01.004.

El Maachi, I., Bilodeau, G., & Bouachir, W. (2020). Deep 1D-convnet for accurate Parkinson disease detection and severity prediction from gait. *Expert Systems with Applications, 143*, 113075. https://doi.org/10.1016/j.eswa.2019.113075.

El-Hasnony, I. M., Barakat, S. I., & Mostafa, R. R. (2020). Optimized ANFIS model using hybrid metaheuristic algorithms for Parkinson's disease prediction in IoT environment. *IEEE Access, 8*, 119252–119270. https://doi.org/10.1109/ACCESS.2020.3005614.

Fister, I., Perc, M., Kamal, S. M., & Fister, I. (2015). A review of chaos-based firefly algorithms: Perspectives and research challenges. *Applied Mathematics and Computation, 252*, 155–165. https://doi.org/10.1016/j.amc.2014.12.006.

Gandomi, A. H., Yang, X.-S., Talatahari, S., & Alavi, A. H. (2013). Firefly algorithm with chaos. *Communications in Nonlinear Science and Numerical Simulation, 18*(1), 89–98. https://doi.org/10.1016/j.cnsns.2012.06.009.

Goldberg, D. E. (1989). *Genetic algorithms in search, optimization and machine learning* (1st ed.). USA: Addison-Wesley Longman Publishing Co., Inc.

Grover, S., Bhartia, S., Akshama, Yadav, A., Abhilasha, Y., & Seeja, K. R. (2018). Predicting severity of Parkinson's disease using deep learning. *Procedia Computer Science, 132*, 1788–1794. https://doi.org/10.1016/j.procs.2018.05.154.

Gunduz, H. (2019). Deep learning-based Parkinson's disease classification using vocal feature sets. *IEEE Access*, 7, 115540–115551.

Gupta, U., Bansal, H., & Joshi, D. (2020). An improved sex-specific and age-dependent classification model for Parkinson's diagnosis using handwriting measurement. *Computer Methods and Programs in Biomedicine, 189*, 105305. https://doi.org/10.1016/j.cmpb.2019.105305.

Haq, A. U., Li, J. P., Memon, M. H., Khan, J., Malik, A., Ahmad, T., et al. (2019). Feature selection based on L1-norm support vector machine and effective recognition system for Parkinson's disease using voice recordings. *IEEE Access*, 7, 37718–37734. https://doi.org/10.1109/ACCESS.2019.2906350.

Helbing, D., Brockmann, D., Chadefaux, T., Donnay, K., Blanke, U., Woolley-Meza, O., et al. (2015). Saving human lives: What complexity science and information systems can contribute. *Journal of Statistical Physics, 158*(3), 735–781. https://doi.org/10.1007/s10955-014-1024-9.

Illner, V., Sovka, P., & Rusz, J. (2020). Validation of freely-available pitch detection algorithms across various noise levels in assessing speech captured by smartphone in Parkinson's disease. *Biomedical Signal Processing and Control, 58*, 101831. https://doi.org/10.1016/j.bspc.2019.101831.

Isler, Y., Narin, A., Ozer, M., & Perc, M. (2019). Multi-stage classification of congestive heart failure based on short-term heart rate variability. *Chaos, Solitons & Fractals, 118*(C), 145–151. https://doi.org/10.1016/j.chaos.2018.11.0.

Karaman, O., Çakın, H., Alhudhaif, A., & Polat, K. (2021). Robust automated Parkinson disease detection based on voice signals with transfer learning. *Expert Systems with Applications, 178*, 115013. https://doi.org/10.1016/j.eswa.2021.115013.

Karapinar, S. Z. (2020). Early diagnosis of Parkinson's disease using machine learning algorithms. *Medical Hypotheses, 138*, 109603. https://doi.org/10.1016/j.mehy.2020.109603.

Kaveh, A., Mahdipour Moghanni, R., & Javadi, S. M. (2019). Optimum design of large steel skeletal structures using chaotic firefly optimization algorithm based on the Gaussian map. *Structural and Multidisciplinary Optimization*, *60*(3), 879–894. https://doi.org/10.1007/s00158-019-02263-1.

Kazem, A., Sharifi, E., Hussain, F. K., Saberi, M., & Hussain, O. K. (2013). Support vector regression with chaos-based firefly algorithm for stock market price forecasting. *Applied Soft Computing*, *13*, 947–958.

Kennedy, J., & Eberhart, R. (1995). Particle swarm optimization. In *Vol. 4. Proceedings of ICNN'95—International conference on neural networks* (pp. 1942–1948). https://doi.org/10.1109/ICNN.1995.488968.

Kuresan, H., Samiappan, D., & Masunda, S. (2019). Fusion of WPT and MFCC feature extraction in Parkinson's disease diagnosis. *Technology and Health Care: Official Journal of the European Society for Engineering and Medicine*, *27*(4), 363–372. https://doi.org/10.3233/THC-181306.

Lacy, S. E., Smith, S. L., & Lones, M. A. (2018). Using echo state networks for classification. *Artificial Intelligence in Medicine*, *86*(C), 53–59. https://doi.org/10.1016/j.artmed.2018.02.002.

Li, X., Zhang, J., & Yin, M. (2014). Animal migration optimization: An optimization algorithm inspired by animal migration behavior. *Neural Computing and Applications*, *24*(7), 1867–1877. https://doi.org/10.1007/s00521-013-1433-8.

Ma, C., Ouyang, J., Chen, H. L., & Zhao, X. H. (2014). An efficient diagnosis system for Parkinson's disease using kernel-based extreme learning machine with subtractive clustering features weighting approach. *Computational and Mathematical Methods in Medicine*, *2014*, 985789. https://doi.org/10.1155/2014/985789.

Maass, F., Michalke, B., Willkommen, D., Leha, A., Schulte, C., Tönges, L., et al. (2020). Elemental fingerprint: Reassessment of a cerebrospinal fluid biomarker for Parkinson's disease. *Neurobiology of Disease*, *134*, 104677. https://doi.org/10.1016/j.nbd.2019.104677.

Mingjuan, X., Su, S., Su, Y., & Xu, M. (2014). *Comparisons of firefly algorithm with chaotic maps.*

Mohamad, A. (2021). *Parkinson's disease diagnosis using deep learning*. arXiv:2101.05631.

Mohammadi, A. G., Mehralian, P., Naseri, A., & Sajedi, H. (2021). Parkinson's disease diagnosis: The effect of autoencoders on extracting features from vocal characteristics. *Array*, *11*, 100079. https://doi.org/10.1016/j.array.2021.100079.

Mohammed, F., He, X., & Lin, Y. (2021). An easy-to-use deep-learning model for highly accurate diagnosis of Parkinson's disease using SPECT images. *Computerized Medical Imaging and Graphics: The Official Journal of the Computerized Medical Imaging Society*, *87*, 101810. https://doi.org/10.1016/j.compmedimag.2020.101810.

Mozhdehfarahbakhsh, A., Chitsazian, S., Chakrabarti, P., Chakrabarti, T., Kateb, B., & Nami, M. (2021). *An MRI-based deep learning model to predict Parkinson\textquoterights disease stages*. medRxiv. https://doi.org/10.1101/2021.02.19.21252081.

Narin, A., Isler, Y., Ozer, M., & Perc, M. (2018). Early prediction of paroxysmal atrial fibrillation based on short-term heart rate variability. *Physica A: Statistical Mechanics and its Applications*, *509*(C), 56–65. https://doi.org/10.1016/j.physa.2018.06.0.

NHS. (2021). *"Diagnosis Parkinson's disease" NHS choices*. Available online: https://www.nhs.uk/conditions/parkinsons-disease/diagnosis/. (Accessed 26 May 2021).

Nunes, A., Silva, G., Duque, C., Januário, C., Santana, I., Ambrósio, A. F., et al. (2019). Retinal texture biomarkers may help to discriminate between Alzheimer's, Parkinson's, and healthy controls. *PLoS One*, *14*(6), e0218826. https://doi.org/10.1371/journal.pone.0218826.

Nuvoli, S., Spanu, A., Fravolini, M. L., Bianconi, F., Cascianelli, S., Madeddu, G., et al. (2020). [123I]Metaiodobenzylguanidine (MIBG) cardiac scintigraphy and automated classification techniques in parkinsonian disorders. *Molecular Imaging and Biology*, *22*(3), 703–710. https://doi.org/10.1007/s11307-019-01406-6.

Osaba, E., Yang, X.-S., Diaz, F., Lopez-Garcia, P., & Carballedo, R. (2016). An improved discrete bat algorithm for symmetric and asymmetric traveling salesman problems. *Engineering Applications of Artificial Intelligence, 48*(C), 59–71. https://doi.org/10.1016/j.engappai.2015.10.006.

Pereira, C. R., Pereira, D. R., Rosa, G. H., Albuquerque, V., Weber, S., Hook, C., et al. (2018). Handwritten dynamics assessment through convolutional neural networks: An application to Parkinson's disease identification. *Artificial Intelligence in Medicine, 87*, 67–77. https://doi.org/10.1016/j.artmed.2018.04.001.

Pham, T. D., & Yan, H. (2018). Tensor decomposition of gait dynamics in Parkinson's disease. *IEEE Transactions on Bio-Medical Engineering, 65*(8), 1820–1827. https://doi.org/10.1109/TBME.2017.2779884.

Quan, C., Ren, K., & Luo, Z. (2021). A deep learning based method for Parkinson's disease detection using dynamic features of speech. *IEEE Access, 9*, 10239–10252.

Rana, A. Q., Ahmed, U. S., Chaudry, Z. M., & Vasan, S. (2015). Parkinson's disease: A review of non-motor symptoms. *Expert Review of Neurotherapeutics, 15*, 549–562.

Ranjan, A., & Swetapadma, A. (2018). An intelligent computing based approach for Parkinson disease detection. In *Proceedings of the proceedings of 2018 2nd international conference on advances in electronics, computers and communications, ICAECC 2018, Bangalore, India, 9–10 February 2018*.

Sakar, B. E., Isenkul, M. E., Sakar, C. O., Sertbas, A., Gurgen, F., Delil, S., et al. (2013). Collection and analysis of a Parkinson speech dataset with multiple types of sound recordings. *IEEE Journal of Biomedical and Health Informatics, 17*(4), 828–834. https://doi.org/10.1109/JBHI.2013.2245674.

Sakar, C. O., Serbes, G., Gunduz, A., Tunc, H. C., Nizam, H., Sakar, B. E., et al. (2019). A comparative analysis of speech signal processing algorithms for Parkinson's disease classification and the use of the tunable Q-factor wavelet transform. *Applied Soft Computing, 74*, 255–263. https://doi.org/10.1016/j.asoc.2018.10.022.

Sayed, G. I., Hassanien, A. E., & Azar, A. T. (2017). Feature selection via a novel chaotic crow search algorithm. *Neural Computing and Applications, 31*(1), 171–188. https://doi.org/10.1007/s00521-017-2988-6.

Segovia, F., Górriz, J. M., Ramírez, J., Martínez-Murcia, F. J., & Castillo-Barnes, D. (2019). Assisted diagnosis of parkinsonism based on the striatal morphology. *International Journal of Neural Systems, 29*(9), 1950011. https://doi.org/10.1142/S0129065719500114.

Shahbakhi, M., Far, D. T., & Tahami, E. (2014). Speech analysis for diagnosis of Parkinson's disease using genetic algorithm and support vector machine. *Journal of Biomedical Science and Engineering, 07*(04), 147–156. https://doi.org/10.4236/jbise.2014.74019.

Sharma, S., Luhach, A. K., & Jyoti, K. (2016). Research and analysis of advancements in BAT algorithm. In *2016 3rd international conference on computing for sustainable global development (INDIACom)* (pp. 2391–2396).

Singh, N., Pillay, V., & Choonara, Y. E. (2007). Advances in the treatment of Parkinson's disease. *Progress in Neurobiology, 81*(1), 29–44. https://doi.org/10.1016/j.pneurobio.2006.11.009.

Solana-Lavalle, G., Galán-Hernández, J.-C., & Rosas-Romero, R. (2020). Automatic parkinson disease detection at early stages as a pre-diagnosis tool by using classifiers and a small set of vocal features. *Biocybernetics and Biomedical Engineering, 40*(1), 505–516. https://doi.org/10.1016/j.bbe.2020.01.003.

Solana-Lavalle, G., & Rosas-Romero, R. (2021). Classification of PPMI MRI scans with voxel-based morphometry and machine learning to assist in the diagnosis of Parkinson's disease. *Computer Methods and Programs in Biomedicine, 198*, 105793. https://doi.org/10.1016/j.cmpb.2020.105793.

Storn, R., & Price, K. (1997). Differential evolution—A simple and efficient heuristic for global optimization over continuous spaces. *Journal of Global Optimization, 11*(4), 341–359. https://doi.org/10.1023/A:1008202821328.

Sveinbjornsdottir, S. (2016). The clinical symptoms of Parkinson's disease. *Journal of Neurochemistry, 139*(Suppl. 1), 318–324. https://doi.org/10.1111/jnc.13691.

Váradi, C., Nehéz, K., Hornyák, O., Viskolcz, B., & Bones, J. (2019). Serum N-glycosylation in Parkinson's disease: A novel approach for potential alterations. *Molecules, 24*(12). https://doi.org/10.3390/molecules24122220.

Viz Hub. (2021). *"GBD compare" data visualizations.* Available online: https://vizhub.healthdata.org/gbd-compare/. (Accessed 26 May 2021).

Wagner, A., Fixler, N., & Resheff, Y. S. (2017). *A wavelet-based approach to monitoring Parkinson's disease symptoms.* arXiv [stat.AP]. Opgehaal van http://arxiv.org/abs/1701.03161.

Wang, Z., Zhu, X., Adeli, E., Zhu, Y., Nie, F., Munsell, B., et al. (2017). Multi-modal classification of neurodegenerative disease by progressive graph-based transductive learning. *Medical Image Analysis, 39*, 218–230. https://doi.org/10.1016/j.media.2017.05.003.

Woo, Z., Hoon, J., & Loganathan, G. V. (2001). A new heuristic optimization algorithm: Harmony search. *Simulation, 76*(2), 60–68. https://doi.org/10.1177/003754970107600201.

Xiao, G., Liu, H., Guo, Y., & Zhou, Y. (2017). Research on chaotic firefly algorithm and the application in optimal reactive power dispatch. *TELKOMNIKA (Telecommunication Computing Electronics and Control), 15*(1), 93. https://doi.org/10.12928/telkomnika.v15i1.3862.

Yang, X.-S. (2009). Firefly algorithms for multimodal optimization. In O. Watanabe, & T. Zeugmann (Eds.), *Stochastic algorithms: Foundations and applications* (pp. 169–178). Berlin, Heidelberg: Springer Berlin Heidelberg.

Yang, X.-S. (2011). Chaos-enhanced firefly algorithm with automatic parameter tuning. *International Journal of Swarm Intelligence Research (IJSIR), 2*(4), 1–11. Opgehaal van https://ideas.repec.org/a/igg/jsir00/v2y2011i4p1-11.html.

Yang, X. (2014). *Cuckoo search and firefly algorithm: Theory and applications.* Cham: Springer, ISBN:978-3-319-02141-6.

Zawbaa, H. M., Emary, E., Parv, B., & Sharawi, M. (2016). Feature selection approach based on moth-flame optimization algorithm. In *2016 IEEE congress on evolutionary computation (CEC)* (pp. 4612–4617). https://doi.org/10.1109/CEC.2016.7744378.

Further Reading

Al Sayaydeha, O. N., & Mohammad, M. F. (2019). Diagnosis of the Parkinson disease using enhanced fuzzy min-max neural network and OneR attribute evaluation method. In *2019 International Conference on Advanced Science and Engineering (ICOASE)* (pp. 64–69). https://doi.org/10.1109/ICOASE.2019.8723870.

Baykasoğlu, A., & Ozsoydan, F. B. (2015). Adaptive firefly algorithm with chaos for mechanical design optimization problems. *Applied Soft Computing, 36*, 152–164. https://doi.org/10.1016/j.asoc.2015.06.056.

Brajević, I., & Stanimirović, P. (2018). An improved chaotic firefly algorithm for global numerical optimization. *International Journal of Computational Intelligence Systems, 12*, 131–148. https://doi.org/10.2991/ijcis.2018.25905187.

Chou, J.-S., & Ngo, N.-T. (2017). Modified firefly algorithm for multidimensional optimization in structural design problems. *Structural and Multidisciplinary Optimization, 55*(6), 2013–2028. https://doi.org/10.1007/s00158-016-1624-x.

Varrecchia, T., Castiglia, S. F., Ranavolo, A., Conte, C., Tatarelli, A., Coppola, G., et al. (2021). An artificial neural network approach to detect presence and severity of Parkinson's disease via gait parameters. *PLoS One, 16*(2), e0244396. https://doi.org/10.1371/journal.pone.0244396.

Index

Note: Page numbers followed by *f* indicate figures and *t* indicate tables.

C

D

E

I

J

K

L

M

N

O

P

Q

R

FSC
www.fsc.org
MIX
Papier | Fördert
gute Waldnutzung
FSC® C083411

Kolibri 360 GmbH
Ferdinand-Jühlke-Straße 7
99095 Erfurt, Deutschland
produktsicherheit@kolibri360.de